Cloud Computing with Security

Naresh Kumar Sehgal • Pramod Chandra P. Bhatt
John M. Acken

Cloud Computing with Security

Concepts and Practices

Second Edition

 Springer

Naresh Kumar Sehgal
Data Center Group
Intel Corporation
Santa Clara, CA, USA

John M. Acken
Electrical and Computer Engineering
Portland State University
Portland, OR, USA

Pramod Chandra P. Bhatt
Computer Science and Information
Technology Consultant
Retd. Prof. IIT Delhi
Bangalore, India

ISBN 978-3-030-24614-3   ISBN 978-3-030-24612-9   (eBook)
https://doi.org/10.1007/978-3-030-24612-9

This Springer imprint is published by the registered company Springer Nature Switzerland AG
The registered company address is: Gewerbestrasse 11, 6330 Cham, Switzerland

Foreword to the Second Edition

Since the advent of microcomputers, Internet, and open-source operating systems, the world is continuously seeing revolutionary changes every decade in social and economic activities due to digital transformation. This decade clearly belongs to the common slogan "Cloud First, Mobile First" which we all routinely hear.

My three decades of experience in this digital economy world has taught me that disruptions in technologies, solutions, and processes always continue to happen. Knowing the current mega trends, either as a practitioner or as an academic, will help one to be a change agent or a key enabler for the next disruption.

From this aspect, I see the second edition of the "CloudBook" from Dr. Naresh Sehgal and his coauthors as a comprehensive text, which connects previous technological changes that led to the disruptive mega trend Cloud Computing, followed by end-to-end technological and business capabilities that made Cloud Computing a key driver of digital transformation in this decade. It touches on emerging trends like edge computing and deep analytics for data democratization with synthesis and access of data towards the next mega trend that is underway. This book has been substantially enhanced with a security focus for the applications and data protection in Cloud.

Intel Fellow and Intel IT Chief   Shesha Krishnapura
Technology Officer, Intel Corp.
Santa Clara, CA, USA
May 28, 2019

Foreword to the First Edition

The Cloud is a massive advancement in computer architecture. It is transforming the way all digital services are delivered and consumed, accelerating the digitization of all industries. With its high efficiency and scale, Cloud Computing enables the democratization of technology worldwide.

With breakneck pace of technology innovation, broad education of the Cloud lags the adoption. This book drives to close that gap and broaden the knowledge base on what is a critical and foundational capability.

For all those looking to contribute to the digital economy, this book is a highly valuable source of information.

Group President, Intel Corp.   Diane M. Bryant
Santa Clara, CA, USA
November 3, 2017

Foreword to the First Edition

A Google search of "Cloud Computing" resulted in 90 million hits in December 2017. It is thus apparent that Cloud Computing is one of the hottest terms in computing. It is a metaphor for computers connected to the Internet, which are enclosed in a Cloud like boundary in figures that are used in power point presentations. According to a Wikipedia entry (since removed) the term was first used academically in a paper entitled "Intermediaries on Cloud Computing" by Ramnath Chellappa in 1997. A start up called Net Centric (now defunct) applied for a trade mark for "Cloud Computing" in 1999 for educational services that was rejected. The terms became popular after it was used in 2006 by Eric Schmidt (then the CEO of Google). The idea of Cloud Computing as a computing utility, that is, "pay for what you use" computing was foreseen in 1961 by John McCarthy. In 1957 (the mainframe era) MIT had one of the most powerful computers at that time, an IBM 704, that was used in a batch mode with punched cards as input. It was time consuming to write complex programs and make them work on this computer. John McCarthy, who had joined the faculty of MIT in 1959, found it very difficult to use this computer for his research. He wrote, in frustration, a note to the Director of the Computer Centre suggesting that tele-typewriters from the offices of faculty members be connected to the IBM 704 permitting interactive time-shared use of the computer by several persons simultaneously. Both hardware and software technology were not available to implement this idea. Fernando Corbató, who was then the Associate Director of the MIT Computer Center, took up the challenge and led a team that designed the hardware and the software of what is known as the Compatible Time Sharing System (CTSS) on an IBM 709, the successor of IBM 704. The system became operational in 1961 that happened to be the centennial year of MIT. John McCarthy in a lecture during the centennial year celebration said that time-shared computers someday in the future will be organized as a public utility just as a telephone system is a public utility and that such a computing utility could become the basis of a new and important industry. It took over five decades to realize the dream of John McCarthy and now Cloud Computing is an important industry that provides not only as much computing and storage as one wishes from one's desk but also application programs "on tap", on a pay only for what you use basis. A number of technologies had to mature to provide what we now call Cloud architecture. Among these are the emergence of the high-speed Internet, rapid development of high band-width communication at competitive rates, availability of huge storage at throw-away prices, and accessibility of powerful and highly reliable computers at low cost. In addition, developments in software had to mature that include, advances in oper-ating systems, powerful high-level language compilers, service oriented architec-ture, and cryptography. The major features of the Cloud model of computing are: service orientation, on-demand provisioning, virtualization, multi-tenancy, self-healing, and SLA driven. Pay as you use is an added business model, which is the main feature of a utility.

The architecture model of Cloud Computing is a paradigm shift and will be the way data centers will be architected in the future. In the curriculum of undergradu-ates in computer science and engineering a number of discrete subjects such as programming, computer logic design, computer architecture, computer networks, Internet architecture, operating systems, World Wide Web software, and object-oriented design are taught. A synthesis of many of the concepts learnt in these courses is required to design Cloud architecture. Cloud Computing will become a very important course taken by students in their final year of the undergraduate cur-riculum and text books are needed to teach such a course. A number of books on Cloud Computing have appeared since 2013. Many of them are addressed to work-ing professionals and are not comprehensive. The book by Dr. Naresh Sehgal and Professor P.C.P. Bhatt is written primarily as a textbook. A look at the table of con-tents shows that the coverage of the book is comprehensive. It starts from basic principles of Cloud Computing, describes its historical evolution, outlines various models of the Cloud, types of services, how to characterize workloads, customers' expectations, management of Cloud, security issues, problems in migrating applica-tions to a Cloud, analytics, economics, and future trends. A uniform feature of the book is a nice combination of a scholarly exposition with a practical orientation. I particularly appreciated their treatment of workload characterization, management, and monitoring. The authors have conducted experiments on the performance of the providers of Cloud services using diverse workloads and have analyzed the results. They have devoted considerable effort to explain the problems of Cloud security. Every chapter has a large number of references to original sources that will provide the students enough information to explore beyond what is in the text. Exercises are well thought out and there is a project orientation in these so that students can learn by doing.

Dr. Naresh Sehgal has a wealth of experience in the design and development of Intel products, and has been training professionals in the industry. This has resulted in the practical orientation of the book. Professor P.C.P. Bhatt has over 40 years of experience in teaching students in reputed academic institutions internationally and also to professionals in industries. He is one of the pioneer teachers of India. This book has gained a lot by this admirable collaboration of an experienced professional with a reputed academic. I thoroughly enjoyed reading this book and learnt a lot form it. I am confident that both students and working professionals will gain a lot of in-depth knowledge on Cloud Computing by reading this book. It is a valuable addition to the literature on Cloud Computing and I congratulate the authors on their fine effort.

Supercomputer Education and Research Centre   V. Rajaraman
Indian Institute of Science
Bangalore, India

Preface to the Second Edition

Cloud Computing is a fast-evolving field with new developments happening every week. It has been a year since the first edition of this book was published, so the coauthors met to discuss the need for an update. It was exacerbated by a spate of recent cybersecurity attacks and comments by the readers of the book asking for more security in the Cloud. Although there are many books on Cloud Computing and on Security topics separately, there is only one we could find on the Cloud Computing Security, and that too was published over a decade ago. Thus, Naresh spoke with his long-time mentor and ex-Intel manager, Dr. John M. Acken, who is now a security research professor at Portland State University, Oregon, to join our team for a new edition. This brings a new book in your hands.

We have incorporated a total of 35 new sections with Security focus in the previous book and added over 130 new pages to describe the intersection of traditional security practices with the Cloud. We have more than doubled the number of abbreviations and definitions introduced at the beginning of this book. In the later chapters, a few contemporary security solutions such as encryption and voice recognition are discussed in the context of Cloud, but many new topics are introduced to match the rising security threats. Specifically, we describe side channel attacks, an introduction to block chain technology, IoT plug-in devices, multiparty Cloud and software-based hardware security modules, etc. Several new possibilities with real-time decision-making support systems, machine learning in a Public Cloud, hardware as the root of trust, and outsourcing computations using homomorphic encryptions are introduced. Points to ponder at the end of each chapter have been augmented to encourage readers to think beyond the usual textbook topics.

It has been a labor of love for us to bring this book in your hands, and we hope you will find it useful. Like the last time, we recognize that a book has a limited shelf life, so we would like to plan ahead for an update. Thus, we invite the readers to send their inputs, comments, and any feedback to CloudBookAuthors@gmail. com. These will be incorporated in the next edition. Many thanks!!

Santa Clara, CA, USA   Naresh K. Sehgal
Bangalore, India   Pramod Chandra P. Bhatt
Portland, OR, USA   John M. Acken

Preface to the First Edition

The idea for this book started with a call from Prof. PCP Bhatt, whom Naresh has known for over a decade. Prof. Bhatt is the author of a widely used Operating Systems book, which is in its 5th edition in India. He asked Naresh to collaborate on his next edition, but Naresh's expertise is in Cloud Computing, on which he had taught a graduate class at Santa Clara University in California. Thus a new book was conceived.

Cloud has evolved as the next natural step after Internet, connecting mainframe like centralized computing capabilities in a data center with many hand-held and distributed devices on the other end. Thus, computing has taken the form of exercise dumbbell equipment, with Servers on one end, and Clients on the other, joined by the fat pipes of communication in between. Furthermore, it has transformed the way computing happens, which for both authors started by carrying decks of JCL (Job Control Language) cards for feeding an IBM mainframe, culminated with collaboration for this book using the Cloud. This is nothing short of amazing, especially considering that even a smart phone has more memory and compute than was present on the entire Saturn V rocket that carried astronauts to the Moon and back. This implies that more such miracles are on the way, enabled by widely and cheaply available compute power made possible by Cloud Computing.

Objective of this book is to give an exposure to the shift in paradigm caused by Cloud Computing, and unravel the mystery surrounding basic concepts including performance and security. This book is aimed at teaching IT students, and enabling new practitioners with hands-on projects, such as setting a website and a blog. Focus areas include Best Practices for using Dynamic Cloud Infrastructure, Cloud Operations Management and Security.

As a textbook our target audience are students in Computer Science, Information Technology, and MBA students pursuing technology based management studies. The book assumes some background and understanding of basic programming languages, operating systems and some aspects of software engineering. To that extent it would suit senior under graduates or graduates in their early semesters. As a technical manuscript, the book has enough in-depth coverage to interest IT managers and SW developers who need to take a call on migrating to Cloud.

In the initial four chapters the book introduces relevant terms, definitions, concepts, historical evolution and underlying architecture. Midway we cover taxonomy based on the Cloud service models. In particular, we cover the service options that are available as infrastructure and software services. This closes with explanation of the features of public, private and Hybrid Cloud service options. Chapters from Chaps. 5, 6, 7, 8, and 9 dive deeper to explain underlying workload characterization, security and performance related issues. To explain the key question of migration we have chosen EDA. EDA presents a typical dilemma of how much and what could, or should, go into Public Cloud. Chapter 10 onwards we expose the readers to business considerations from effective cost benefit and business perspective. Also presented is a key area of Analytics to reinforce the utility of Cloud Computing. Finally, to keep the user interested we offer glimpses of hands on operations and also project ideas that can be carried forward. To encourage further reading and challenge we have included a few advance topics towards the end of the book.

In our opinion, as a textbook the student's interest would be best served if there is an accompanying weekly laboratory exercise using some Public Cloud or captive university Cloud platform. Public Cloud services are available from Google, Microsoft, Amazon and Oracle under their education support programs.

We recognize that a book has a limited shelf life so would like to plan ahead for an update. Thus, we invite the readers to send their inputs, comments and any feedback to CloudBookAuthors@gmail.com. These will be incorporated in the next edition. Many thanks!!

Santa Clara, CA, USA   Naresh K. Sehgal
Bangalore, India   Pramod Chandra P. Bhatt

Acknowledgments

Pramod would like to acknowledge the indirect contribution of his numerous students. They shared their understanding and thoughts. As a result, he benefited and grew professionally.

Naresh wants to acknowledge the help from several experts, starting with Tejpal Chadha, who had invited him to teach the SCU class in 2015, and his colleagues at Intel who taught him Cloud. He thanks his teaching assistant, Vijaya Lakshmi Gowri, for sharing her project assignments.

We appreciate the help from Mrittika Ganguli, Shesha Krishnapura, Vipul Lal, Ty Tang, and Raghu Yeluri for reviewing this book and giving their technical feedback(s) for its improvement. Vikas Kapur gave his valuable time to suggest many improvements for the first edition, which have improved the quality of this book. The concepts and ideas from several previously published papers with Prof. John M. Acken, Prof. Sohum Sohoni, and their students were used to compile this book, for which we are very grateful. We appreciate the excellent support from Charles Glaser and Brian Halm, along with the staff of Springer Publications, to bring this book in your hands. Various trademarks used in this book are respectively owned by their companies. An approval from Naresh's Intel managers, Tejas Desai and Chris Hotchkiss, was instrumental in bringing this book out for a publication in a timely manner.

Finally, Naresh wants to thank his wife Sunita Sehgal for accommodating this labor of love.

Contents

Abbreviations

2FA	Two-Factor Authentication (also, see MFA)
AES	Advanced Encryption Standard
AMI	Amazon Machine Image
AWS	Amazon Web Service
AZ	Availability Zone
B2B	Business-to-Business
B2C	Business-to-Consumers
BC	Block Chain
C2C	Consumer-to-Consumer
CAD	Computer-Aided Design
CapEx	Capital Expenditure
CAPTCHA	Completely Automated Public Turing Test to Tell Computers and Humans Apart
COMSEC	Communication Security
CPS	Cyber-physical Systems
CSP	Cloud Service Provider
CSRF	Cross-Site Request Forgery
CV	Coefficient of Variation
CVE	Common Vulnerabilities and Exposures
CVSS	Common Vulnerability Scoring System
DARPA	Defense Advanced Research Projects Agency
DBMS	Database Management System
DDOS	Distributed Denial of Service
DES	Data Encryption Standard
DHS	Department of Homeland Security (in the USA)
DOS	Denial of Service
DR	Disaster Recovery
DRM	Digital Rights Management
DSA	Digital Signature Algorithm
DSS	Decision Support System
DUK	Device Unique Key

EC2	Elastic Compute Cloud
EKE	Encrypted Key Exchange
EMR	Elastic Map Reduce
ERP	Enterprise Resource Planning
ESB	Enterprise Service Bus
FaaS	Function as a Service
FIPS	Federal Information Processing Standard
FIT	Failure in Time
FMC	Follow-ME Cloud
FMR	False Match Rate
FNMR	False Non-match Rate
FRR	False Rejection Rate (see FNMR)
GDP	Gross Domestic Product
HE	Homomorphic Encryption
HIP	Human Interaction Proof
HPC	High-Performance Computing
HSM	Hardware Security Module
HTML	Hypertext Markup Language
HTTP	Hypertext Transfer Protocol
HTTPS	HTTP Secure
IaaS	Infrastructure as a Service
IB	In-Band
IETF	Internet Engineering Task Force
INFOSEC	Information Security
IoT	Internet of Things
IP	Intellectual Property
IP	Internet Protocol
LAN	Local Area Network
LFSR	Linear-Feedback Shift Register
MFA	Multifactor Authentication (also, see 2FA)
ML	Machine Learning
MTBF	Mean Time Between Failure
MTTF	Mean Time to Failure
MTTR	Mean Time to Repair
NFV	Network Function Virtualization
NIST	National Institute of Standards and Technology
NSA	National Security Agency
OEM	Original Equipment Manufacturer
OOB	Out of Band
OpEx	Operational Expenditure
OTP	One True Password
OWASP	Open Web Application Security Project
PaaS	Platform as a Service
PIN	Personal Identification Number
PK	Public Key

PoC	Proof of Concept
QoS	Quality of Service
REST	Representational State Transfer
RFID	Radio-Frequency Identification
RoT	Root of Trust
RSA	Rivest–Shamir–Adleman
SaaS	Software as a Service
SDDC	Software-Defined Data Center
SDN	Software-Defined Networking
SEC	Security
SGX	Software Guard Extensions
SHA3	Secure Hash Algorithm 3
SLA	Service-Level Agreement
SMB	Small and Medium Business
SMPC	Secure Multiparty Computation
SNIA	Storage Networking Industry Association
SOA	Service-Oriented Architecture
SOAP	Simple Object Access Protocol
TCB	Trusted Compute Boundary
TCP/IP	Transmission Control Protocol/Internet Protocol
TEE	Trusted Execution Environment
TRNG	True Random Number Generator
UDDI	Universal Description, Discovery, and Integration
US-CERT	United States Computer Emergency Readiness Team
VLSI	Very Large-Scale Integration
VM	Virtual Machine
VMM	Virtual Machine Monitor
VPC	Virtual Private Cloud
VPN	Virtual Private Network
WAN	Wide Area Network
WSDL	Web Services Description Language
WWW	World Wide Web
XML	Extensible Markup Language
XSS	Cross-Site Scripting

Definitions

Acceptability	Acceptability indicates the willingness of users to perform a particular action. It is used for evaluating factors acceptable for authentication.
Access Control	Access control is a mechanism that determines whether to allow or deny access to the requested data, equipment, or facility.
Asymmetric Encryption	An encryption algorithm has a different key for encryption than decryption; thus, the encryption key is made public so anyone can send a secret, but only the possessor of the second key (private) can read the secret. It is also known as public key encryption. An example is RSA.
Attack Surface	Attack surface for a system refers to the set of access points, which can be used to compromise security of the system.
Attestation	Attestation is a process of certification. It enhances the trust level of remote users that they are communicating with a specific trusted device in the Cloud.
Authentication Factor	An authentication factor is data or a measurement used to validate an identity claim.
Auto-scaling	Auto-scaling is used by AWS to automatically add new servers to support an existing service, when more customers are using it and service may be slowing down.

Block Chain	It is a growing sequence of records, each called a block. Each block contains a cryptographic signature of the previous record. Each new transaction record has a time stamp and new encrypted content.
Botnet	A collection of systems successfully infected by an attacker to launch Web-based robot (aka Wobot) attack such as Denial of Service (DOS).
Breach	A successful attack on a secured system.
Cache	A cache is a high-speed memory included in a CPU (central processing unit) package, for storing data items that are frequently accessed, to speed up a program. There may be multiple levels of cache memories.
Cache Side-Channel Attack	A cache side-channel attack uses cache behavior to statistically discover a secret, for example, measuring memory access time for a targeted memory location showing a hit or miss for a particular value. Also, see Side-Channel Attack.
Check Pointing	Check pointing is a practice of taking frequent snapshots of a database or virtual machine memory to restore operations in case of a machine crash or failure.
Checksum	See hash checksum.
Ciphertext	Ciphertext is an encrypted message.
Clickjacking	Clickjacking is a security attack to hijack a user's click on a Website or link, and direct it to an unintended target, to trick them into giving away private information.
Client-Server	Client-server is a computing model where a single server provides services to many client devices.
Cloud Bursting	Cloud bursting is a process to request and access resources beyond an enterprise's boundary, typically reaching into a Public Cloud from a Private Cloud, when user load increases.
CloudWatch	CloudWatch is an AWS service to monitor performance of a user's virtual

machine. It typically reports CPU utilization, and if that exceeds beyond a preset limit, then it may indicate a slow-down, thus starting an action such as auto-scaling (see "Access Control" above).

Cluster
Cluster refers to a group of intercon-nected servers, within a rack or combin-ing a few racks, to perform a certain function such as supporting a large data-base or to support a large workload.

Container
A container is the packaging of applica-tions, including their entire runtime files.

Controllability
Controllability is the measure that indi-cates the ease with which state of a sys-tem can be controlled or regulated.

Cookies
Cookie is a small piece of data sent by a Website and stored locally on a user's computer. It is used by Websites to remember state information, such as items added in the shopping cart in an online store.

CSRF
Cross-site request forgery, also known as one-click attack or session riding and abbreviated as CSRF (sometimes pro-nounced sea-surf) or XSRF. It is a type of malicious exploit of a Website where unauthorized commands are transmitted from a user that the Web application trusts. It is used to perform unauthorized actions such as money transfer on a bank's Website.

Cyber-Physical System
A cyber-physical system is a physical system with an embedded computer control. It contains various physical devices being controlled by system soft-ware and measurement devices.

Data Provenance
To track and record the original sources of shared data objects in a distributed Cloud environment.

Decryption
Decryption is the algorithmic modifica-tion of a ciphertext using a key to recover the plaintext content of the message.

Denial of Service Attack	A denial of service (DoS) attack is flooding a server with repeated service requests, thereby denying its services to other requestors.
Detection Error Trade-Off	A detection error trade-off (DET) curve is used for identity authentication. It relates false match rate to the false non-match rate.
Digital watermarking	Digital watermarking refers to embedding information (such as ownership) within an image's data. It may be imperceptible when viewing the image. Such information is extractable from the raw digital data of the image.
Distributed Denial-of-Service Attack	A distributed denial-of-service (DDoS) attack uses multiple clients on the Internet to repeatedly send excessive numbers of requests to a server. It prevents the recipient from providing services to other requestors. Frequently, this causes a Website to completely crash.
Eavesdropping	A security attack where the attacker (Eve) intercepts data exchange on the network between the sender (Alice) and the recipient (Bob) but does not alter the data. This is also called a passive attack.
Edge Computing	Edge computing refers to the notion of having compute ability at the edge of a network with local storage and decision-making abilities. Edge denotes the end point of a network.
Elasticity	Elasticity is a property of computing and storage facilities in a Cloud to expand in view of a growing need and shrink when the need goes away.
Electronic Signature	An electronic signature is a secure hash of a message with its author's identification. This electronic hash is often used as a replacement of a physical signature on a paper document.
Encryption	Encryption is the algorithmic modification of a message using a key to convert plaintext to ciphertext. This makes it difficult for an attacker to read the content of the message except with a key.

Enterprise Service Bus (ESB)	ESB is a communications system between mutually interacting software applications in a service-oriented architecture (SOA).
Equal Error Rate	An equal error rate (EER) occurs when the false non-match rate equals the false match rate, used for identity authentication.
Error Checking Hash	Error checking hash is the calculation of a short fixed sized number based on the content of a message. It is used to check whether the integrity of a data or message has been compromised.
Fab	Fab refers to a fabrication plant for manufacturing silicon chips.
Failure in Time	Failure in time (FIT) is the failure rate measured as the number of failures per unit time. It is usually defined as a failure rate of 1 per billion hours. A component having a failure rate of 1 FIT is equivalent to having an MTBF of 1 billion hours. FIT is the inverse of MTBF. See also MTBF, MTTR, and MTTF.
False Match Rate	False match rate (FMR) is the rate at which authentication evaluation allows an imposter access because measured data incorrectly matched expected data.
False Non-match Rate	False non-match rate (FNMR) is the rate at which the authentication evaluation prevents an authorized user access because the measured data didn't match the expected data.
Fault Tolerance	Fault tolerance is the property of a computer system to keep functioning properly in the presence of a hardware or software fault.
Follow-ME Cloud	Refers to migration of an ongoing IP (Internet Protocol) service from one data center to another, in response to the physical movement of a user's equipment, without any disruption in the service.
Frame Busting	Frame busting is a technique used by Web applications to prevent their Web pages from being displayed within a frame to prevent malicious attacks.

Frame	Frame is a portion of Web browser's window.
Gateway	A connection that allows access to a network or between networks. It also refers to a point that enables access to a secure environment or between multiple secure environments.
Grid Computing	Combining compute servers, storage, and network resources to make them available on a dynamic basis for specific applications. Grid computing is a form of utility computing. Also, see Utility Computing.
Hadoop	Apache Hadoop is an open-source software framework used for distributed storage and processing of dataset of big data using the MapReduce programming.
Hardware Security Module	Hardware security module (HSM) is a physical computing device acting as a vault to hold and manage digital keys.
Hash Checksum	Hashing for error checking or authentication produces a checksum. This can be a simple count or parity of the data stream or be calculated using a linear-feedback shift register (LFSR).
Hash Digest	A hash digest is short fixed size number created using a secure hash function such as SHA-3.
Hashing	Hashing is the calculation of a short fixed size number based upon the content of a message. Hashing can be used for error checking, authentication, and integrity checking.
Homomorphic Encryption (HE)	Homomorphic encryption is such that a certain form of computation on both plaintext and ciphertext results in identical results, in plaintext and encrypted forms, respectively.
Hybrid Cloud	Hybrid cloud is a Cloud-computing environment, which uses a mix of on-premises, Private Cloud, and third-party, Public Cloud services with orchestration between the two platforms.

Identity Authentication	Identity authentication is determination that a claimant is who it claims to be. It uses a process or algorithm for access control to evaluate whether to grant or deny access.
Imposter	An imposter is a person or program that presents itself as someone or something else in order to gain access, circumvent authentication, trick a user into revealing secrets, or utilize someone else's Cloud service. See also Masquerader.
Interoperability	Interoperability is the ability of different information technology systems and software applications to communicate, exchange data, and use the information that has been exchanged.
Key Distribution	The method or algorithm to solve the problem of getting one or more keys to only the desired participants. These are used in a secure system that uses encryption.
Key Generation	The method or algorithm to generate the keys in a secure system that uses encryption.
Key Management	The problem and solutions involving key generation, key distribution, key protection, and key updates.
Latency	Latency refers to time delay in transmission of network packets or in accessing data from a memory.
Machine Learning	Machine learning refers to the ability of computers to learn without being explicitly programmed.
Malleability	Malleability refers to a class of weak encryption algorithms. Given an encryption of plaintext p and a known function f, malleable cryptography generates another ciphertext, which decrypts to $f(p)$, without knowing p. This is generally not desirable but useful for homomorphic encryption.
Malware	Malware is a program (such as virus, worm, Trojan horse, or other code-based malicious entity infecting a host) that is covertly inserted into a system with the

intent of compromising the confidentiality, integrity, or availability of the victim's data, applications, or operating system.

Map-Reduce	Map-Reduce is a programming model and implementation for processing and generating big datasets with a parallel, distributed algorithm on a cluster.
Masquerader	Pretending to be someone else in order to gain access, circumvent authentication, or trick a user into revealing secrets. See also Imposter.
Mean Time Between Failures	Mean time between failures is the downtime or basically mean time to failure plus the mean time to repair. See also FIT, MTTF, and MTTR.
Mean Time to Failure	Mean time to failure is the average time between failures. See also FIT, MTTR, and MTBF.
Mean Time to Repair	Mean time to repair is the average time from when a failure occurs until the failure is repaired. See also FIT, MTTF, and MTBF.
Meltdown	Meltdown is a side-channel attack that takes advantage of some speculative execution implementation details.
Multifactor Authentication	Multifactor authentication (MFA) uses more than one factor to determine authorization. Commonly used factors are information, biometrics, physical device, and location. Example factors are password, fingerprints, car key, and GPS coordinates. Also, see Single-Factor Authentication.
Network Function Virtualization (NFV)	It refers to packaging networking services in virtual machines (VMs) to run on general-purpose servers. NFV services include routing, load balancing, and firewalls. NFV replaces dedicated hardware devices and offers flexibility along with reduced costs.
Noisy Neighbors	Noisy neighbor is a phrase used to describe a Cloud Computing infrastructure co-tenant that monopolizes bandwidth, disk I/O, CPU, and other

resources and can negatively affect other users' Cloud performance.

Observability
Observability is a measure of how well internal states of a system can be inferred from knowledge of its external outputs.

OpenStack
OpenStack is a Cloud operating system that controls large pools of compute, storage, and networking resources throughout a data center.

Optimizations
In computing, optimization is the process of modifying a system to make some features of it work more efficiently or use fewer resources.

Phishing
Phishing is an attempt to fraudulently obtain personal information such as passwords, bank or credit card details, by posing as a trustworthy entity.

Plaintext
Plaintext is data to be encrypted to yield ciphertext.

Power Analysis Side-Channel Attack
Power analysis side-channel attack uses power supply measurements to statistically discover a secret, for example, measuring power usage during decryption for an unknown key and checking the correlation to power usage for known keys.

Private Cloud
Private Cloud is dedicated to a single organization.

Public Cloud
A Public Cloud offers its services to a full range of customers. The computing environment is shared with multiple tenants, on a free or pay-per-usage model.

Quality of Service
The overall performance for a Cloud service, as documented in a service-level agreement between a user and a service provider. The performance properties may include uptime, throughput (bandwidth), transit delay (latency), error rates, priority, and security.

Reliability
Reliability is the quality of performing consistently well.

Secure Hash Function
A function that generates a relatively small unique digest from a very large set of data used to verify that the large set of data is authentic and unaltered. An example is SHA3.

Self-Service	Self-service Cloud Computing is a facility where customers can provision servers, storage, and launch applications without going through an IT person.
Service-Oriented Architecture (SOA)	A service-oriented architecture (SOA) is a style of software design, where services are provided by application components using communication protocol over a network.
Side-Channel Attack	An attempt to create a security breach by indirectly attacking the secured information, such as guessing a secret by measuring power supply current. Also, see Cache Side Channel Attack.
Single-Factor Authentication	Single-factor authentication uses only one factor to determine authorization. Example factors are password, fingerprints, car key, or GPS coordinates. Also, see Multifactor Authentication.
Software-Defined Data Center (SDDC)	A software-defined data center (SDDC) uses virtualization and provides IT managers with software tools to monitor customers' usage patterns, control allocation of networking bandwidth, and prevent a malicious attacker from causing a denial of service (DOS) attack.
Speaker Identification	Speaker identification (or voice recognition) uses audio signals to identify a speaker. One use of speaker identification is for authentication.
Spectre	Spectre is a side-channel attack that takes advantage of some speculative execution and indexed addressing implementation details.
Spoofing	Spoofing is the act of creating false and seemingly believable environment with an intent to cheat.
Streaming	Streaming is a technique for transferring data so that it can be processed as a steady and continuous stream. Streaming technologies are becoming important because most users do not have fast enough access to download large datasets. With streaming, the client browser

	or plug-in can start displaying the data before the entire file has been transmitted.
Symmetric Encryption	An encryption algorithm that has the same key for encryption and decryption, thus, the key is kept private and must be privately communicated between the sender and receiver. It is also known as private key encryption. An example is AES.
Test Side-Channel Attack	A test side-channel attack uses the test access port (IEEE 1149.1 standard test access port) to read internal values and find secrets.
Thick and Thin Clients	Thick client is a full-featured computer that may be connected to a network. A thin client is a lightweight computer built to connect to a server from a remote location. All features typically found on a thick client, such as software applications, data, memory, etc. are stored on a remote data center server when using a thin client.
Threat Model	Threat modeling is an approach for analyzing the security of an application. It is a structured approach to identify, quantify, and address the security risks associated with an application.
Time Machine	When "digital" twin model of a real machine is used to mimic its behavior on a Cloud platform, it simulates the machine health conditions in real time. This is the concept of "time machine" used to detect anomalies in the real machine's operation. Also, see Twin Model below.
Trojan Horse	A computer program that appears to have a useful function but also has a hidden and potentially malicious function that evades security mechanisms, sometimes by exploiting legitimate authorizations of a system entity that invokes the program.
True Match Rate	True match rate (TMR) is the rate at which the authentication evaluation

allows an authorized user access because the measured data correctly matched expected data.

True Non-match Rate
True non-match rate (TNMR) is the rate at which the authentication evaluation prevents an imposter access because measured data didn't match the expected data.

Trust Anchor
A certification authority that is used to validate the first certificate in a sequence of certificates. The trust anchor's public key is used to verify the signature on a certificate issued by a trust anchor certification authority. A trust anchor is an established point of trust from which an entity begins the validation of an authorized process or authorized (signed) package. See Trusted Arbitrator.

Trust List
A structured collection of trusted certificates used to authenticate other certificates. See Trusted Arbitrator.

Trusted Agent
Entity authorized to act as a representative of an agency in confirming subscriber identification during the registration process. Trusted agents might not have automated interfaces with certification authorities. See Trusted Arbitrator.

Trusted Arbitrator
A person or entity that certifies the level of trust for a third party. This term covers trust anchor, trust list, trusted agents, trusted certificate, and the process of checking for trust levels.

Trusted Certificate
A certificate that is trusted by the relying party (usually a third party) on the basis of secure and authenticated delivery. The public keys included in trusted certificates are used to start certification paths. See also Trust Anchor and Trusted Arbitrator.

Trusted Compute Boundary
Trusted compute boundary (TCB), or trust boundary, refers to a region that consists of devices and systems on a platform that are considered safe. Sources of data outside of TCB are untrusted and thus checked for security violations,

such as network ports. In general, TCB on a system should small to minimize the possibility of security attacks.

Trusted Computing   Trusted computing refers to a situation where users trust the manufacturer of hardware or software in a remote computer and are willing to put their sensitive data in a secure container hosted on that computer.

Twin model   Twin model refers to the representation of a physical machine in a virtual domain. It enables a physical operation to be coupled with a virtual operation by means of an intelligent reasoning agent.

Usability   Usability is a metric of user experience. It represents ease of use or how easily a product can be used to achieve specific goals with effectiveness, efficiency, and satisfaction for typical usage.

Utility Computing   Delivering compute resources to users, who pay for these as a metered service on a need basis. Also, see Grid Computing.

Verification   Verification is the process of establishing the truth, accuracy, or validity of something.

Virtual Machine Monitor   A virtual machine monitor (VMM) enables users to simultaneously run different operating systems, each in a different VM, on a server.

Virtual Machines   A virtual machine (VM) is an emulation of a computer system.

Virtual Private Cloud   A virtual Private Cloud (VPC) is an on-demand configurable pool of shared computing resources allocated within a Public Cloud environment, providing a certain level of isolation between the different organizations.

Virus   A computer program that can copy itself and infect a computer without permission or knowledge of the user. A virus might corrupt or delete data on a computer, use resident programs to copy itself to other computers, or even erase everything on a hard disk.

Vulnerability	Vulnerability refers to the inability of a system to withstand the effects of a hostile environment. A window of vulnerability (WoV) is a time frame within which defensive measures are diminished and security of the system is compromised.
Web Service	Web service is a standardized way of integrating and providing Web-based applications using the XML, SOAP, WSDL, and UDDI open standards over an Internet Protocol backbone.
Workload	In enterprise and Cloud Computing, workload is the amount of work that the computer system has been given to do at a given time. Different types of tasks may stress different parts of a system, e.g., CPU-bound or memory-bound workloads.
Worm	A self-replicating, self-propagating, self-contained program that uses networking mechanisms to spread itself. See Virus.
Zombie	A computer process that is consuming resources without doing any useful work.

Chapter 1
Introduction

1.1 Motivation

The Internet has brought revolutionary changes in the way we use computers and computing services. The paradigm mostly used is a client-server architecture supported over Internet. Client refers to computers that are generally used by individuals and include desktops, laptops, handheld smartphones, or other similar devices. A server refers to bigger computers that can simultaneously support multiple users, typically using many multicore processors, larger memories, and bigger storage capacity. Historically, mainframes fit this description, and lately smaller compute servers have been pooled together to meet the needs of several users at once. These servers are placed in a rack, also known as a rack of servers networked together, and housed in a building called a data center (DC) along with storage devices. The power, cooling, and physical access to a data center is tightly controlled for operational and security reasons.

Recently, networking and storage technologies have increased in complexity. As a consequence the task to manage them has become harder. IT (information technology) professionals manage server infrastructure management in a DC to ensure that hardware and software are working properly. Hence, a fully operational DC requires large amounts of capital and operational expenditures, but may not be fully utilized at all hours of a day, over a week, or a year. This has led some people to consider sharing the resources of a DC with other users from other organizations while maintaining data security, application level protection, and user isolation. This has been enabled by operating systems in the past and most recently with hardware-based virtualization. The overall net effect of such technologies has been to eliminate the direct capital and operational expenditure of a DC for users on a shared basis and for the DC operator to spread their costs over many users. This is akin to airlines industry, where buying a commercial ticket for occasional travel is much cheaper than the cost of owning an airplane and hiring staff to operate it.

© Springer Nature Switzerland AG 2020
N. K. Sehgal et al., *Cloud Computing with Security*,
https://doi.org/10.1007/978-3-030-24612-9_1

1.2   Cloud Computing Definitions

Cloud Computing refers to providing IT services, applications, and data using dynamically scalable pool(s), possibly residing remotely, such that users do not need to consider the physical location of server or storage that supports their needs. According to NIST, the definition of Cloud Computing is still evolving [1]. Their current definition for Cloud Computing is "a model for enabling convenient, on-demand network access to a shared pool of configurable computing resources (e.g., networks, servers, storage, applications, and services) that can be rapidly provisioned and released with minimal management effort or service provider interaction." Armbrust provides another, similar, definition of Cloud Computing as "the applications delivered as services over Internet and the hardware and system in the data centers that provide these services". The services themselves have long been referred to as Software as a Service (SaaS) [2]. The Cloud can include Infrastructure as a Service (IaaS) and a Platform as a Service (PaaS). These service models are defined, starting from the top of a symbolic pyramid, as follows:

1. *Software as a Service (SaaS):* is focused on end users of Cloud, to provide them with application access, such that multiple users can execute the same application binary in their own virtual machine or server instance. These application sessions may be running on the same or different underlying hardware, and SaaS enables application providers to upgrade or patch their binaries in a seamless manner. Examples of SaaS providers are Salesforce.com providing CRM (customer relationship management), Google.com serving docs and gmail, etc., all of which are hosted in Cloud.
2. *Platform as a Service (PaaS):* is focused on application developers with varying computing needs according to their project stages. These are met by servers that can vary in number of CPU cores, memory, and storage at will. Such servers are called elastic servers. Their services can auto scale, i.e., new virtual machines can start for load balancing with a minimal administrative overhead. Examples of PaaS providers are Google's App Engine, Microsoft's Azure, Red Hat's Makara, AWS (Amazon Web Services) Elastic Beanstalk, AWS Cloud Formation, etc. These cloud service providers (CSPs) have the capability to support different operating systems on the same physical server.
3. *Infrastructure as a Service (IaaS):* is the bottom-most layer in a Cloud stack, providing with direct access to virtualized or containerized hardware. In this model, servers with given specification of CPUs, memory, and storage are made available over a network. Examples of IaaS providers are AWS EC2 (Elastic Compute Cloud), OpenStack, Eucalyptus, Rackspace's CloudFiles, etc.

Different types of Cloud Computing providers [2, 3] defined as below:

1. *Public Cloud:* A Public Cloud offers its services to a full range of customers. The public nature of this model of a Cloud is similar to Internet, i.e., users and services can be anywhere on the World Wide Web. The computing environment is shared with multiple tenants, on a free or pay-per-usage model.

2. *Private Cloud: Private Cloud* restrict their users to a select subset, usually to a specific organization within a given company. The Private Cloud is similar to an intranet, i.e., their services are provided internally via an organization's internal network.
3. *Hybrid Cloud: Hybrid Cloud* providers offer their services to a narrowly defined range of private users, which if needed, can expand to reside on a Public Cloud infrastructure. Alternatively, a public service provider can remotely manage part of the infrastructure in a private organization and use Cloud for backups.

An example of a Hybrid Cloud is Microsoft Azure Stack, deployed in an enterprise but is managed externally, and if the computing requirements increase, then some tasks are migrated to an external Public Cloud. This process is often called cloud bursting. Clearly Public Cloud pose the greatest security challenges due to their wider open access. In general, when a particular model is not specified, the term Cloud Computing refers to Public Cloud. In all models, a client's usage is independent of the source and location of the service provider.

Cloud customers often choose a Cloud service on the basis of their needs:

1. Workload attributes:

    (a) Performance
    (b) Security
    (c) Integration complexity
    (d) Data size

2. Business needs:

    (a) Time to deployment
    (b) Compliance regulatory
    (c) Geographical reach
    (d) SLA (service-level agreements)

3. Ecosystem available:

    (a) Maturity of SaaS offerings
    (b) Viability of alternate services
    (c) Availability of resellers/system integrators

1.3   Cloud Computing Operational Characteristics

In the extreme case, Cloud Computing can be defined as a service-oriented architecture (SOA) that provides any type of computing component [3]. An example is Internet-based email providers, where the content for each user resides in the Cloud and the primary interface is via a browser. A user who may be travelling can access her email from any location in the world by simply typing a URL, without caring about the location of a service provider's database. However, the response time may vary depending on the physical distance and latency between the user and service provider's locations. To overcome this limitation, some international service providers use a distributed database replicating each user's email information among multiple data centers in different locations. One of these is picked by intelligent routers to provide the shortest response time. At any given time, the user doesn't know which physical location is providing her email service and thus considers it to be residing in a Cloud. Another advantage of Cloud Computing is the economy of scale. The utilization of servers inside an enterprise, a small-medium business, or even on a home computer can vary widely but rarely reaching a near 100% level at all times. In fact, averaged over 24 hours, and 7 days a week, an IEEE study recently showed that most servers will show CPU utilization between 10% and 15%, and the same is true of network bandwidth [4]. Storage is also underutilized. Combining the usage of several such customers served from a single large data center enables the operators to spread their capital investment and operational costs over a large set of customers. The sharing of resources can drive the utilization higher. Such higher utilization meets the need to fill the installed capacity with new users by allowing a flexible Cloud environment. Amazon uses this concept for their Elastic Compute Cloud [5]. This allows them to rent their servers with a reduced total cost of ownership (TCO) for their customers. This is the broadest definition of Cloud Computing, that information is stored and processed on computers somewhere else, "in the Cloud," and results are then brought back to the end users' computer [6].

The trend toward Cloud Computing continues due to financial benefits accrued to both the users and providers. As the trend continues, Cloud-specific security issues are added to the existing security issues [7–9]. In a Cloud, services are delivered to the end users via the public Internet or via enterprise networks in case of a Private Cloud, without any user intervention. Private Cloud are deployed behind a firewall for an organization's internal use and enable IT capabilities to be delivered as a service. Companies often resort to using Private Cloud as they do not want to trust the Public Cloud with their confidential or proprietary information. Some key characteristics [10] of a Public or Private Cloud include:

- *Automated provisioning and data migration*: Cloud Computing eliminates the users' need to worry about buying a new server, loading an OS, and copying data to or from it when scalability is needed. Advance capacity planning is important in a traditional environment: users need to forecast their needs in advance, outlay capital budget to procure necessary hardware and software, then wait for days and weeks before systems are brought online. In case a user under-forecasts her needs, their applications will not be available or may run slow while over-forecasting results in wasted money. In contrast to the traditional computing environments, with Cloud Computing, users can order and get new capacity almost immediately. An analogy for fluctuating user demand and remotely installed capacity is with water and electricity utility companies.
- *Seamless scaling:* Pay as you go, instead of making an upfront investment in hardware and software, some of may be partially used, above two features allow customers to get on-demand applications and services and then pay for only what

they use [5]. This is similar to households paying for electricity and water utility bills based on their monthly consumption. In fact, several services on the public Internet Cloud are available free to the end users, such as email and search capabilities, and a fee is charged to the advertisers. In this model, each user is promised some space to store their mail on the Cloud, but multiple users share storage and computers at the same time, often referred to as multi-tenancy on a server.

- *Increased multi-tenancy:* With the advent of multicore computers and virtualization, it is possible for several customers to share a single server [11] with each being isolated in a separate virtual machine (VM). However, the topic of virtualization is orthogonal to Cloud Computing as some leading operators are currently not using virtualization in their data centers, but when used together, the benefits of these two technologies can multiply. Virtualization is used by some Cloud operators to provide immediate capacity by creating a new VM. Multiple VMs are consolidated on a single physical server to improve HW utilization. When demand decreases, any unused servers are shut down to save electricity and air-conditioning costs in a data center.

Cloud Computing is based on many prior innovations, e.g., the ability to do task consolidation and VM migration using virtualization for dynamic load balancing. It enables a service provider to optimize their data center usage by booting new servers and migrating some of the existing users to new machines to meet the expected response time. However, this also opens up potential security threats as mission critical applications and confidential data from different users are co-located on the same hardware. Current methods of disk encryption are no longer deemed sufficient if one can get physical access to a competitor's servers in a third-party data center. Physical access controls are placed on the employees working in such a data center, but there is no line of defense if one can gain access to the contents of main memory on a server. An example is the launch plan for next-generation product kept behind the firewalls, lest some competitors can access it, while sharing the same server in a Public Cloud. Someone with physical access to a shared server can simply copy the files, and in a virtualized environment, even though the VMs are isolated, their runtime images are backed up at regular intervals to enable recovery or migration of a session. Hence, the back-up images also need to be protected.

The benefits and some potential risks of Cloud Computing may be classified as follows:

1.3.1 Cloud Computing Benefits

1. Shared infrastructure reduces cost.
2. Pay as you go or only pay for what you use.
3. On-demand elasticity, from a large pool of available resources.
4. Increased focus on the application layer.
5. Let someone else worry about the hardware.

1.3.2   Cloud Computing Potential Risks

1. You lose direct knowledge and control of underlying hardware.
2. Noisy neighbors or other VMs sharing the same hardware can affect your performance.
3. Hard to diagnose performance issues, due to limited visibility and virtualization.
4. Potential security risks of placing your mission critical data on remote servers.
5. Vendor lock-in, means getting stuck with a Cloud provider who has your data.

1.4   Cloud Computing Trends

Cloud Computing is rapidly evolving from providing end-user services (e.g., search, email, social networking) to support mission critical business and commercial grade applications as shown in Fig. 1.1. Some services for some companies are already based entirely in Cloud, e.g., Salesforce.com, and coming years will see new emerging trends, such as:

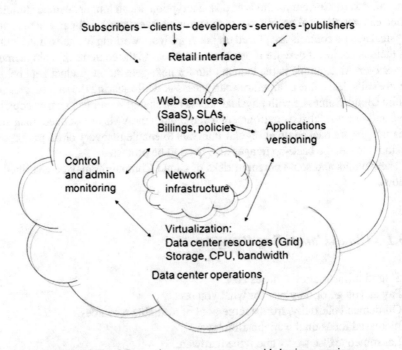

Fig. 1.1 Interaction of Cloud Computing components to provide business services

1.4.1   Trend #1: Abstraction of Network, Storage, Database, Security, and Computing Infrastructure

- This helps with software applications and data migration between Cloud.
- Offering image of on-demand, virtual data center with flexibility implied in scalability and agility.

1.4.2   Trend #2: A Pricing Model that Is Retail in Its Conception

- E.g., Pennies per gigabyte, massive CPU cycles, and bandwidth, which in turn will make computing more affordable beyond Moore's or Metcalfe's law can predict. This is due to a higher utilization of infrastructure in a Cloud data center.

1.4.3   Trend #3: Service-Level Agreements (SLAs)

Increasingly, SLAs are being used for the following purposes:

- Data persistence, system reliability, and business continuity as individual consumers may be patient to wait for their search or email results, but businesses need predictability to meet their goals and deliver products in a timely manner.
- SLAs imply that Cloud service providers will need systems in place to ensure redundancy and security of their customers' information.
- SLAs also need to cover performance aspects of the Cloud services being provided in terms of the required computing speed and network bandwidth.

1.5   Cloud Computing Needs

Cloud Computing is still in its nascent stages during the first two decades of the twenty-first century. With the present implementations, two Cloud can be quite different from each other. A company's cluster of servers and switches can be termed as a Private Cloud, accessible only to the people inside an enterprise. Some companies have started to outsource their IT services to external Cloud providers for economic reasons, which gives rise to the expansion of Hybrid Cloud. A company can also offer both internal and external computing services. As an example, Google has internal infrastructure for its email and search products and also has a Public Cloud offering. The following areas need to be addressed in order for Cloud services to be competitive with internally dedicated servers:

- *Network latency:* Customers are demanding high bandwidth to overcome the latency issues, but in a data-flow chain, delay is determined by the weakest link. Thus service providers are often using local caching near a customer's location. This further expands the attack surface, as multiple copies exist with the same confidential information.
- *Fine-grained migration and provisioning:* Solutions are needed to avoid copying gigabytes of data when a physical server needs to be brought down for planned or unplanned maintenance, e.g., to change its memory or fans. In this case, regular snapshots are taken of a VM to help with quick migration, but one needs to ensure that these memory snapshots are protected to avoid compromising a user's information in the Cloud.
- *Standards and open-source solutions for Cloud SW* infrastructure support [12] are needed as currently most commercial grade Cloud providers use their internal data representations. In case a provider declares bankruptcy or the customer wishes to migrate her data to another provider, it is important that the information is stored in a manner that can be read by others with necessary security keys.
- *Offline vs. online synchronization of data* is needed as Internet or electric service is rarely guaranteed 24x7 in the emerging markets. Thus users want to use devices that go beyond Internet browsers running on thin clients, which can store local copies of application and data that can run using an uninterrupted power supply and upon restoration of public utility service can sync up with the Cloud providers' database. An example of this is an accountant's office which wants to continue work on client's tax return even if the Cloud service is not available. However, they will also want the customers' data to be protected.

Cloud Computing has found widespread adoption among public consumers with personal emails and photo storage. However, enterprises have hesitated to move their internal email services, financial data, or product design databases to external Cloud due to real or perceived security concerns. This has given rise to internal or Private Cloud within enterprises, tucked behind corporate firewalls. For Cloud Computing to become truly universal, these real or perceived security concerns need to be outlined as described in this chapter. Until then, economic benefits of broadly shared Cloud Computing infrastructure will be slow to reduce the IT costs of enterprises.

1.6   Information Security Needs

It is said that data is the new oil of the twenty-first century. Cloud aggregates data at a few central points, which also makes it more susceptible to leakage and ensuring damage. Security of information has three aspects: *confidentiality, integrity,* and *availability* (CIA of security policies and objectives). Sometimes, security is shortened to SEC, and information security is shortened to INFOSEC.

1. Confidentiality refers to access control for the data, lets it falls in the wrong hands. The protection of private data includes storage devices, processing units, and even cache memory.
2. Integrity refers to maintaining the fidelity or correctness of the data, so it is not accidentally or maliciously altered.
3. Availability refers to the servers, storage, and networks in the Cloud remaining available on a 24 × 7 basis.

An example of lack of availability is DOS (denial of service) attack, which can potentially bring down one or more elements in a Cloud. Using a Cloud also requires secure communication between an end user and the Cloud. This topic will be discussed in more details in Chap. 7.

1.7   Edge Computing and IoT Trends

Edge denotes end point of a network. Internet of Things (IoT) expands the definition of a computer network to include sensors and items found in our daily usage, such as household devices and industrial machines. Edge computing refers to the notion of having compute ability at the edge of a network with local storage and decision-making abilities. Use cases include ability to control your garage door or refrigerator settings remotely from a smartphone, or using vibration sensors in a factory to predict impending failure of an electrical motor, etc. By 2020, spending on edge infrastructure will reach up to 18% of total IoT infrastructure spending [13]. The IoT data deluge requires a massive investment. This is due to large number of sensors on the edges of a network, generating enormous amounts of data. While much of the data can be processed locally, a lot can be aggregated remotely for real-time decision-making. This also can be utilized for off-line analytics. We shall visit this topic in details in the later chapters. The edge computing has a side effect, as it expands the attack surface for potential security attacks. This opens up new threat models, such as someone hijacking Internet-connected devices. For example, surveillance cameras at unsuspecting homes were used to launch a distributed denial-of-service (DDOS) attack on targeted victim Websites [14].

1.8   This Book's Organization

This book is organized in 15 chapters and 3 appendices. In the first chapter, we began with a brief overview of the Cloud Computing domain, and then the second one delves deeper in the technologies that laid down the foundations of Cloud over the preceding half-century. It also introduces the basic information security concepts, an example of an attack and Cloud software security requirements. Then next two Chaps. 3 and 4, review the taxonomy of Cloud, followed by a classification of compute workloads that is suitable for Cloud Computing in Chap. 5. We review Cloud Management and Monitoring topics in Chap. 6 and an idea of Follow-Me Cloud. In Chap. 7, the information security technologies are introduced followed by topics on the use of encryption, side channel attacks, and an introduction to block chain. Chapter 8 starts by looking at industries that have well adopted the Cloud, including a start-up effort using IoT devices, multi-party Cloud, and software-based hardware security modules. While Chap. 9 studies an industry that has struggled with the concept of Cloud Computing and has now started to migrate to Cloud. Along with new applications for block chain, it should serve as a case study for any new users who want to experiment with the Cloud. No discussion of Cloud can be complete without the economic aspects, so we visit the billing topics in Chap. 10, followed by additional security considerations in Chap. 11, including novel topics of access control using speaker identification and real-time control of cyber-physical systems. Analytical models and future trends are discussed in Chaps. 12 and 13, respectively, including machine learning in a public cloud and new sections related to hardware as a root of trust, multi-party Cloud, and security solutions for edge computing. Finally, we wrap up the book with a short quiz, some project ideas, and appendices to cover topics that supplement the main text of this book. This includes Linux containers, the usage of which is rising rapidly in the Cloud Computing. There is a new appendix D, with introduction of trust models for IoT devices.

Our recommended way to read this book is serially starting with Chap. 1 onward, but if some readers are already familiar with the basics of the Cloud, then they can take the test of Chap. 13 to judge their proficiency. After basics are understood, then Chaps. 5, 6, and 7 can be reviewed to appreciate the performance and security issues related to Cloud. If an IT manager is looking to migrate complex workloads to Cloud, then they will benefit from considerations of Chap. 9. New research ideas and directions can be found in Chap. 13, and hands-on exercises in Chap. 15 will give an appreciation of actual usage scenarios. Our earnest hope is that you, our readers, will enjoy reading this book as much as we enjoyed writing it.

1.9   Points to Ponder

1. Cloud Computing has enjoyed double digit yearly growth over the last decade, driven mainly by economics of Public Cloud vs. enterprise IT operations. Can you think of three economic reasons why Public Cloud Computing is cheaper than an enterprise DC?
2. Concentration of data from various sources in a few data centers may increase the risk of data corruption or security breaches. How can this possibility be minimized?
3. Under what circumstances an enterprise should opt to use a Public Cloud vs. its own Private Cloud facilities?

4. Why does a Public Cloud provider want to support a Hybrid Cloud, instead of asking enterprises to use its public facilities?
5. Is there a scenario when Private Cloud may be cheaper than the Public Cloud?
6. Can you think of scenarios where a Cloud user favors (a) PaaS over IaaS and (b) SaaS over PaaS?
7. Think of ways in which you can detect if a DOS attack is underway?

References

1. Mell, P., & Grance, T. (2009). *The NIST definition of Cloud Computing (Version 15 ed.)*. Available: http://csrc.nist.gov/groups/SNS/Cloud-computing/Cloud-def-v15.doc.
2. Armbrust, M., Fox, A., Griffith, R., Joseph, A. D., Katz, R., Konwinski, A., Lee, G., Patterson, D., Rabkin, A., Stoica, I., & Zaharia, M. (2010). A view of Cloud Computing. *ACM Communications, 53*, 50–58.
3. Ramgovind, S., Eloff, M.M., & Smith, E.. (2010). The management of security in Cloud computing. *Information Security for South Africa (ISSA)*, Sandton, South Africa, pp. 1–7.
4. Vasan, A., Sivasubramaniam, A., Shimpi, V., Sivabalan, T., & Subbiah, R. (2010). Worth their Watts? – an empirical study of datacenter servers. In *High Performance Computer Architecture (HPCA), IEEE 16th International Symposium*, Bangalore, India, pp. 1–10.
5. *OProfile*. Available: http://oprofile.sourceforge.net.
6. Fowler, G. A., & Worthen, B. (2009). Internet *industry is on a Cloud — whatever that may mean*. Available: http://online.wsj.com/article/SB123802623665542725.html.
7. Kaufman, L. M. (2010). Can public-cloud security meet its unique challenges? *IEEE Security & Privacy, 8*, 55–57.
8. Anthes, G. (2010). Security in the Cloud. *ACM Communications, 53*, pp. 16–18.
9. Christodorescu, M., Sailer, R., Schales, D. L., Sgandurra, D., & Zamboni, D. (2009). Cloud Security is not (just) virtualization security: A short chapter. *Proceedings of the 2009 ACM workshop on Cloud Computing Security*, pp. 97–102. Chicago.
10. Soundararajan, G. & Amza, C. (2005). Online data migration for autonomic provisioning of databases in dynamic content web servers. *Proceedings of the 2005 conference of the Centre for Advanced Studies on Collaborative Research*, Toronto, pp. 268–282.
11. Nicolas, P. *Cloud Multi-tenancy*. Available: http://www.slideshare.net/pnicolas/Cloudmulti-tenancy.
12. Bun, F. S. (2009). Introduction to Cloud Computing, presented at the Grid Asia.
13. https://www.i-scoop.eu/internet-of-things-guide/edge-computing-iot/
14. https://www.forbes.com/sites/thomasbrewster/2016/09/25/brian-krebs-overwatch-ovh-smashed-by-largest-ddos-attacks-ever/

Chapter 2
Foundations of Cloud Computing and Information Security

2.1 Historical Evolution

During the past half-century, computing technologies have evolved in several phases as described below. There appears to be a cyclic relationship between them, but in reality computing has grown as an outward going spiral.

1. *Phase 1:* This was the era of large mainframe systems in the backrooms connected to multiple users via dumb terminals. These terminals were electronics, or electromechanical hardware devices, using separate devices for entering data and displaying data. They had no local data processing capabilities. Even before the keyboard or display capabilities were card-punching systems with JCL (job control language), which was an early form of scripting language to give instructions to mainframe computers [1].

   The main takeaway of this era is the concept of multiple users sharing the same large machine in a backroom and each unaware of other users. At an abstract level, this is similar to Cloud Computing with users on thin clients connected to racks of servers in the backend data centers. INFOSEC relied entirely upon restriction of physical access to the computational machinery.

2. *Phase 2:* This was the era starting in the 1980s with personal computers (PCs), many of which were stand-alone or connected via slow modems [2] [3]. Each user interacted with a PC on a one-on-one basis, with a keyboard, a mouse, and a display terminal. All the storage, computing power, and memory were contained within a PC box. Any needed software was installed via floppy disks with limited storage capacity to run on the PCs. These PCs evolved in the early 1990s to laptops with integration of display, keyboard, mouse, and computing in a single unit. There was also an attempt in the early 1990s to create a network computer, which was diskless and connected to more powerful computers in the back end, but perhaps the idea was before its time as networks were still slow with 28.8 kbit/s dial-up modems [3]. A more practical solution in terms of client-server solution

© Springer Nature Switzerland AG 2020
N. K. Sehgal et al., *Cloud Computing with Security*,
https://doi.org/10.1007/978-3-030-24612-9_2

[4], which enabled remote devices with a little computing power to connect with servers over corporate Ethernet networks, became prevalent.

The main takeaway of this era was the birth of an online desktop to emulate the desk of working professionals. It represented multiple tasks that were simultaneously kept in an open state on the PC. With GUI and OS utilities, it was possible to create the notion of a desktop on a computer, to go from a single user-single job model, to single user-multiple jobs running simultaneously. This caused user interactions to move from command prompts to mouse-driven clicks.

3. *Phase 3:* In the mid-1990s, this was the era of Web browsers [5], which is a software application to retrieve, present, and traverse information resources on the World Wide Web (WWW). These came out of a research project, but became popular with everyday computer users to access information located on other computers and servers. Information often contained hyperlinks, clicking which enabled users to traverse to other Websites and locations on the WWW. These browsers needed full-fledged PCs to run on, which were more than dumb terminals so the evolution cycle was not yet complete.

   The main takeaway of this era was the birth of WWW, with PC forming a gateway for connecting users to the Internet. It used Internet infrastructure to connect with other devices through TCP/IP protocol, for accessing a large set of unreliable resources and get a reliable output.

4. *Phase 4:* New century heralded an era of full-fledged Internet browsing with PCs, and a mobility revolution with cell phones abound. It was inevitable that the twains shall meet launching innovative mobile applications running on cell phones. Cell phones created a yet another gateway to Cloud. This enabled users to book hotels, rent rooms, and buy goods on the move.

   This main takeaway of this era was the birth of mobile clients, similar to client-server model, except with limited compute power in small form factors. Thousands of powerful servers were located in large, far-flung data centers. It may be noted that this represented one revolution of spiral akin to mainframes, as depicted in Fig. 2.1.

5. *Phase 5:* When companies discovered that the population of the world limits number of smartphones and mobile devices they can sell, they started to look for new business opportunities. These came in the form of Internet of Things (IoT), which enables everyday common objects such as a television, a refrigerator, or even a light bulb to have an IP (Internet Protocol) address. This gives rise to new usage models, for example, to conserve energy, IoT objects can be remotely monitored and turned on or off. This also includes consumer services for transportation and a utilitarian phase for user interactions with appliances, leading to higher productivity and improved quality of life. Computing reach extended to a better-informed decision-making and into social relationships. Hence, the information security considerations have also became critical and will be briefly discussed in Sect. 2.16.

This era completes the virtual cycle of computing evolution, with many front-end dumb devices connected to powerful servers on the back end, as shown in Fig. 2.1.

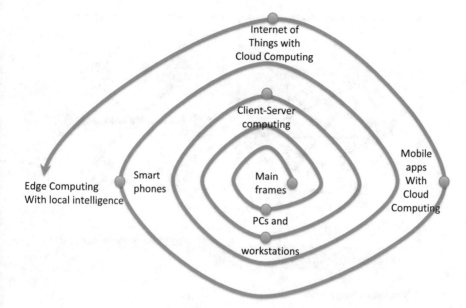

Fig. 2.1 Evolution of computing models in a spiral of growing computing needs

Another way to look at the computing transitions is by imagining a pendulum that oscillates between centralized servers on one end and client devices on the other. In phase 1, computing power was concentrated on mainframe side, while in the next phase, it shifted to the PC side. Such oscillations continued across different phases of the spiral shown in Fig. 2.1.

2.2   Different Network Protocols

Two models prevailed in the networking domain: peer to peer and client-server. In the former, each computer that wants to talk to another computer needs a dedicated line, so N machines will need N^2 connections. This while being fast was obviously not scalable; also when no data is being transferred, then network capacity is unutilized. Meanwhile, the client-server model with a central server supporting multiple clients was economical and scalable for the enterprises, as data and files could be easily shared between multiple users as shown in Fig. 2.2:

On networked systems with a client-server model, a computer system acting as a client makes request for a service, while another computer acting as a server provides a response when a request is received, as shown in Fig. 2.3. The greatest advantage of the networking model was its scalability, i.e., one could add any number of servers and clients. It offered an opportunity to create open-source software that offered interoperability. The network protocol was based on TCP/IP, and going forward this growth led to several interconnecting networks, eventually into the

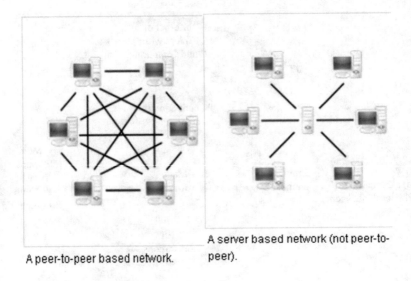

A peer-to-peer based network.   A server based network (not peer-to-peer).

Fig. 2.2 Two networking models

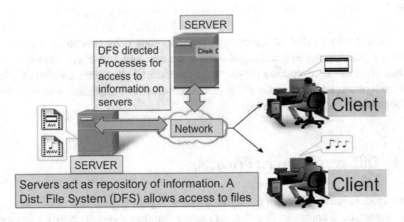

Fig. 2.3 The networking model

Internet using HTTP protocol. The protocol-based control meant that one could operate without a vendor lock-in, i.e., not be dependent on any single supplier. Networks within an enterprise grew as LAN (local area network) and WAN (wide area network). LAN connected computers in confined areas, such as within a single office building, or there could be several LANs, e.g., one for each floor. WAN connected computers spread over larger geographical areas, such as across the cities. Several LAN standards were developed over time, such as IEEE 802.2, Ethernet or IEEE 802.3, IEEE 802.3u for faster Ethernet, IEEE 802.3z and 802.3ab for Gigabit Ethernet, IEEE 802.5 for token rings, and IEEE 802.12 for 100VG on Any LAN. The Internet in turn is simply a net of networks, using TCP/IP suite of protocols. These were mostly packet switching protocols, with each packet having a structure offering a virtual connection, and no single dedicated path committed to maintain the connection between a source and its destination. Paths were temporarily created to maintain connections as needed between users and/or machines. Sometimes, time division multiplexing was used to share the same physical line between different pairs of users, and every pair was given a short time slot. As an example, if a time unit has ten slots, then ten pairs can communicate simultaneously, oblivious of the presence of others. Connections are made over a network of switched nodes, such that nodes are not concerned with content of data. The end devices on these networks are stations such as a computer terminal, a phone, or any other communicating device. Circuit switching operates in three phases, to establish a connection, transfer data, and then disconnect. Networking routers used an intelligent algorithm to determine optimal path that takes the least time or to minimize cost to establish a path of communication.

TCP/IP is a fault-tolerant protocol, so if a packet is lost or corrupted, then the connection is retried and packet is sent again. It was originally designed by DARPA (Defense Advanced Research Projects Agency) during the Cold War era to survive a nuclear attack. TCP/IP is a layered protocol where TCP provides reliability of a connection by attempting to send the same packet again if the previous transmission failed, while IP provides routability between the communicating nodes by finding an optimal path.

When computing needs exceed beyond what can be reasonably supported by a single server, an effort is made to share the workload among multiple servers. Note that in sharing a workload, the latencies on servers that are loosely connected shall be determined by TCP/IP suite of protocols. The concern often is that these

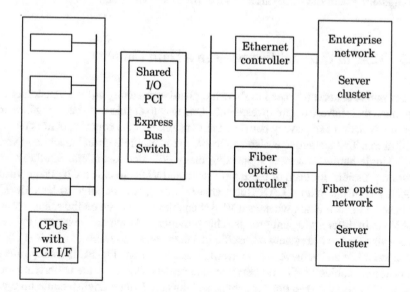

Fig. 2.4 Connecting multiple servers in a cluster

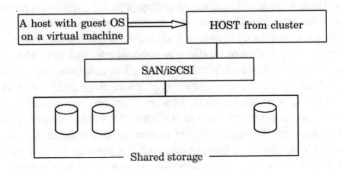

Fig. 2.5 Connecting multiple storage elements in a cluster

latencies would be in far excess of what may be acceptable. Therefore, an alternate approach is to use server clusters. The latency in tightly bound server clusters is far less than the networked servers. One of the techniques for clustering is by sharing I/O switches between the servers [6]. The switches connect individual class of controllers such as Ethernet controller in an enterprise network, as shown in Fig. 2.4.

Clustering ensures that large applications and services are available whenever customers and employees need them. This enables IT managers to achieve high availability and scalability for mission critical applications, such as corporate databases, email, Web bases services, and support external facing retail Websites. Clustering forms the backbone of enterprise IT's servers and storage elements. Clustering may operate with SAN (storage area network) or iSCSI (Internet Small Computer System Interface) architectures [6] as shown in Fig. 2.5.

2.3   Role of Internet Protocols in a Data Center

Many public data centers need to share their hardware among different customers, to optimize their infrastructure investments. One way to enable this sharing, while preserving isolation and privacy between their customers, data center operators use virtualization. This technique is also applicable to run multiple virtual machines (VMs) on a single hardware server, supporting many different customers. Similarly, networking sessions for each of these multi-tenanted VMs are isolated by using virtual LANs (VLANs). With a LAN (local area network), it is possible to have devices connected to each other, whereas a VLAN enables hosts to be on the same physical LAN yet in different logical groups. This provides isolation and security. However, it can also lead to heavy network traffic in a data center, so Fibre Channels (FC) are used to isolate and achieve high-performance interactions [7]. FC typically runs on optical fiber cables within and between data centers. However, the underlying technology can also run over copper cables, and thus the British English name fibre was

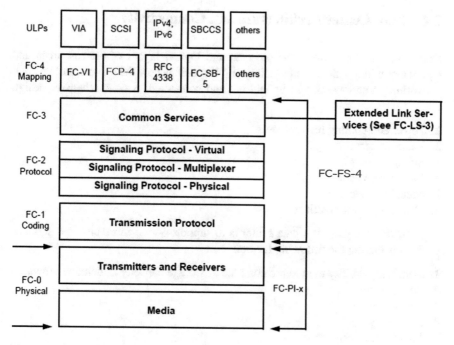

Fig. 2.6 Layers of Fibre Channel technology implementation

used to define the standard, to avoid the confusion with fiber optical cables [7]. Fibre Channel networks provide in-order and lossless delivery of raw block data.

Fibre Channel does not follow the OSI model layers and, as shown in Fig. 2.6, is split into five layers:

FC-4 – Protocol-mapping layer, in which upper-level protocols are grouped into Information Units (IUs) for delivery to FC-2.

FC-3 – Common services layer, a thin layer that implements functions like encryption or RAID redundancy algorithms; multiport connections.

FC-2 – Signaling Protocol consists of the low-level Fibre Channel protocols; port-to-port connections.

FC-1 – Transmission Protocol, which implements line coding of signals.

FC-0 – PHY includes cabling, connectors, etc.

The main goal of Fibre Channel is to create a storage area network (SAN) to connect servers to storage. A SAN refers to a dedicated network that enables multiple servers to access data from one or more storage devices. Such devices may be flash memories, disk arrays, tape libraries, and other backup media. SANs are often designed with dual fabrics to increase fault tolerance.

2.4   Data Center Architecture and Connectivity

Data center architecture is the physical and logical layout of the resources and equipment within a data center facility [8]. It specifies how the servers, storage, and networking components will be laid out and connected. A comprehensive design takes into account:

1. Power and cooling requirements
2. Racks of servers
3. Storage arrays
4. Network switches
5. Security aspects
6. IT management practices

Typically, the goal of a data center is to support one or more business services while considering the following factors:

1. Scalability: ability to increase compute or storage capacity as demand grows
2. Performance
3. Flexibility
4. Resiliency
5. Maintenance aspects

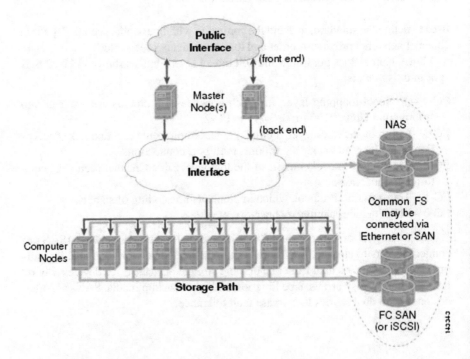

Fig. 2.7 Logical view of a server cluster [8]

Servers are often clustered together to handle many jobs at ones. Storage elements, such as network-attached storage (NAS), are grouped to house a file system (FS), which connects to the servers via an Ethernet or storage area network (SAN) protocol, as shown in Fig. 2.7.

2.5 Evolution of Enterprise IT

Enterprise computing refers to business-oriented information technology that is critical to a company's day-to-day operations. This includes various types of activities, such as product engineering, accounting, human resources, and customer services using software for database management and other service specific modules. One key aspect of enterprise computing is its high availability, as crash of a hardware or software component can lead to loss of revenue and customer base. Thus multiple redundant computers exist for mission critical applications, as well as regular data backups are taken to ensure that there is no single point of failure within an enterprise.

IT has evolved over the decades to closely align with business units (BUs) within a large corporation. A BU is typically responsible for providing a product or service to the external world, whereas the IT department has mostly internal interfaces to ensure a smooth flow of information and data. IT is the lifeblood of a successful enterprise. Any problems in IT can bring a corporation to halt, such as email disruptions, but worse yet security vulnerability in an IT system can cause loss of valuable information. As an example, Target's security and payment system was hacked, leading to loss of Target's customers' information [9]. Hackers came through the HVAC (heating, ventilation, and air conditioning) system and installed malware, which targeted "40 million credit card numbers—and 70 million addresses, phone numbers, and other pieces of personal information." About 6 months before this happened, Target invested $1.6 million to install a malware detection tool made by FireEye; their security product is also used by the CIA. FireEye's tool is reasonably reliable software and it spotted the malware, but it was not stopped at any level in Target's security department. Target's unresponsiveness to the alerts resulted in the exposure of confidential information of one in three US consumers.

Much less dramatic examples of other IT failures include lack of scalability. For example, in August 2016, a computer outage at the Delta Airlines headquarters in Atlanta prompted the airline to cancel about 2300 flights, delaying the journeys of hundreds of thousands of passengers and prompting three days of chaos [10]. Delta executives said a malfunction in an aging piece of equipment at its data center had caused a fire that knocked out its primary and backup systems. The carrier said the system failure reduced revenue by $100 million in the month of August 2016.

Examples abound in other key economic sectors also. At the NYSE (New York Stock Exchange), trading was suspended for nearly 4 hours on July 8, 2015, after a botched technology upgrade, freezing one of the world's biggest financial markets. The NYSE blamed a technical "configuration problem" that was subsequently fixed [11].

One way to deal with an unavoidable hardware or software failure is the rise of distributed computing. In such a system, components are located on networked computers and interact with each other to achieve a common goal. The goal is to provide service with an acceptable level of reliability and timeliness. This became possible with the advent of service-oriented architecture in enterprises, which we will study in a subsequent section.

2.6 Evolution of Web Services

A Web service is an interface described by some form of service from a remote provider. These are based on "open" and "standard" technologies, as HTTP, HTTPS, XML, REST, etc. These evolved from client-server and distributed computing concepts, to offer a "Web of Services," such that distributed applications can be assembled from a Web of software services in the same way that Websites are assembled from a Web of HTML pages [12].

The advent of the Internet is based on a set of open standards, some of which are:

1. *TCP/IP:* Transmission Control Protocol/Internet Protocol, for network applications to exchange data packets in real time. Data is packed in byte packages, ranging up to 64 K (65,535 bytes). These are sent and acknowledged and, if not acknowledged, then sent again with multiple retries, until each packet arrives at the destination. These packets are then reassembled to create a complete copy of the information content in whole.
2. *RPC:* Remote Procedure Call allows functions written in C, Java, or any other procedural languages to involve each other across a network, allowing software services to reach other servers across the network.
3. *HTTP:* Hypertext Transport Protocol, for sharing data between machines based on top of TCP/IP protocol.
4. *HTML:* Hypertext Markup Language, the format for representing data in a browser-friendly manner.
5. *XML:* It stands for Extensible Markup Language. It enables any data to be represented in a simple and portable way. Users can define their own customized markup language, to display documents on the Internet.
6. *SOAP:* Service-Oriented Architecture Protocol, for connecting computers. It allows message passing between endpoints and may be used for RPC or document transfer. These messages are represented using XML and can be sent over a transport layer, e.g., HTTP or SMTP. An example of communication using SOAP and XML is shown in Fig. 2.8.
7. *UDDI:* Universal Description, Discovery, and Integration is an XML-based registry for businesses to list themselves on the Internet. It can streamline online transactions by enabling companies to find one another on the Web and make their systems interoperable for e-commerce.

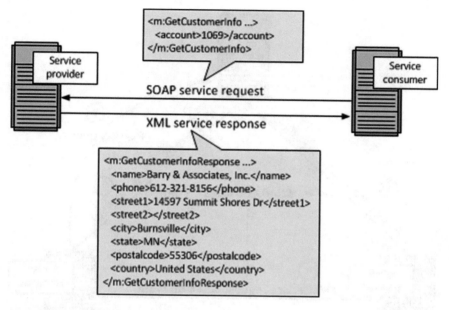

Fig. 2.8 Interacting using SOAP and XML [13]

8. *WSDL:* Web Services Description Language is used to describe a Web service in a document. It uses an XML format for describing network services as a set of end points operating on messages. These can contain either document-oriented or procedure-oriented information. The operations and messages are described abstractly and then bound to a concrete network protocol and message format to define an end point. The Web Services Description Language (WSDL) forms the basis for the original Web Services specification. Figure 2.9 illustrates the use of WSDL. At the left is a service provider. At the right is a service consumer. The steps involved in providing and consuming a service are described below:

   I. A service provider describes its services using WSDL. This definition is published to a repository of services. The repository could use Universal Description, Discovery, and Integration (UDDI) as a way to publish and discover information about Web services. UDDI is a SOAP-based protocol that defines how other UDDI clients communicate with registries. Other forms of directories could also be used.
   II. A service consumer issues one or more queries to the repository to locate a service and determine how to communicate with that service.
   III. Part of the WSDL provided by the service provider is passed to the service consumer. This tells the service consumer what the requests and responses are for the service provider.
   IV. The service consumer uses the WSDL to send a request to the service provider.
   V. The service provider provides the expected response to the service consumer.

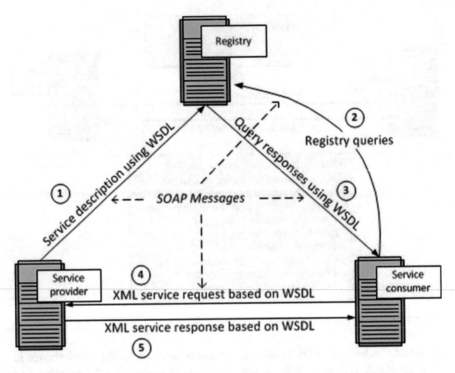

Fig. 2.9 An example of using WDSL [13]

9. *REST:* Representational State Transfer is a protocol used to create and communicate with the Web services. REST is language independent. Developers prefer
   REST due to a simpler style that makes it easier to use than SOAP. It is less
   verbose so less data wrappers are sent when communicating. An interaction is
   illustrated in Fig. 2.10.
10. *JSON:* JavaScript Object Notation uses a subset of JavaScript. An example is
   shown in Fig. 2.11. It uses name/value pairs and is similar to tags used by
   XML. Also, like XML, JSON provides resilience to changes and avoids the brit-
   tleness of fixed record formats. These pairs do not need to be any specific order.
11. *DCB:* Datacenter Bridging (DCB) is a set of enhancements [14] to the Ethernet
   protocol for use with clustering and storage area networks. Ethernet was originally designed to be a best-effort-based network, which can experience packet
   loss when the network or devices are preoccupied. TCP/IP adds end-to-end reliability to Ethernet, but lacks the finer granularity to control the bandwidth allocation. This is especially required with a move to 10 Gbit/sec and even faster
   transmission rates, as such a network pipe can't be utilized fully by TCP/IP. DCB
   eliminates loss due to queue overflows (hence, called lossless Ethernet). It also
   includes a Priority-based Flow Control (PFC), an IEEE 802.1 standard that provides a link-level control mechanism.

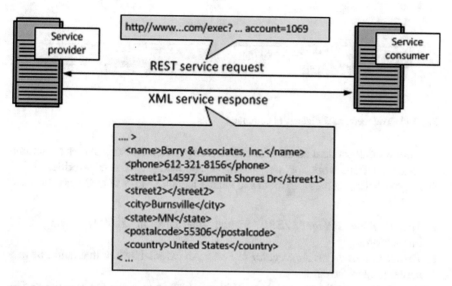

Fig. 2.10 Interaction of two computers using REST [13]

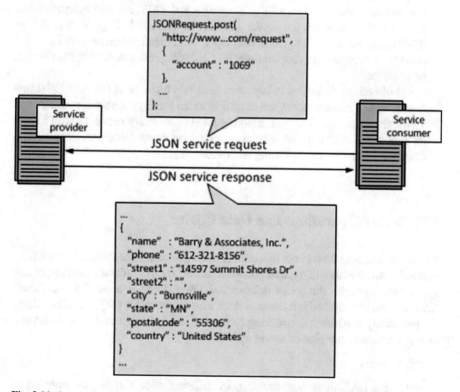

Fig. 2.11 Interaction of two computers using JSON [13]

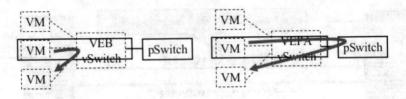

Fig. 2.12 Two methods for inter-VM communications

It allows independent control for each frame priority. It also provides a common management framework for assignment of bandwidth to frame priorities. IEEE 802.1Qbg-2012 specifies two modes to handle local VM-to-VM traffic, as shown in Fig. 2.12:

1. *Virtual Edge Bridge (VEB):* Switch internally in a VMM using CPU instructions.
2. *Virtual Ethernet Port Aggregator (VEPA):* An external switch that could be in a network interface card.

VEPA relays all traffic to an external bridge, which in turn can forward this traffic in a mode, known as "Hairpin Mode." This allows visibility of VM-to-VM traffic for policy enforcement. This also gives better performance than a simple vSwitch that puts additional load on a CPU. Both VEB and VEPA can be implemented on the same NIC in the same server and can be cascaded. Edge Virtual Bridge (EVB) management can be done with Open Virtualization Format (OVF), which provides port profiles to support resource allocation, resource capability, vSwitch profiles, etc.

The goal of both mechanisms is to ensure zero loss under congestion in DCB networks. These enhancements are crucial to make Ethernet viable for storage and server cluster traffic. However, enabling DCB on arbitrary networks with irregular topologies and without special routing may cause large buffering delays, deadlocks, head of line blocking, and overall unfairness.

2.7   Server Operations in a Data Center

The server stores or has access to a database, for which clients initiate queries from terminals. As an example, type of software called "DBMS" (database management system) manages the storage of information, often stored as a set of tables (files). Access to retrieve desired information is via a query processor (SQL queries). Using the previously explained networking technology and client-server model, the following are some examples of server operations:

- Print servers
  - Over a LAN or remote printing, by different clients to one or more set of printers

- FTP (File Transfer Protocols):
  - Client may download or upload.
  - 7-bit ASCII character set.
  - Text or images.
  - Commands and responses.

- DNS (Domain Name Server): for resolving IP addresses
- Mail: below are various protocol processing services offered by mail servers:
  - SMTP
  - POP
  - IMAP
  - POP3

- Media
  - Streaming and buffering

- Security servers: for issuing and maintaining encryption keys, digital certificates, user transaction, monitoring logs, etc.

Development of GUI-based browsers using HTTP led to an infrastructure that we currently call World Wide Web (WWW). This led to Web services and various applications, such as search engines, e-commerce, banking apps, etc.

To support many servers in one place, in turn serving many users spread across a large area, one needs a lot of servers, routers, and switches. These server clusters are also called data centers (DCs). The sheer density of equipment means the DCs need special cooling equipment to dissipate the heat generated by the equipment. Typically, $1 spent to power a server also needs another $1 to cool it. A data center can be hierarchically organized, as shown in Fig. 2.13.

Note that different layering of switches is done for efficiency reasons. At layer 2, one single tree spans the access of local server racks to prevent looping and ignores alternate paths. At layer 3, aggregation is done with shortest path routing between

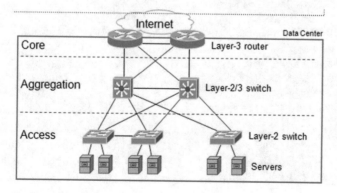

Fig. 2.13 Organization of a data center

source and destination, with best-effort delivery of packets. A failure at this level can be catastrophic so backups are provided with the appropriate trade-off between cost and provisioning services.

These servers need to have massive storage, connectivity, and traffic management capability using switches. Of these, storage is a key capability required to provide large amounts of data needed by remote users, for which SNIA (Storage Networking Industry Association) has recommended storage architecture, as depicted in Fig. 2.14. It consists of a layered architecture, starting from the bottom going upward as:

1. Storage devices
2. Block aggregation layer with three functional placements:

    (a) Device
    (b) Network
    (c) Host

3. File/record layer

    (a) File system
    (b) Database

4. Applications at the top to use the storage

Key advantages of the SNIA-proposed architecture include scalability and 24 x 7 connectivity. The former is needed to support easy addition of storage volumes as data needs to grow, while the latter is needed to provide reliable, distributed data sharing with high performance and no single point of failure.

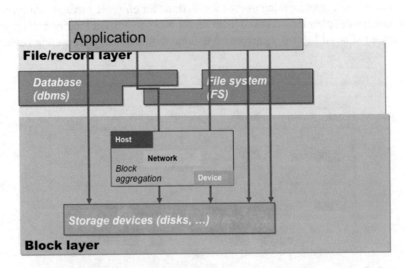

Fig. 2.14 SNIA recommended storage architecture for large data centers

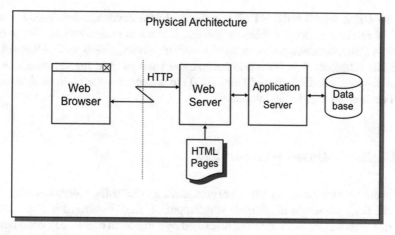

Fig. 2.15 A representation of how clients can communicate with servers

Web services are provisioned from a client's browser to access applications and data resident on remote servers in a remote data center. Figure 2.15 represents this linkage with the following architecture, often using thin clients that:

- Offer reduced energy consumption.
- May not have any native SW application, other than a browser.
- May not support a file system.
- Hence no SW distribution licenses are required on the client side.
- Applications running on the server side.

Cross-platform compatibility is provided via browsers across OSes' such as Linux and Windows with:

- Support for standard networking protocols
- HTTP (for browser communications)
- HTML (for rendering)

Clients select the type of information needed, access it remotely on servers, and then display it locally. The net result of this setup is that at an airport, one can get flight info, such as timing, reservation status, etc., using a smartphone application or browser. Similarly, terminals at public libraries allow patrons to browse services offered by the library. This is often done with thin clients, which are minimally networked computers using:

- Thin Client Java Viewer
- Browser
- Embedded ActiveX Control

However, a thin client has several limitations such as reduced computing, so a server will need to validate the inputs, slowing the server down. Also, with no native software, client-side GUI design limits most server interactions to be of pull nature and displays the incoming data as is, unable to do any computations or processing on the client side. Lastly, HTTP is a packet-based connectionless protocol with limited reliability. Despite these limitations, thin client usages abound due to their cheaper costs, such as Chrome netbooks, mobile phones, tablets, etc. As we will see in a later chapter, these have also became the basis of even thinner clients in the form of sensors and Internet of Things (IoT) to connect regular household or industrial devices directly to the Internet.

2.8 Server-Based Web Services

An agent running on server hardware provides a service using a Web-based app and fits the basic paradigm of client-server computing, as re-illustrated in Fig. 2.16. In short, a Web service is an interface described by some form of service from a remote provider. The actual phrase, "Web Services," came from Microsoft in the year 2000 and was adopted by W3C. The idea was to offer services using "open" and "standard" technologies.

Various service components can be bound together to offer an integrated service, an example of which is a vacation planner. Given a date, desired location, and hotel requirements, the Web agent can search and come back with options for transportation and hotel arrangements. To enable this various airlines and hotels from different areas can register with the vacation planner, to offer an integrated services package.

A Web app is the software aimed at an end-user and can be used from a remote client, providing a dedicated program-to-program communications-based service. It operates with formatted messages and instructions. Examples of Web applications include Internet banking or the Uber service app. These allow users to communicate from different operating systems such as Windows or Linux, using different devices such as a personal computer or a smartphone. The open standards include HTTP or HTTPS for transportation of HTML or XML messages. In these scenarios, the client side is running a Web browser, such as Chrome, whereas the server side is running a Web server such as Apache, enabling multiple users to interact with one or more servers in a remote data center. The size of the attack target in a Web service is quite large, as the attacker can compromise the browser running on a client device, do a passive or active attack on the network in-between, or target the server in a data center. We will examine each of these in the next section.

Fig. 2.16 Web services based on a client-server paradigm

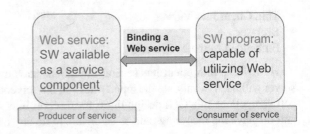

2.9   Evolution of Service-Oriented Architecture

A service-oriented architecture (SOA) is a style of software design where services are provided to the other components by application components, through a communication protocol over a network [15]. SOA is composed of loosely coupled set of services with well-defined interfaces. These services communicate with each other and are used for building higher-level applications. The basic principles of service-oriented architecture are independent of vendors, products, and technologies. A service is a discrete unit of functionality that can be accessed remotely and acted upon and updated independently, such as retrieving a credit card statement online.

A service has four properties according to one of the many definitions of SOA:

1. It logically represents a business activity with a specified outcome.
2. It is self-contained.
3. It is a black box for its consumers.
4. It may consist of other underlying services.

Different services can be used in conjunction to provide the functionality of a large software application. So far, the definition could be a definition of modular programming in the 1970s. Service-oriented architecture is about how to compose an application by integration of distributed, separately maintained, and deployed software components, as shown in Fig. 2.17. It is enabled by technologies and standards that make it easier for components to communicate and cooperate over a network.

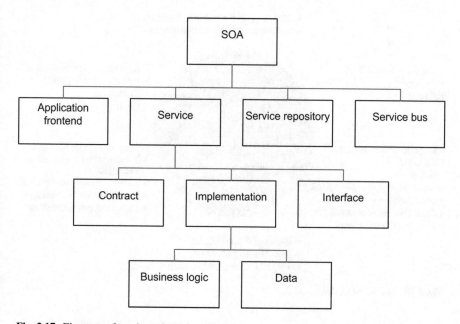

Fig. 2.17 Elements of service-oriented architecture

Thus, SOA helps to organize discrete functions contained in enterprise applications into interoperable, standards-based services that can be combined and reused quickly to meet business needs [16]. By organizing enterprise IT around services instead of around applications, SOA provides the following key benefits:
- Improves productivity, agility, and speed for both business and IT
- Enables IT to deliver services faster and align closer with business
- Allows the business to respond quicker and deliver optimal user experience

Six SOA domains are depicted in Fig. 2.18 and enumerated below:

1. Business strategy and process: By linking an organization's business strategy with IT management and measurement, SOA enables continuous process improvements.
2. Architecture: By providing an IT environment based on standards, SOA enables integration of functionality at enterprise level.
3. Building blocks: By reusing implementation and core infrastructure, SOA provides consistency and repeatability of IT's success.
4. Projects and applications: By cataloging and categorizing functionality of applications and systems across an enterprise, SOA drives out redundancy and promotes consistency in business execution.
5. Organization and governance: By standardizing delivery of IT services, SOA ensures maximal reuse of any developed functionality.
6. Costs and benefits: By planning and reusing functionality, SOA ensures that existing IT investments are leveraged for a sustainable enterprise value.

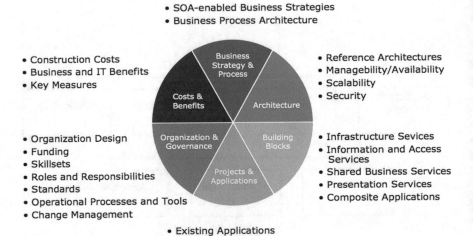

Fig. 2.18 The six SOA domains [15]

Business objectives, such as a policy for retail banking, financing, individual loans, etc., are achieved by SoA regardless of technology and infrastructure. The specific solution architecture may include:

1. Retail banking solution
2. Financing solution
3. Personal loans management system

Technical architecture for SoA would cover data abstraction, messaging services, process orchestration, event management or monitoring, etc. A service comprises of a stand-alone unit of functionality available only via a formally defined interface. A mature rollout of SOA effectively defines the API of an organization. Reasons for treating the implementation of services as separate projects from larger projects include:

1. Separation promotes the decoupling of services from consuming projects.
2. Encourages good design as the service is designed without knowing the consumers.
3. Agility fosters business innovations and speeds up time-to-market.
4. Documentation and test artifacts of the service are not embedded within the detail of the larger project.
5. Enables reuse of components later on.

SOA also promises to simplify testing indirectly. Services are autonomous, stateless, with fully documented interfaces, and separate from the concerns of the implementation. A full set of regression tests, scripts, data, and responses is also captured for the service. The service can be tested as a "black box" using existing stubs corresponding to the services it calls. Each interface is fully documented with its own full set of regression test documentation. It becomes simple to identify problems in test services.

However, SOA suffers from drawbacks, as "state-full" services (i.e., where the memory state of previous transactions needs to be preserved) require both the consumer and the provider to share the same consumer-specific context. This could reduce the overall scalability of the service provider if the service provider needs to retain the shared context for each consumer. It also increases the coupling between a service provider and a consumer and makes switching service providers more difficult.

2.10   Transition from SOA to Cloud Computing

Any action tends to produce an equal and opposite reaction. SOA has not been immune to this universal law, and there has some resistance in the enterprises to adoption of SOA. However, customers of IT have been demanding faster services, cost cutting, and more control on their procurement processes. For example, just a decade ago, it took weeks or even months in an organization to acquire and deploy a new server or storage capacity. This hindered the timeliness of business, which led to automation of ordering and deployment processes. Furthermore, as enterprises discovered unused capacity due to the inevitable up and down nature of business cycles, there was a desire to share the unused resources and monetize them through other users within the organization. SOA accelerates and supports Cloud Computing through the following vectors [17]:

1. Faster application development
2. Reuse of services
3. Reduction of application maintenance
4. Reduced integration costs
5. Support of application portfolio consolidation

The above benefits have led many organizations to transition faster into Cloud Computing by leveraging SOA governance disciplines, shared infrastructure service, shared data services, and well-defined and layered enterprise architectures. A case in point is Amazon, which first used its IT infrastructure to manage and sell books, then expanded into selling into other consumer merchandise, and finally evolved into renting compute capacity via Amazon Web Services (AWS). In the early 2000s, Amazon was expanding rapidly as an e-commerce company, struggling with scaling problems. It required the company to set up internal systems to cope up with its hyper growth, which led to the creation of a new strategy that laid the foundations for AWS. This included a set of well-documented APIs that third-party merchants could use to sell books on Amazon's Website. Its processes were decoupled and applications were built to handle bursts of high traffic. Amazon's internal teams shared a set of common infrastructure services that different applications could access, such as compute and storage components. Since the e-commerce load varies over the year, especially high during the shopping season from Thanksgiving to Christmas holidays, the underlying infrastructure was not always fully utilized. That led Amazon to launch AWS as its Cloud infrastructure services on a rental basis to public in August 2006. Increasing revenues and profits attracted many other competitors, such as Google and Microsoft; however, Amazon still leads with the highest Public Cloud market share. We shall examine AWS Public Cloud capabilities in more details in the later chapters.

2.11   Building an Enterprise SOA Solution

Applications built using SOA style can deliver functionality as services. These can be used or reused when building applications or integrating services within the enterprise or trading partners. These services are the building blocks of business flows, as shown in Fig. 2.19. A service may be simple, such as "get me a person's address," or complex as "process a check and make the payment," etc. It ensures that business function is executed consistently and within the quality of service parameters, to return predictable results.

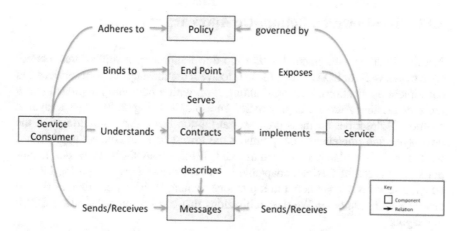

Fig. 2.19 An SOA service provisioning with contract

SOA allows integration by using open standards, which ensure compatibility across enterprise boundaries. Consumer of data is expected to provide only the stated data on the interface definition, and service handles all processing, including exceptions, e.g., if an account has insufficient funds to pay in exchange of a check.

SOA services are stateless, e.g., they do not maintain any memory between invocations. Each service takes the input parameters, executes a previously identified process, and returns the results. If a transaction is involved and executes properly, then data is saved to the database (e.g., checking amount is updated after a payment). Users of the service do not need to worry about the implementation details for accessing the service. This is all done with loosely coupled building blocks with well-defined integration points, such that any changes can be made to evolve the service. An example is to allow ATM and electronic transactions on a bank account in addition to paper check processing. SOA service is carried from end-to-end (E2E), and no context is maintained between different service invocations. Rollbacks are implemented if a particular transaction fails, e.g., once a check is deposited in an ATM, it is processed through the depositor's account, and a message is sent to the originator's bank. Even if a temporary credit is given to the depositor, most of it is on hold for a few days until the check clears. If the check writer doesn't have sufficient funds, then the check is returned and any temporary credit is rolled back. Similarly, a withdrawal process from an ATM machine first checks if the account holder has sufficient funds, deducts the amount, and then dispenses the cash. If for any inventory or mechanical reasons, cash cannot be given out, then the transaction rolls back and the account is credited resulting in no withdrawal. This ensures that each transaction is complete and all possible failure conditions are accounted for, leaving the database (in this case bank account) in a consistent and correct integrity state for the next transaction to take place.

2.12   Top-Down vs. Bottom-Up Approach

Just like buildings are planned with a top-down approach, a similar architectural style works well for SOA. The purpose is to avoid redundancy between services and constituent applications, although mixing it up with a bottom-up approach and a reuse of existing structure is pragmatic. An example in Fig. 2.20 shows a layered solution. It describes business processes at the top layer, service interfaces at the next layer, and implementation points at the lowest layer to connect applications written in different languages, such as .NET, J2EE, legacy C language, etc. It supports automation of flexible, adaptable business processes by composing loosely coupled services. As we noted in a previous section, Web service frameworks are evolving with numerous WS-∗ specifications that can be composed using SOAP messaging.

This multilayered architecture demonstrates the decoupling of an end-user on a front-end terminal from the backend servers in a Cloud. The presentation layer in the Cloud formulates and sends HTML messages to a browser on a client device, often without knowing the type or make of the client machine. This allows the browser to update only specific frames, such as weather information or a share's price, while keeping other display parts constant such as the name of the cities or number of shares, etc. It minimizes the traffic across distant Internet connection while speeding up the necessary communications. On the Cloud side, the motivation to decouple presentation and business layers is to simplify the task of rendering and

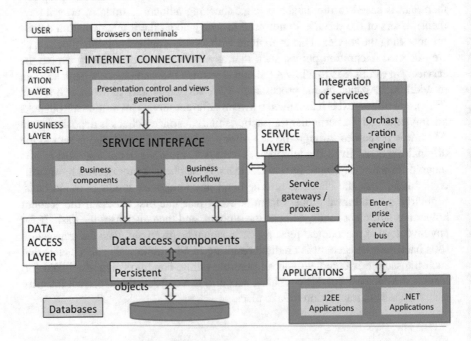

Fig. 2.20 A multilayered reference model for an enterprise

serving information, from the rules of who can access which kind of information. An example is need-to-know-based access, e.g., for medical records. A patient can read their own records but not modify them, whereas a caregiver can add but not delete any notes, while a doctor may be able to edit the medical records to alter the medical treatment direction. Data is often kept on a different server, with provisions to create backups and for access by multiple users at the same time. This is useful for serving content of common interest from one to many, such as news and movies. It enables scalability when the number of users increases by adding more servers in the business logic or database layers. SOA defines the tasks for each layer and protocols between them as indicated by bi-directional busses.

A top-down design starts with inventory analysis. Just like for a house, the blueprint SOA is established. An SOA design can cross enterprise and Internet boundaries to provide compounded services, supported by well-defined service-level agreements (SLAs) between the organizations. The next step is the design of Service Contracts, which includes inputs and outputs between different entities involved. It is followed by Service Logic, which specifies tasks performed by each individual service block. Then each service is further developed, and it may be composed of sub-services. An example is shown in Fig. 2.21, which describes orchestration of custom services for a travel agency's portal with multiple service providers and consumers.

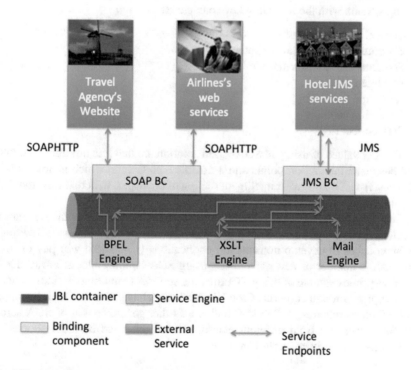

Fig. 2.21 A travel agency conceptual architecture

It integrates services [15] from airlines and hotels, offering a one-stop shop to make travel and staying arrangements for a business or vacation trip. Input-output specification for a service block allows it to be tested in a stand-alone manner before integration in the SOA flow. After each block has been confirmed to work per its specification, the service can be deployed. Lastly, Service governance is performed to ensure that overall SOA delivery is happening per the contractual terms. Complex business processes are defined using a Business Process Execution Language (BPEL) in a portable XML format. This describes how services interact to form complex business process, fault handling, and transaction rollback. A top-down architectural approach reduces the overall burden of subsequent service governance, because services were modeled as a part of an inventory.

2.13 Enterprise Service Bus (ESB)

In its simplest form, an ESB delivers a message from one point to another. It is middleware for connecting heterogeneous services and orchestrating them. This is done with loose coupling, a location transparency in a transport neutral manner. The messages can be XML document, sent using SOAP, as shown in Fig. 2.22.

ESB can also be thought of as an abstraction layer on top of an enterprise messaging system, with the following key characteristics:

- Streamlines development
- Supports multiple binding strategies
- Performs data transformation
- Intelligent routing
- Real-time monitoring
- Exception handling
- Service security

The key values of using an SOA-based solution are that it is not tightly coupled and has clean integration points and a flexible architecture, which is amenable to changes. An ESB allows for intelligent routing of messages, with real-time monitoring and service security.

Benefits of SOA initiatives should be measured over years rather than months or quarters. As shown in Fig. 2.23, initial investment may be higher than the traditional IT approaches to develop nonstandard applications [15]. These will pay off over time as the number of new capabilities using SOA building blocks grows due to reuse. A prime example of this is IT providing services to multiple business units in an enterprise through corporate Cloud, instead of each business unit maintaining its own IT infrastructures. A Public Cloud is a further generalization of SOA across various consumers from different enterprises and unrelated business entities, as long as their security concerns can be met.

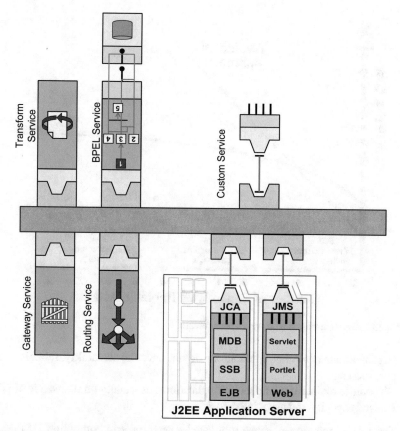

Fig. 2.22 An Enterprise Service Bus with compounded applications

2.14   Enterprise Implementation on Private Cloud

Increased adoption of client-server architecture and Internet computing infrastructure has resulted in a large number of servers scattered across different divisions of an enterprise. However, IT experts soon learnt that each organization works in a silo and sharing of servers across groups was minimal. Furthermore, due to time zone differences, it is possible to share the hardware and software licenses across different divisions. It is further enabled by the use of virtual machines. Hence, the idea of creating an enterprise-wide data center, where jobs and tasks from different divisions can be scheduled to run, seemed very attractive for cost savings and run-time efficiencies as powerful servers can be purchased with pooled resources.

The need for a Private Cloud, which is a virtual data center backed by one or more physical data centers, was driven by the following factors:

- Need for rapid scaling of applications when additional capacity is required.
- Inability to predict capacity and performance trends.

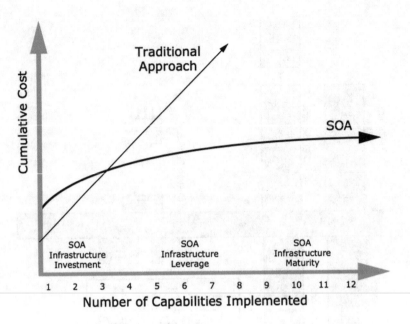

Fig. 2.23 SOA vs. traditional delivery cost structure

- Organizations grow, but applications do not have the scalability, reliability, and performance.
- Difficult to maintain client/server applications, especially on thousands of PCs.
- Unused capacity and associated costs.

The factors mentioned above motivated growth of grid computing [18] during the early 2000s, which has been largely subsumed by Cloud Computing in subsequent years. With grid computing, several inexpensive servers are combined to function as a large server. Thus IT departments no longer had to acquire large and expensive servers to run existing or future workloads. Moreover, new capacity could be added to existing infrastructure by simply adding new servers and systems. These are generally stacked in racks of servers, and at the top of rack is a networking switch to monitor the loading of each server and load it appropriately. Each department contributes to the purchase and maintenance of servers, and run-time costs of these servers, based on the extent of its usage. This is a basic tenant of Cloud Computing, except that the cost has to be paid upfront for IT department to buy the servers. Any financial adjustments are made during the lifetime of that server in terms of credits and debits. Furthermore, enterprise customers do not know exactly which servers their jobs are running on and have no direct control of these machines. However, they trust the IT managers who are the employees of the same enterprise to run their jobs in Private Cloud [19].

2.15   Enterprise Implementation on Hybrid Cloud

A general contrast between Amazon's AWS and Microsoft's Azure has been latter's ability to tap into enterprise segment. IT managers tend to be conservative and want to maintain a semblance of control on the servers used by their corporate customers. Thus they are reluctant to adopt Public Cloud and migrate their corporate data into unknown server locations. This gave rise to computing environments, which are a mix of on-premises, Private Cloud and third-party, Public Cloud servers with orchestration between the two sets of platforms.

A Hybrid Cloud [20] enables enterprise IT managers to combine one or more Public Cloud providers such as AWS and Google Cloud Platform (GCP), with a Private Cloud platform. Using encrypted connection technology, data and application can be ported between private and Public Cloud components. One of the usage models is Cloud bursting, i.e., when a number of enterprise jobs exceed the Private Cloud capacity, some of these are seamlessly migrated to an external Public Cloud server. Later on, if and when the internal demand for computing decreases, external jobs can be brought back in to run on the internal servers. This in summary is how Cloud bursting works.

Microsoft has further extended this model with an Azure Stack [21] that enables IT managers to deliver Azure services from an enterprise data center. Azure Stack is an integrated system, ranging in size from 4 to 12 server nodes deployed on the premise, and jointly supported by a hardware partner and Microsoft. Azure Stack uses Public Cloud for any bursting needs as well as backup of production workloads to assure 24x7 business continuity. Additional use cases are still evolving.

2.16   Information Security Basic Concepts

The previously defined Cloud Computing business models and implementation architectures have extended access to a wide variety of capabilities. Consequently, its security needs have also extended beyond the basic information security issues. However, basic security concepts still apply. Information security or INFOSEC begins with access control. Access control includes several abilities including issuing control commands, sending information, reading information, and physical access to locations or machinery. Access control is based upon identity authentication. The entity to be identified can be a person, a device, or a computational process. There are four basic factors of identity authentication: information, physical item, biological or physical characteristics, and location. These factors involve answering the four key questions of identity authentication:

1. What you have?
2. What you know?
3. What you are?
4. Where you are?

Examples of the information factor are username and password, birthdate, or mother's maiden name. Examples of the physical item factor include car key, credit card, debit card, or employee badge. What you are as a person is called biometrics. Examples of what you are include fingerprints, blood, DNA, eye scans, and voice patterns. Examples of location metrics are GPS coordinates, city or state, current time, current temperature, and altitude. The most common form of access control is based upon the single factor of what you know, specifically username and password. Another common form of authentication is based upon the single factor of what you have (a credit card) or sometimes what you know (your credit card number). A common two-factor authentication (2FA) is a debit card with a pin. The card is the physical device you have, and the pin is the information you know. In general, access control is improved with added factors for authentication, that is, by using multi-factor authentication (MFA). There are many trade-offs for identity authentication and access control. The trade-offs include level of security, speed of performance, usability, and acceptability of method.

The next category of security for Cloud Computing is protecting information both during transmission and during storage. Protection of information includes keeping secrets and private data away from unauthorized entities, preventing changes by unauthorized entities, and detection of attempts at tampering with the data. Separate from security is the detection of errors due to transmission noise or equipment problems. While the methods for this are related to security methods, they are not sufficient in the presence of malicious participants. The primary basic technique for preventing unauthorized reading of data is encryption. The originator (person A) encrypts the data using a key to convert the original plaintext to ciphertext. The ciphertext is then stored or transmitted. The recipient or reader (person B) uses his (her) key to decrypt the ciphertext to obtain the original information. An unauthorized person (person E) only sees the ciphertext, and upon this idea rests the secrecy of the message. There are two basic classes of encryption, symmetric encryption and asymmetric encryption. In symmetric encryption the same key is used to encrypt and decrypt the message. For asymmetric encryption a different key is used to encrypt a message from the key to decrypt the ciphertext. Symmetric encryption is also called private key encryption because the key must be kept secret. Asymmetric encryption is also called public key encryption because the encryption key can be made public and only the decryption key needs to be kept secure for the message to be secure. Whereas encryption hides information, hashing is used to detect data tampering. Specifically, changes in information (such as the terms of a contract) must be reliably detected. This involves creating a digest of the data that can be used to check for tampering. Calculating the digest for the tampered data can be compared to the digest of the original data to detect (but not identify) changed data. The common information protection techniques in today's Cloud environment and the Internet are Advanced Encryption Standard (AES) for symmetric encryption, RSA (Rivest-Shamir-Adleman) for asymmetric encryption, and SHA-2 (Secure Hash Algorithm 2) or SHA-3 (Secure Hash Algorithm 3) for information hashing.

While access control and information protection are required for preventing security breaches, some security attacks will occur. The detection of positional attacks and an appropriate response mechanism is required. For example, a person trying to login and repeated getting the password wrong is suspicious. Hence many systems limit the number of failed attempts (detection of an attack) and then close the login access (response to suspected attack). This is a straightforward approach for access control. However, most attacks are of type denial-of-service (DOS). The goal is not about getting or changing information by accessing the resources, but about preventing others from utilizing the resources. Here a device or several devices repeatedly and rapidly attempt to perform a normally permitted activity at such a volume and rate with a view to prevent other entities from accessing the resources. A common example is many rapid Website inquires resulting in the server crashing. The trade-off here is that techniques for the detection of malicious activity slow down the performance of legitimate normal activities.

The fundamental information security concepts will be described in more detail in later chapters, especially as they apply to Cloud Computing. Information security traditionally identifies the security boundary, and that leads to identifying potential security attack scenarios. The systems are then designed to defend against those attacks. The first thing to note is that in many information security breaches, the problem was not the theoretical security of its system, but the implementation. The second thing to note is that with Cloud Computing, the identification of the security boundary is difficult and always changing.

2.17 An Example of a Security Attack

In a previous section, we mentioned how DNS is used to resolve IP addresses. It allows users to type a human-readable address, such as www.cnn.com for the Cable News Network (CNN) site, and translate it to an IP address. Attackers have found ways to manipulate the DNS records [22] located on a DNS.

Hackers can replace a legitimate IP address with a booby-trapped address and then carry out some malicious activities, such as harvesting users' login information. Furthermore, attackers can cover their tracks by substituting the site's security certificates. The Internet uses Transport Layer Security (TLS) and Secure Sockets Layer (SSL) protocols to provide privacy and integrity [23]. A Certificate Authority issues TLS/SSL certificates. These are digital files, also called root certificates that contain the keys, which are trusted by browsers. This attack happens in the following five steps:

1. The attacker sets up a fake site, resembling the real target name.
2. Then somehow the attacker compromises login credentials of a DNS provider server. The attacker changes the IP address of a targeted domain, such as a bank's Website, to the fake one. The attacker also generates a new valid TLS certificate for the malicious Website.

3. The victim inadvertently approaches the DNS provider looking for real site.
4. Using DNS record of the compromised site, the user is directed to the fake site.
5. The fake site asks for victim's user credentials and records this information in the attacker's database. Later it is harnessed to access the victim's records from the target site.

This process is illustrated in Fig. 2.24, which may lead unsuspecting users to think that they are on a legitimate Website, and then enter their passwords, such as for a bank account. The attackers can later on use this information to steal money from the actual bank account.

The root cause of this attack is an improperly secured DNS server and ability to generate valid digital certificates for a new target site. Like other attacks that we shall study in this book, understanding the root cause is a key to prevent these attacks.

2.18 Cloud Software Security Requirements

Traditionally, software has to meet functional and performance requirements. However, security needs focus on minimizing the attack surface and vulnerabilities. It is also required that even under an attack, software will perform as specified [24].

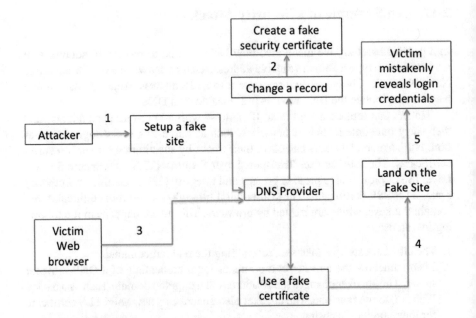

Fig. 2.24 Process of redirecting incoming IP traffic to a different Website

US Department of Defense's Cyber Security and Information Analysis Center (CSIAC) [25] has specified that all software must meet the following three security needs:

1. *Dependability:* Software should be dependable under anticipated operating conditions and remain fairly dependable under hostile operating conditions.
2. *Trustworthy:* Software should be trustworthy in its own behavior and robust. That is its inability to be compromised by an attacker through exploitation of vulnerabilities or insertion of malicious code.
3. *Resilience:* Software should be resilient enough to recover quickly to full operational capability with a minimum of damage to itself, the resources and data it handles, and the external components with which it interacts.

This means that security needs should be considered during all phases of development, starting with architecture, implementation, validation, and deployment. From a Cloud user's point of view, security is subtle, invisible, and almost taken for granted. However, Cloud developers and operators need to observe the following practices to assure security:

1. *Language options:* Start by considering strengths and weaknesses of available options, preferring a language with strong type checking and built-in security measures. For example, because C is a weakly typed high-level language, therefore it is unable to detect or prevent improper memory allocation, resulting in buffer overflows. So, a program will need to check for boundary limits. Whereas Java is a strongly typed high-level language, it has intrinsic security mechanisms based on trusted byte code interpretation, which prevent the use of uninitialized variables and language constructs to mitigate buffer overflows.
2. *Secure coding:* Adopt coding practices that eliminate incidents of buffer overflows, strings overwrites, and pointer manipulations, preferably by adding checks on the array sizes, string lengths, and pointer overrides, respectively.
3. *Data handling:* Separate sensitive or confidential data, and identify methods to securely handle it, e.g., using encryption during storage and transmission and at run time.
4. *Input validation:* Add additional checks to ensure that the range and type of data entered by users are correct, before passing it to the downstream APIs.
5. *Physical security:* All equipment connected to Cloud servers should have restricted physical access. System logs must be maintained and reviewed to trace back the root cause of attacks.

Overall, ensuring security is a matter of following the "trust but verify" approach, based on a quote by the late President Ronald Reagan, which is even truer for Cloud Computing.

2.19   Rising Security Threats

Security is no longer just a theoretical concern, but costing our world's economy more than half a trillion dollars per year, as shown in Table 2.1.

Table 2.1 Cost of security breaches [26]

Security incidents	Results
Cost of security breaches/year globally	$600B
Insiders contributing to security incidents in 2017	46%
Security experts expecting a major attack in the next 90 days	30%

Table 2.2 Biggest data breaches of this century, so far [27]

Year	Enterprise compromised	Number of people impacted
2018	Marriott	500 M
2017	Equifax	143 M
2016	Adult Friend Finder	412.2 M
2014	eBay	145 M
2013	Yahoo	3B

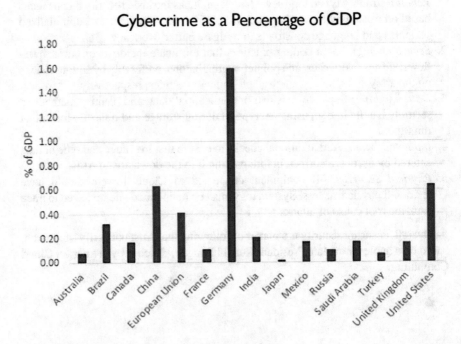

Fig. 2.25 Drag of cybercrime on national economies [28]

Each year, prominent breaches are happening and affecting hundreds of millions of people. Some of the largest impacts during the past decade are shown in Table 2.2.

Lastly, cybercrime is starting to drag down gross domestic product (GDP) of many national economies, as shown in Fig. 2.25. Beyond these figures, there is a multiplier effect on the economy, as productivity slows down. It is due to additional security checks that need to be implemented as preventive measures.

There is a continuous battle to outdo each other between the hackers and security professionals. In the latter chapters of this book, we will study methods that will help the latter.

2.20   Summary

Cloud Computing emerged only a decade ago. It is based on basic technologies that have been around and under development for more than half a century. These technologies were hardened in other environments, such as defense, personal computing, etc. Note that as we advance into a Web application domain, it also expands the hacking attack surface. Business models and relationships have to be completely reevaluated due the fundamental technologies of Cloud Computing. Threat agents can have multiple entry points on the client-side browsers, network in-between, and server-side hardware or software. Each risk needs to be evaluated and mitigation strategies developed in advance to prevent harm to the users of Web applications.

2.21   Points to Ponder

1. What led to the desire of people with PCs to connect with one another?
2. What led to the growth of thin clients? How thick should thick clients be and how thin should thin clients be? Which use cases suit each category (e.g., an information panel at the airport vs. an enterprise handheld computer)? Or given a usage class which client they should use? Besides usage, software update frequency and security are also a consideration.
3. How has SOA helped with the evolution of Cloud Computing?
4. Note that Cloud Computing is a natural evolutionary step as communication links grew in capacity and reliability.
5. Even though cookies pose security risks, why do browsers allow them and what's the downside for a user to not accept cookies?
6. What's the minimal precaution a public Website's user should take?
7. What are the trade-offs of securing information during transmission?

References

1. https://searchdatacenter.techtarget.com/definition/JCL
2. https://www.britannica.com/technology/personal-computer
3. https://en.wikipedia.org/wiki/Dial-up_Internet_access
4. https://condor.depaul.edu/elliott/513/projects-archive/DS513Spring99/figment/CSCONC. HTM
5. https://www.techopedia.com/definition/288/Web-browser
6. Bhatt, Pramod Chandra P, "An Introduction to Operating Systems Concepts and Practice (GNU/Linux)", PHI Learning Pvt Ltd, Jan 1, 2014.
7. https://en.wikipedia.org/wiki/Fibre_Channel
8. https://www.cisco.com/c/en/us/td/docs/solutions/Enterprise/Data_Center/DC_Infra2_5/ DCInfra_1.html
9. Riley, M., Elgin, B., Lawrence, D., & Matlack, C. (2014). Missed Alarms and 40 Million Stolen Credit Card Numbers: How Target Blew It. Bloomberg Businessweek. Retrieved from http://www.businessweek.com/articles/2014-03-13/target-missed-alarms-in-epic-hack-of-credit-card-data.
10. https://www.charlotteobserver.com/news/nation-world/national/article94453227.html
11. https://www.datacenterknowledge.com/archives/2015/07/08/technical-issue-halts-trading-on-nyse
12. https://www.w3.org/standards/Webofservices/
13. https://www.service-architecture.com/articles/Web-services/Web_services_explained.html
14. https://en.wikipedia.org/wiki/Data_center_bridging;
15. https://en.wikipedia.org/wiki/Service-oriented_architecture
16. http://www.soablueprint.com/yahoo_site_admin/assets/docs/BEA_SOA_Domains_ WP.290214359.pdf
17. http://www.agile-path.com/_media/whitepapers/AgilePath-SOA-Cloud-Computing-Transition.pdf
18. http://www.gridcomputing.com
19. https://www.techopedia.com/2/29245/trends/Cloud-computing/building-a-private-Cloud
20. http://www.zdnet.com/article/hybrid-Cloud-what-it-is-why-it-matters/
21. https://docs.microsoft.com/en-us/azure/azure-stack/
22. https://arstechnica.com/information-technology/2019/01/a-dns-hijacking-wave-is-targeting-companies-at-an-almost-unprecedented-scale/
23. https://www.acunetix.com/blog/articles/tls-security-what-is-tls-ssl-part-1/
24. Krutz, R. L., & Vines, R. D. (2010). Cloud Security: A Comprehensive Guide to Secure Cloud Computing. Indianapolis: Wiley Publishing.
25. https://www.csiac.org/
26. https://www.vmware.com/radius/rising-costs-cybersecurity-breaches/
27. https://www.csoonline.com/article/2130877/the-biggest-data-breaches-of-the-21st-century. html
28. Https://www.financialsense.com/contributors/guild/cybercrime-s-global-costs

Chapter 3
Cloud Computing Pyramid

3.1 Roots of Cloud Computing

"Cloud Computing" term became popular about two decades ago. However, its roots extend at least half a century back when users sat in front of blinking terminals far away from mainframe computers connected via cables. Telecommunication engineers about a century ago used Cloud concepts. Historically, telecommunications companies only offered single dedicated point-to-point data connections [1]. In the 1990s, they started offering virtualized private network (VPN) connections, with the same QoS (quality of service) as their dedicated services but at a reduced cost. Instead of building out physical infrastructure to allow for more users to have their own connections, telecommunications companies were now able to provide users with shared access to the same physical infrastructure. Later on, notions of utility computing, server farms, and corporate data centers formed the foundation of current generation of large Cloud data centers. The same theme underlies the evolution of Cloud Computing, which started with mainframes [2] and now has evolved to 24 × 7 services (24 hours service available 7 days a week, i.e., no downtime). The following list briefly explains the evolution of Cloud Computing:

- *Grid computing:* Solving large problems using parallelized solutions, e.g., in a server farm
- *Utility computing:* Computing resources offered as a metered service
- *SaaS:* Network-based subscriptions to applications
- *Cloud Computing:* "Anytime, anywhere" access to IT resources delivered dynamically as a service

Server farms didn't have APIs (Applications Programming Interface). Each user was made to think that he or she had full access and control of a server, but in reality time-sharing and virtual machine isolation kept each user's processes independent of others. Let us consider an example: assume a pharmaceutical company called P. Also, assume there is a Hollywood movie production studio called H. P needs 1000

© Springer Nature Switzerland AG 2020
N. K. Sehgal et al., *Cloud Computing with Security*,
https://doi.org/10.1007/978-3-030-24612-9_3

machines to run a drug trial on a continuous 24×7 basis for a whole month. *H* also needs 1000 machines on a weekend to render the scenes shot during the week. If only 1000 physical servers exist in the data center, on an exclusive basis, these can be loaned to one or the other user. In practice, if each user is running their respective tasks in a VM, then additional 1000 VMs can be spawned on the weekend, which may slow down *P's* jobs, but not impact their confidentiality or integrity. Each user thinks that he or she has a full access to 1000 server "platforms." In this context, the term "platform" is used to represent a collection of hardware and software components such as CPU, memory, and storage in an entity such as a server. Several servers are stacked to form a rack, and racks of servers constitute a data center that can be accessed via the Internet.

Each platform hides the complexity and details of the underlying infrastructure from users and applications by providing a Web-based graphical interface or APIs. These server platforms provide "always-on demand services" and are infrequently rebooted for planned maintenance such as HW upgrades or low-level BIOS patches. Meanwhile, load balancers and routers direct user traffic to other such servers, to give users an impression of "always-on" Cloud services. The hardware- and software-based Cloud services are made available to general public, enterprises, and SMBs (small and medium businesses) by the CSPs (Cloud service providers).

The NIST (National Institute of Standards and Technology) in the USA has defined Cloud Computing [3] as "Cloud Computing is a model for enabling ubiquitous, convenient, on-demand network access to a shared pool of configurable computing resources (e.g., networks, servers, storage, applications, and services) that can be rapidly provisioned and released with minimal management effort or service provider interaction." This Cloud model is composed of five essential characteristics, three service models, and four deployment models. The three service models mentioned in the NIST definition are described as follows:

1. *Software as a Service (SaaS):* allows the consumer to use the provider's applications running on a Cloud infrastructure. A user need not worry about the Operating system, CPU types, Memory or any storage required for these applications. Also the software developers are able to roll out new version and patches without worry about CDs or other distribution models, charging users on a subscription basis, much like cable TV services that charge for channel bundles based on number of Television sets at home (Fig. 3.1).
2. *Platform as a Service (PaaS):* allows the consumer to deploy onto the Cloud infrastructure consumer-created or acquired applications created using programming languages, libraries, services, and tools supported by the provider. This is one level lower than SaaS, but user doesn't need to worry about the underlying hardware, such as the memory, storage, or CPU capabilities.
3. *Infrastructure as a Service (IaaS):* allows the consumer to provision CPU processing, storage, networks, and other computing resources at the lowest abstraction levels, with an ability to deploy and run software, which can include operating systems and applications.

Fig. 3.1 User view of
Cloud Computing services

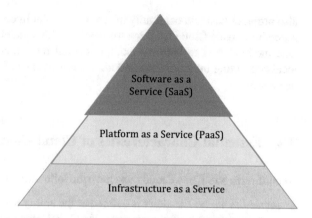

Business models differentiate Cloud providers in the following five deployment models:

1. *Public Cloud:* Cloud infrastructure is made available to the general public. An example is Amazon Web Services (AWS) offering Elastic Compute Cloud (EC2) to anyone with a credit card. Typically, server rental prices are charged on a per-hour rental basis.
2. *Private Cloud:* Cloud infrastructure is operated solely for an organization. An example is an enterprise that offers Cloud-based services behind corporate fire-walls to its employees located at different sites. Internal departments contribute or pay to build such a Cloud, typically maintained by the corporate IT services.
3. *Hybrid Cloud:* Cloud infrastructure is composed of two or more Cloud that inter-operate or federate through technology. An example is a company with an inter-nal data center, which further uses a Public Cloud for certain non-mission critical tasks or during overage demand periods. Central IT service determines the pol-icy on when to use internal vs. external resources, generally based on the usage patterns, or confidentially of data being shared and task urgency.
4. *Community Cloud:* Cloud infrastructure is shared by several organizations to supporting a specific community. An example is a large housing complex with Cloud-based community services exclusively for its residents. All residents, similar to backup power generation or water facilities in a closed community, use and share the cost of such services.
5. *Virtual Private Cloud (VPC):* This simulates a Private Cloud experience but located within a Public Cloud infrastructure cluster, such as a health mainte-nance organization (HMO) offering private services to its members, but hosting their data and applications on segregated servers within AWS. Due to the nature of data and tasks being handled, a VPC cluster may be physically isolated in terms of storage, networking, etc., which means that VPC users are charged a fixed cost and then a variable amount on the basis on their usage patterns.

Hybrid Cloud is the fastest-growing segment as it enables enterprise ITs and even SMB (small and medium business) owners to keep onsite computing capabilities. It also provides business continuity in case of external Internet disruptions, and enable access to outside Cloud resource for dealing with sudden compute demand surges. One motivation is to leverage existing capital investments in an already existing local computing infrastructure and alleviate wait times for additional resources on need basis.

3.2   Essential Characteristics of Cloud Computing

According to NIST, any Cloud must have the following five characteristics (Fig. 3.2):

1. *Rapid Elasticity:* Elasticity is defined as the ability to scale resources both up and down as needed. To the consumers, the Cloud appears to be infinite, and they can purchase as much or as little computing power as they need. This is one of the essential characteristics of Cloud Computing in the NIST definition.
2. *Measured Service:* In a measured service, aspects of the Cloud service are controlled and monitored by the Cloud provider. This is crucial for billing, access control, resource optimization, capacity planning, and other tasks.
3. *On-Demand Self-Service:* The on-demand and self-service aspects of Cloud Computing mean that a consumer can use Cloud services as needed without any human interaction with the Cloud provider.
4. *Ubiquitous Network Access:* Ubiquitous network access means that the Cloud provider's capabilities are available over the network and can be accessed through standard mechanisms by both thick and thin clients.
5. *Resource Pooling:* Resource pooling allows a Cloud provider to serve its consumers via a multi-tenant model. Physical and virtual resources are assigned and reassigned according to consumers' demand. There is a sense of location independence in that the customers generally have no control or knowledge over the exact location of the provided resources but may be able to specify a geographical location (e.g., country, state, or data center).

Fig. 3.2 Five essential characteristics of Cloud Computing

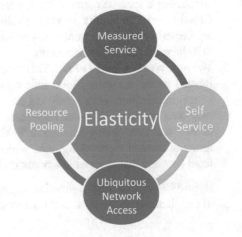

3.3 Role of Virtualization

Data center servers are powerful machines. No single application can possibly utilize their computational power on a 24 × 7 basis over long periods of time. Hence, there is a good business case to share a data center's resources among many different users. The enabling technology is called virtualization, which creates a layer of abstraction between the applications and the underlying hardware or a host operating system, as shown in Fig. 3.3.

Virtualization [14] refers to the act of creating a virtual image of the computer hardware, including CPU, memory, storage, and network elements. It is done to isolate the software stack from the underlying hardware. This isolation is the form of a virtual machine (VM), in which the user jobs run, isolating them from other such VMs. A layer of system software, called virtual machine monitor (VMM) or hypervisor, resides between the VMs and the platform hardware. It plays a role similar to an operating system (OS), managing various VMs, just as the OS manages the user processes. Only difference is that each VM may belong to a different and independent user, known as a guest or tenant. A VM may contain its own OS, enabling Windows and Linux to coexist on the same hardware platform.

Virtualization makes each user feel that he or she has full access and control of the machine. In reality there is time-sharing between different users on the same machine. With multi-cores, it is possible to allocate each user to a dedicated core, but still some I/O resources will need to be shared between the users. This may result in occasional performance bottlenecks.

Virtualization is essential to enabling a Public Cloud. Multiple users on the same machine pay a fractional cost to rent the system. It is similar to public airlines with many paid passengers, each buying a ticket for the journey, and the cost of an expensive airplane is amortized over many such trips. The only key difference is that in a Public Cloud, the users are not aware of each other nor can see the content of other

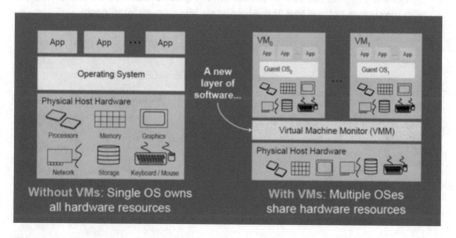

Fig. 3.3 An OS vs. a virtualization stack

VMs. Now the virtualization technology is extending beyond the server hardware, giving IT managers' tools to manage a data center's resources. This is called a software-defined data center (SDDC), providing an ability to monitor customers' usage patterns, controlling allocation of precious networking bandwidth, and preventing a malicious attacker from causing a denial-of-service (DOS) attack.

3.4   Cloud Players and Their Concerns

Fig. 3.4 shows a consumer-producer chain for Cloud services, starting with the users requesting services from their accounts located in a Cloud portal, e.g., an email service hosted in the Cloud. A Cloud user can start the request by opening a Web page and logging into the user account for which all the incoming mails are stored somewhere in the Cloud. Users may not care about the actual location of Cloud servers as long as they can get a reasonable response time and some assurance of secured, continuous service. The Cloud infrastructure may be located in a remote data center or spread across many, depending on the location and mobility of users. The IT manager in charge of Cloud operations manages the quality of service (QoS) by monitoring several metrics, e.g., average wait and response times for users, as well as the application performance, etc. Oftentimes, the Cloud service is hosted in a third-party data center, and here, the hardware usage monitoring and maintenance is done by the IT service providers, whereas the application performance monitoring is done by the Cloud service providers. In any event, Cloud users do not care about who is doing what as long as they get their service.

In a typical Public Cloud, the following actors are involved:

1. *Facility Managers:* are responsible for the Cloud infrastructure and its operations, i.e., efficient operations of a data center (DC), including the servers, storage, power supplies, and air-conditioning equipment. Their concern is to run DC with maximum uptime and at the lowest possible operational cost.

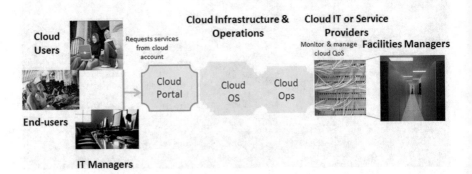

Fig. 3.4 Consumer-producer chain in the Cloud [11]

2. *Cloud IT Managers or Service Providers:* are responsible for scheduling tasks in one or more DCs. Their goal is to maximize utilization of equipment. This refers to maximal loading while leaving sufficient headroom for any usage spikes, monitoring infrastructure performance, planning future capacity growth, and compliance to ensure secure and efficient DC operations.
3. *Cloud Users:* are the customers of Cloud services and may be different than the applications' end-users, e.g., someone who is renting machines in a Cloud to host an application or provide a service, such that end-users may not know where their jobs are running. A Cloud user's concern is to provide a quality of service that the end-user expects at the lowest possible cost. An example is Netflix using Amazon's AWS to host movies.
4. *End-Users:* are the people or businesses at the other end of a Cloud. They may be using mobile or other devices to consume services that are hosted in the Cloud. The end-users' concern is the protection of their data and QoS.
5. *IT Managers:* are in charge of computing infrastructure operation. If the Cloud user is a business, it typically has an IT manager for the internal or enterprise data center, but also using a Cloud for backups, dealing with occasional work-load spikes, or certain type of non-mission critical business services.

Figure 3.5 outlines pain points in Cloud, starting with Cloud users who must specify the type of services they will need in advance. If their future requests exceed the forecast, then level of service will depend on the elasticity of service providers' capacity, e.g., since the Cloud hardware is shared between many tenants, it may be fully utilized at a given time. In this case, level of service will deteriorate for all users on a given server, e.g., if one of the users starts to do excessive I/O operations, it will slow down I/O requests from other users. This is referred to as a noisy neighbor VM (virtual machine) problem such as caused by placing a streaming media task next to the database-accessing job on the same server. A Cloud user may care only for his or her own QoS, while an IT manager needs to worry about satisfying all users. The facilities manager needs to ensure that hardware is being fully utilized and the data center is receiving adequate power and cooling, etc. There is no silver bullet to solve all Cloud actors' problems. An orchestration of actions is needed in response to dynamic monitoring of usage situation at various levels. This will be further explained in later chapters.

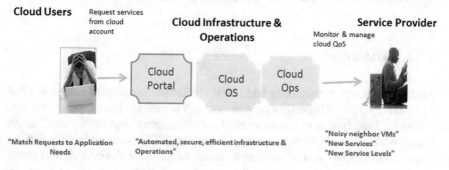

Fig. 3.5 Pain points in the Cloud

3.5   Considerations for Cloud Data Centers

A major consideration in selecting a Cloud service provider is the service-level agreements (SLA). Several researchers have targeted the problem of meeting SLA requirements and optimizing resource allocation for Cloud users, because SLAs define measurable considerations, such as listed below:

1. Response time
2. Output bandwidth
3. Number of active servers to monitor for SLA violations
4. Changes in the environment and
5. Responding appropriately to guarantee quality of service

Appleby et al. [4] developed the Oceano framework to match Cloud resource allocation to users' requirements. Emeakaroha et al. [5] presented LoM2HiS, a framework that uses resource usage metrics such as network bandwidth availability, data transfer, and down−/uptime to prevent SLA violations. Others perform statistical analysis and employ heuristics to improve resource allocation in autonomous compute environments such as a Cloud infrastructure. Ardagna et al. [5] employed heuristics with local search algorithm to better match physical servers with application sets that are currently using them. Bennani and Menasce [6] used analytic performance models to better match application environments with physical servers.

These concepts of relating a workload to hardware utilization can be applied to the Cloud by monitoring the virtual machines (VMs) and utilizing a load balancing mechanism. An example of a simplistic approach is the switch hierarchy within a standard data center. A standard data center design involves a switch on top of each rack connecting to higher-level switches forming a tree, with a load balancer at the rack level between the two lowest-level aggregate switches. The load balancer simply reallocates jobs when one VM, computer, or server gets too full. This is analogous to pouring drinks to guests but only switching glasses when the liquid overflows out of the cup onto the table. According to the work done by Khan et al. [7], monitoring a single VM over time appears only as noise, but observing a cluster of VMs over time reveals a pattern, showing that if a moderately accurate prediction about the workload could be made, the load balancing would be much more effective.

3.5.1   Migration

Another challenge is migrating traditional workloads that require computer clusters such as HPC (high-performance computing) workloads from captive data centers to a Cloud. The best-known example of commercial support for HPC in the Cloud is Amazon's Cluster Compute Instances (CCI). Amazon's CCI tries to address the upfront costs of running a cluster by allowing the customer to rent a scalable number of instances by the hour. For onetime HPC jobs or low utilization HPC needs, this solution appears cost-effective. However, for highly utilized clusters, the cost of running HPC jobs on Amazon CCI can be significantly higher. Carlyle et al. [8] show that current pricing on Amazon CCI is prohibitive when compared to the total cost incurred by their university maintained cluster. The relatively high utilization ratio on their university cluster resources amortizes the upfront and maintenance costs of owning the cluster.

3.5.2   Performance

Besides cost, the performance of HPC in a Cloud is of significant concern. Zhai et al. [9] evaluate Amazon CCI for tightly coupled Message Passing Interface (MPI)-based parallel jobs. They compare jobs run on CCI and an InfiniBand backed local cluster and found that performance in the Cloud is significantly reduced due to the lower bandwidth provided by the 10 gigabit network on CCI as compared to the InfiniBand network on their dedicated cluster. They also found significant performance variability with storage provided by Amazon's virtual storage infrastructure, often with periods of very low disk performance lasting for hours. Evangelinos and Hill [10] also reached the same finding and illustrated as much as two orders of magnitude higher latency in internode communication as compared to dedicated high-performance clusters. Most recently, IBM has started to use Watson as their HPC solution for advanced analytics [12] and several real-life problems such as medicine, etc.

3.5.3   Security

A Public Cloud is open to anybody with a credit card and needs to access its compute resources. This opens up possibility of attacks to access other customers' data or to launch a denial-of-service attack. Such attack has to be on a large scale, as Cloud by definition has massive amounts of resources to offer. Any malicious actor will need to consume them well before other users feel any significant performance impact. Also, since most Cloud service providers (CSPs) do not inspect incoming virtual machines to preserve their customers' confidentiality, they have to rely on indirect measures to detect if an attack is underway. This is done by statistically monitoring a VMs resource consumption behavior to detect anomalies and using resource throttling to mitigate the ensuing threats [13].

With a view to improve critical infrastructure security, the National Institute of Standards and Technology (NIST) has issued a draft framework [15]. Its purpose is to help organizations manage their cyber security risks in the nation's critical infrastructure, such as electrical grid and bridges, etc. It contains a common vocabulary for the actions that need to be taken to identify, protect, detect, respond, and recover from the security threats (Fig. 3.6).

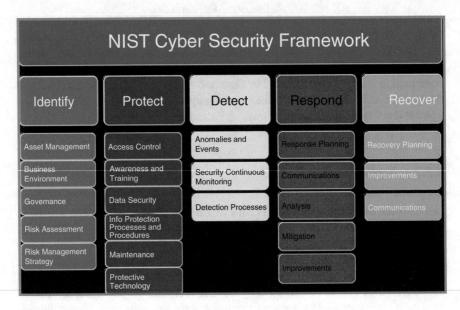

Fig. 3.6 NIST cyber security framework [15]

Five functions of NIST cyber security framework include:

1. *Identify:* Develop an organizational understanding to manage cyber security risk to systems, people, assets, data, and capabilities.
2. *Protect:* Develop and implement appropriate safeguards to ensure delivery of critical services.
3. *Detect:* Develop and implement appropriate activities to identify the occurrence of a cyber security event.
4. *Respond:* Develop and implement appropriate activities to take action regarding a detected cyber security incident.
5. *Recover:* Develop and implement appropriate activities to maintain plans for resilience and to restore any capabilities or services that were impaired due to a cyber security incident.

This framework has been widely adopted by many organizations across the USA and other countries. This is applicable to both the Private and Public Cloud as well.

3.6   Points to Ponder

1. Small and medium business (SMB) users lack financial muscle to negotiate individual SLAs with Cloud service providers. So what are their options?
2. Some potential Cloud users are on the fence due to vendor lock-in concerns. How can these be addressed?

3. What are the new business opportunities that enable movement of users and data between different Cloud providers?
4. What are some key differences between the concerns of a private vs. Public Cloud's IT managers?
5. Why is virtualization important for Public Cloud? Would it also help in Private Cloud?
6. Is workload migration a viable solution for a large Public Cloud service provider?
7. Why is NIST framework applicable to Public Cloud?

References

1. Donovon and Magi. https://www.amazon.com/Operating-Systems-Computer-Science-Madnick/dp/0070394555.
2. https://www.ibm.com/blogs/Cloud-computing/2014/03/a-brief-history-of-Cloud-computing-3/
3. Mell, P., & Grance, T. (2011). The NIST definition of Cloud Computing (draft). *NIST Special Publication, 800*, 145.
4. Appleby, K., Fakhouri, S., Fong, L., Goldszmidt, G., Kalantar, M., Krishnakumar, S., Pazel, D. P., Pershing, J., & Rochwerger, B.. (2001) Oceano-SLA based management of a computing utility. *Integrated network management proceedings, 2001 IEEE/IFIP international symposium on*, pp. 855–868.
5. Emeakaroha, V. C., Brandic, I., Maurer, M., & Dustdar, S.. (2010). Low-level metrics to high-level SLAs-LoM2HiS framework: Bridging the gap between monitored metrics and SLA parameters in Cloud environments. *High performance computing and simulation (HPCS), 2010 international conference on*, pp. 48–54.
6. Bennani, M. N. & Menasce, D. A. (2005). Resource allocation for autonomic data centers using analytic performance models. *Autonomic computing, 2005. ICAC 2005. Proceedings. Second international conference on*, pp. 229–240.
7. Khan, A., Yan, X., Tao, S., & Anerousis, N. (2012). Workload characterization and prediction in the Cloud: A multiple time series approach. *Network Operations and Management Symposium (NOMS), 2012 IEEE*, pp. 1287–1294.
8. Carlyle, A. G., Harrell, S. L., & Smith, P. M. (2010) Cost-effective HPC: The community or the Cloud?. *Cloud Computing technology and science (CloudCom), 2010 IEEE second international conference on*, pp. 169–176.
9. Zhai, Y., Liu, M., Zhai, J., Ma, X., & Chen, W. (2011). Cloud versus in-house cluster: Evaluating amazon cluster compute instances for running mpi applications. *State of the Practice Reports*, p. 11.
10. Evangelinos, C., & Hill, C. (2008). Cloud Computing for parallel scientific HPC applications: Feasibility of running coupled atmosphere-ocean climate models on Amazon's EC2. *Ratio, 2*, 2–34.
11. Mulia, W. D., Sehgal, N., Sohoni, S., Acken, J. M., Stanberry, C. L., & Fritz, D. J. (2013). Cloud workload characterization. *IETE Technical Review (Institution of Electronics and Telecommunication Engineers, India), 30*(5), 382–397.
12. https://www.ibm.com/analytics/watson-analytics/us-en/
13. http://palms.ee.princeton.edu/system/files/thesis.pdf
14. Bhatt, P. C. P. *An introduction to operating systems concepts and practice (GNU/LINUX)* (4th ed., pp. 305–311). Prentice Hall India Pvt Ltd, New Delhi, Jan 2014 pp. 558–562, and pp. 681.
15. https://www.nist.gov/news-events/news/2017/01/nist-releases-update-cybersecurity-framework

Chapter 4
Features of Private and Public Cloud

4.1 Customer Expectations of Cloud Computing

While economics is the main driver of Public Cloud Computing, its customers want the same performance and security aspects that they enjoy on a private server. This is a classic case of wanting to have your cake and eat it too. Specifically, Public Cloud users want full observability and controllability for their workloads. This refers to the applications and data they deploy on a remote server. They also expect a Cloud server to be secure as if it was on their own premises behind a firewall. These requirements are well captured in NIST's five essential characteristics that we listed in the previous chapter. However, there are some ambiguities; for example, "on-demand self-service" requires that a consumer can unilaterally provision computing capabilities, without requiring human interaction with a Cloud service provider. This implies automation on the Cloud service provider's site, but doesn't specify anything on the user side. In reality, for a user to provision hundreds of jobs at a moment's notice or to monitor them requires some automation. Furthermore, if a particular application is not behaving well, then the user will need some diagnostics to identify the root cause of problems and be able to migrate the job to another Cloud server. This implies availability of suitable monitoring and alerting tools. Automation of actions is such that a user and Cloud provider's environments work in unison. This will help to realize the full potential of Cloud Computing. The reality lies somewhere in between, as shown in Fig. 4.1.

Most enterprises already own some servers and storage facilities in their private datacenters. For customers with confidential or performance sensitive workloads, Public Cloud providers offer a Virtual Private Cloud, which can be thought of as a hosted service, or a group of dedicated servers cordoned off for such a customer. These offer dedicated facility but at a higher cost since the Cloud service provider can't share this infrastructure with other customers. This is preferable for some Cloud users who do not wish to maintain their own IT services but want the assurance of privacy.

© Springer Nature Switzerland AG 2020
N. K. Sehgal et al., *Cloud Computing with Security*,
https://doi.org/10.1007/978-3-030-24612-9_4

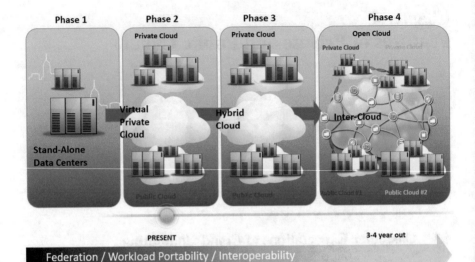

Fig. 4.1 A phased approach to adoption of Cloud Computing

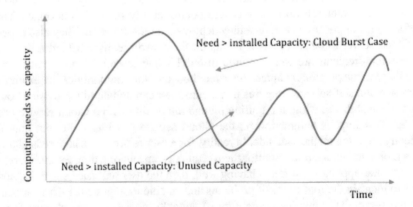

Fig. 4.2 Variations in computing needs of an organization

Next phase is a Hybrid Cloud, where users experience large variations in the Cloud consumption profiles, some of which exceeds their internally installed server base. Then they have two choices, either to buy more servers or let some computing demand remain unsatisfied. The former requires more capital investment and operational costs, as these servers may remain idle during off-peak times. The second choice has an implication of lost business opportunities as user tasks will need to wait in a queue, or worst yet customers will go somewhere else to meet their need. This is particularly true for online retailers who can't respond to their shoppers in a timely manner. This dilemma is depicted in Fig. 4.2, where one way to meet the unsatisfied demand is via migration of computing tasks to a Public Cloud.

"Cloud bursting" is a term used to define the jobs that were running in an internal data centers but are moved out to a Public Cloud at peak usage and then return back to a Private Cloud when the internal capacity is available. Since Public Cloud charge on a pay-peruse basis, this proposition is attractive. However, as we will see in later chapters, this is nontrivial because computing jobs typically require a setup, which includes an operating system environment, sometimes with a plethora of supporting tools and often a large amount of data. Such jobs can't be easily migrated between Private and Public Cloud at a moment's notice, which means that computing environments on both sides need to be kept in synchronization. One way to do this is by data mirroring between the data centers, so only computing programs need to migrate, while associated databases always stay in synchronization. However, this adds to the operational cost to keep the Public Cloud environment always ready to go at a moment's notice.

4.2   Interoperability of Cloud Computing

The last phase in computing evolution curve, as shown in Fig. 4.1, is interoperability, which refers to data formats and interfaces, not just between Private and Public Cloud but among various Public Cloud players, so ideally users can decide when and where to go on need basis. This is far from the current reality, as each Cloud provider uses different set of tools, formats, and environments, so customers tend to get locked-in with a Public Cloud vendor, which hampers the growth and adoption of Cloud by mainstream computing users.

4.3   System Failures, Diagnostics, and Recovery

Cloud Computing relies heavily upon systems composed of many components. The reliability of each component has inherent variability. The overall reliability of the Cloud is a function of the various reliabilities of the components as well as separability of these components. The separability of components allows overall Cloud Computing system to continue providing services with some parts of the Cloud, while other parts are experiencing a failure. Providing Cloud Computing services in the event of a failure relies upon detection of the failure, diagnosis of the cause of a failure, and subsequent repair of the failure.

The reliability of a system is measured in terms of statistics of failures over time. One measure is how often failures occur during a unit of time. Failure in time (FIT) is the failure rate measured as the number of failures per unit time. FIT is usually defined as the occurrence of 1 failure per billion hours. Another measure of reliability is the average time elapsed between failures. The mean time between failures (MTBF) is the measure of the average amount of time from a failure until the next failure occurs. A component having a failure rate of 1 FIT is equivalent to having an

MTBF of 1 billion hours. FIT is the inverse of MTBF. When a system fails, it needs to be repaired. The average time to fix a component is the mean time to repair (MTTR). The average time until the next failure is the mean time to failure (MTTF). The average time between failures is the average time until the next failure plus the average time to repair. Hence, MTBF = MTTF + MTTR. Note that MTTR includes time to diagnose the problem.

Extending the component concepts of reliability to systems and then to the Cloud is more complicated than initially imagined. This is due to the fact that traditional reliability of a system does not include the Cloud Computing concept of separability. Let us begin with a straightforward model of system reliability. The system reliability is dependent upon the reliability of each of its components and how many of those components there are in series or parallel. So consider a computer system that has a CPU, ten memory chips, four disk drives, and an IO controller. The computer system will fail if any of these components fails. So failure rate is the sum of the failure rates for each of the components times the number of each of the components. For our example consider that the MTTF for the CPU is 1 billion hours, the MTTF for each memory chip is 700 million hours, the MTTF for each of the disk drives is 500 million hours, and the MTTF for the IO controller is 700 million hours. The system failure rate (FIT) is equal to the sum of the failure rates of the components. Therefore the system

$$\text{FIT} = (1/10^9 + 10 * (1/7 \times 10^8) + 4 * (1/5 \times 10^8) + (1/7 \times 10^8)$$
$$= 24.7 \text{failures per billion hours.}$$

The overall system MTTF is about 40 million hours. And then one must add the repair time for the system to be a viable again. Suppose 2 days for repair then for MTBF = MTTF + MTTR. We see the repair is relatively insignificant for this highly reliable system. Reliability can be further improved by putting critical components in parallel, so any single failure will not impact the system uptime. An example of this is the cooling fans inside a server or duplicate storage elements in a Cloud environment.

In Cloud Computing, number of computer systems is very large, as many as 100,000 servers in one data center. Also there are thousands of other components connecting the servers such as routers and storage elements. Clearly one server does not cause the entire data center to fail, let alone the entire Cloud. However, the specific cause of each failure must be detected and diagnosed for the Cloud to return to full performance and reliability. The diagnosis requires analyzing the data remotely as it is impractically to climb through racks of 100,000 servers to physically measure the functioning of each component. This is where testing and diagnosis comes in conflict with security and privacy. To test and diagnose the system, we need access to vast quantities of data. However, to maintain the security and privacy, we must limit access to actual system management data. Hence, testing is often carried out not on the actual but using a representative set of data.

4.4 Reliability of Cloud Computing

Generally, each Cloud provider has a SLA (service-level agreement), which is a contract between a customer and the service provider. The service provider is required to execute service requests from the customer with negotiated quantity and quality ranges for a given price. Due to variable load, as shown in Fig. 4.2, dynamic provisioning of computing resources is needed to meet an SLA. Optimum resource utilization on the Cloud provider side is not an easy task. Hence, large Cloud service providers have fixed SLAs that are offered on a take-it-or-leave-it basis, unless the user is a large agency with high computing needs and ability to pay for these needs. In such cases, AWS is known to offer dedicated compute resources on an exclusive basis, which may solve the availability problem but only for large customers.

Availability is defined as the ratio of uptime/total time, where uptime is total time minus downtime. Availability ratio in percentage is measured by 9 s, which is number of digits shown in Table 4.1. So 90% has one 9, while 99% has two 9 s, etc. A good high availability (HA) package, even with substandard hardware, can offer up to three 9 s, while enterprise class hardware with a stable Linux kernel has 5+ nines.

An example is a system composed of two servers in series, such that each server has an expected availability of 99%, so the expected availability of such a system is 99%*99% = 98.01%, because if either server fails, it will bring the system down. Now consider a 24 × 7 e-commerce site with lots of single points of failures, e.g., the following eight components with their respective availability, as shown in Table 4.2 below:

Now the expected availability of such a site would be less than 60%, as calculated by simply multiplying the availability of each row, i.e.,

$$\rightarrow 85\% * 90\% * 99.9\% * 98\% * 85\% * 99\% * 99.99\% * 95\% = 59.87\%.$$

This is not acceptable, since the Website may be down for half the time.

One way to improve this situation is by using individual components with higher reliability, but these are more expensive. Another way is to put two low availability systems in parallel, to improve the overall component-level reliability. An example is data stored on parallel disks or two servers running in parallel, one of which can be used if the other is down. The availability of a parallel system with two components, each of which has an availability of Ac, can be computed as

Table 4.1 Example availability of various computing systems

9 s	Availability	Downtime per year	Examples
1	90.0%	36 days 12 hours	Personal computers
2	99.0%	87 hours 36 minutes	Entry-level business
3	99.9%	8 hours 45.6 minutes	ISPs, mainstream business
4	99.99%	52 minutes 33.6 seconds	Data centers
5	99.999%	5 minutes 15.4 seconds	Banking, medical
6	99.9999%	31.5 seconds	Military defense

Table 4.2 Example availability of Cloud components

	Component	Availability
1	Web server	85%
2	Application	90%
3	Database	99.9%
4	DNS	98%
5	Firewall	85%
6	Switch	99%
7	Data center	99.99%
8	ISP	95%

$$As = Ac + \left(\left(1 - \left(Ac / 100 \right) \right) * Ac \right)$$

Hence, two Web servers, each with only 85% availability, in parallel will have availability of 85 + ((1–85/100)*85) = 97.75%. Another way to think is that Server A is down 15% of the time (with 85% reliability). Server B is also down 15% (with 85% reliability). Therefore, the chance of both Server A and Server B down at the SAME time is 0.15 × 0.15 = 0.0225 or 2.25%. So, the uptime of A and B is 100–2.25 = 97.75%. Just this one change alone will improve the overall system-level reliability to nearly 80%, as shown below:

$$\rightarrow 97.75\% * 90\% * 99.9\% * 98\% * 97.75\% * 99\% * 99.99\% * 95\% = 79.10\%.$$

This is better than 60% before, so let us add a second ISP to improve the overall reliability to 94.3%, which represents 500 hrs of downtime/year vs. 3506 hrs/yr with 60% downtime.

$$\rightarrow 97.75\% * 99\% * 99.9999\% * 99.96\% * 97.75\% * 99.99\% * 99.99\% * 99.75\% = 94.3\%.$$

However, improving it further requires more work and becomes expensive.

4.5 Performance of Cloud Computing

Cloud Computing combines the needs of several users and serves them out of one or more large data centers. Characterization of computer workloads has been extensively studied with many practical applications. Benchmark suites such as SPEC and PARSEC are often used to evaluate improvements to the microarchitecture such as cache designs. Jackson et al. [1] evaluated the performance of HPC applications on Amazon's EC2 to understand the trade-offs in migrating HPC workloads to Public Cloud. Their focus is on evaluating the performance of a suite of benchmarks to capture the entire spectrum of applications that comprise a workload for a typical HPC center. Through their Integrated Performance Monitoring (IPM) [2] framework, they provide an analysis of the underlying characteristics of the application and quantitatively identify the major performance bottlenecks and resource constraints for the EC2 Cloud. Another study by Ferdman et al. [3] looks specifically at emerging scale-out workloads that require extensive amounts of computational resources. They introduce a benchmark suite, Cloudsuite, to capture these workloads and characterize their performance via performance counters available on modern processors. Their high-level conclusion is that modern superscalar out-of-order processors are not well suited to the needs of scale-out workloads. While the goal of these studies was to do detailed performance characterization of individual servers, our work below aims to demonstrate the Cloud variability at the user application level and examine its causes.

There is a 2–3× of performance difference observed between different Cloud vendors [4] or in the different data centers owned by the same Cloud vendor. Even for the same Cloud data center location, there is a 30–50% performance variation across the same class of Cloud servers. Such a wide variation poses serious problems for the customers of Public Cloud, especially when they host end-user facing commercial applications or run mission critical tasks. In the first case, end-users expect real-time response and may take their business elsewhere. In the second case, batch mode applications may take longer time impacting deadlines and violating service-level agreements.

4.6 A Sample Study

Performance tests were conducted with 24 users, 6 Public Cloud vendors, and the megaApp application from Meghafind Inc. These experiments lasted for a full quarter during Oct to Dec 2015, with 351 samples collected and analyzed for the purpose of this book, as outlined in Table 4.3.

A user-level application (called "megaApp") collects the metrics, using a real-time clock provided by the operating system. The application creates measurement threads and schedules them to run on all CPU cores on a server. These scores are dependent on the CPU, memory speed and data access architecture, storage technology (speed, hard disk vs. SSD), operating system, threads scheduling and management, virtualization machine, virtualization guest OS, and any other applications running when the measurements are performed. The following components were measured:

Table 4.3 Attributes of our study

Attribute	Value
Number of users	24
Number of samples	351
Number of Cloud providers	6
Number of guest operating systems	2 (Linux and Windows)
Number of VMMs or host operating systems	6

1. *CPU score* is a measure of how many seconds do certain thousands of CPU instructions take to complete. It is measured in real time by performing integer, floating-point, and string operations. Lower duration scores are better than the higher duration scores.
2. *Memory score* is a measure of how many seconds do certain thousands of memory instructions take to complete. It is measured in real time by allocating blocks of memory and reading from and writing to the allocated memory blocks. Lower duration scores are better than the higher duration scores.
3. *Storage score* is a measure of how many seconds do certain thousands of file creation, deletion, and input and output operations take to complete. It is measured in real time by creating and deleting files and reading and writing to the created files. Lower duration scores are better than the higher duration scores.
4. *Total score* is a simple addition of CPU, memory, and storage scores. A higher value of this score indicates that the server is running slower.

We have used coefficient of variation (CV) to compare performance of servers across various Cloud and over different days of a week. CV quantifies the performance variability of the servers and incorporates the metrics described above. CV is computed as a ratio of standard deviation with average values of total score, i.e.,

$$CV = \frac{\sigma(\text{Total Score})}{\mu(\text{Total Score})}$$. Lower values of CV are desirable as these indicate minimal

perturbations across the observed scores. For baseline comparison purposes, megaApp was run on a stand-alone machine, as shown in Fig. 4.3 over a period of 3 months. As expected, these show minimal CV if the loading conditions remain consistent.

However, megaApp results in Cloud showed up to 14× higher coefficient of variation over the same period of time. This indicates large performance variability for

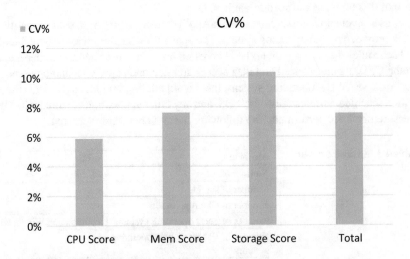

Fig. 4.3 Variations by type of scores on a stand-alone machine, lower is better

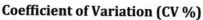

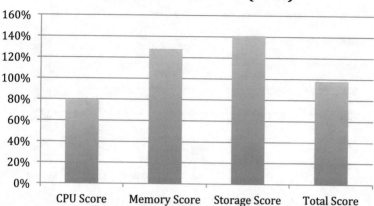

Fig. 4.4 Variations by type of scores in Public Cloud, lower is better

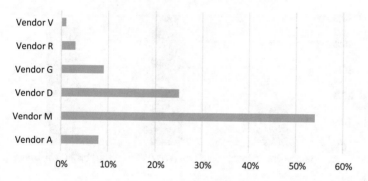

Fig. 4.5 Total score variation by Cloud vendor, lower is better

Public Cloud users. As shown in Fig. 4.4, using aggregated benchmarks across all Cloud providers, CPU scores had the least variations overall. This may be due to multi-cores servers being used by various Cloud providers. Since individual cores are allocated to different users, interference between users tends to be minimal. However, memory and storage read-write traffic for all users on a physical server passes through shared controllers, thus giving opportunity for noisy neighbors [5].

Total score variation across different Cloud vendors is shown in Fig. 4.5, indicating that choice of a Cloud Vendor is a key consideration if run-time consistency is desired.

Different vendors use varied server-sharing schemes, so it was interesting to observe a wide gap between normalized total scores across different VMMs (virtual machine monitors) in Fig. 4.6.

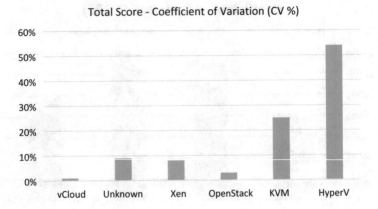

Fig. 4.6 Total score variation by VMM (i.e., host OS), lower is better

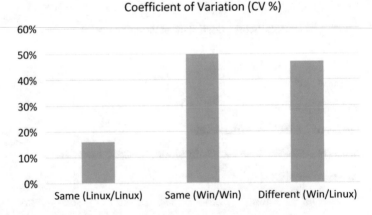

Fig. 4.7 Total score variation given a host/guest combination, lower is better

An IT manager in the Cloud data center replaces only a fraction of her servers when a new processor is launched, to smoothen out capital expenditure investments and minimize any perceived risks associated with any new hardware migrations. This results in a mix of various generations of servers and processors, with differing performance in a large data center. This heterogeneity is masked by a virtualization layer, but underlying performance differences can't be hidden and affect all user jobs running on this mix of servers. Furthermore, different users' jobs cause different loading conditions, resulting in differing performance, even on the same generation of hardware servers. Performance also depends on the host and guest OS combination, as shown in Fig. 4.7.

"Day of the week" variations were surprising in that Saturday had the worst CPU and storage scores and Monday had the worst memory score. Overall, Thursday and Friday were most predictable days to get a consistent performance on the Cloud, as seen in Fig. 4.8.

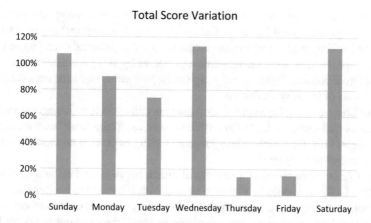

Fig. 4.8 Total score variation on a given Cloud server, lower is better

This is quite different than the results observed previously [6], where several patterns of performance variation in Public Cloud were observed, but no clear impact of time of the day or the day of the week was found on a Cloud's performance. We were able to detect variations as our benchmarking process depended on the execution time of individual key instructions of certain type, rather than an end-user application in the Cloud.

Our observed variation may represent human users' scheduling behaviors in the Cloud. Collectively, people may like to run their tasks so to finish them by early Thursday, so the last 2 days of the working week are used to report results, and then launch a new set of tasks before going home for the weekend on Friday. If multiple users exhibit the same behavior collectively as a group, its impact on server performance will cascade due to noisy neighbor effect, as documented previously [7], thus further slowing down the observed performance.

4.7   Security

Besides performance variability, security is another big concern among the Public Cloud users. Since a VM user doesn't know which other tenants are sharing the same hardware infrastructure, there is a perception of vulnerability. Following is a summary of top five threats [8]:

1. *Data Breaches:* A data breach is a security incident in which information is accessed without authorization. It can hurt businesses and consumers, as the data may be sensitive, proprietary, or confidential such as healthcare, industrial trade secrets or credit card information, etc. In 2017 there were 791 data breaches reported [9], a 29% increase over the same period in 2016. A single breach in Equifax, a credit card agency, exposed sensitive personal data of more than 145 million people in the USA.

2. *Inadequate Access Management:* Who can access which information is critical to manage the Cloud business. Data thieves use common techniques like password hacking or phishing to gain access to valuable personal data. Organizations can improve their access management protocols by using multifactor authentication. Additional techniques requiring strong passwords and time period enforced renewals of security credentials.
3. *Side Channel Attacks:* Sophisticated hackers can often use side access methods, such as malicious code, to read system contents. There have been some recent incidents [10], for which industry reacted swiftly to provide necessary patches so no damage was reported.
4. *Data Loss:* Sometimes if a successful attack happens and critical data is lost, then a business should be prepared to deal with it by having backups and ability to deal with it quickly. An example is after a loss of credit card numbers, can the business issue new cards and deactivate or block the old ones quickly? Public Cloud users should review contracts carefully for indemnity clauses to see who pays for the loss.
5. *Denial of Service (DoS) Attacks:* A DoS attack can disable a machine or network, rendering it inaccessible to the intended users. A Public Cloud operator needs to guard against such an attack coming from any user or internal source and plan to have redundant capacity to keep the business running.

4.8   Summary

Cloud Computing performance has inherent variability. The study involved running megaApp scouts to take performance samples over time and observed large performance variations. More than 350 samples were collected over a quarter, on different days and times, to minimize the temporal dislocations. Wide variations were seen across the same type of machines in a Cloud for the same vendor and even on the same machine over time. This study demonstrates how an end-user can measure Cloud Computing performance, especially exposing the performance variability.

4.9   Points to Ponder

1. Multi-tenancy drives load variations in a Public Cloud. Different users running different types of workloads on the same server can cause performance variations. How can you predict and minimize the undesirable effects of performance variability?
2. How can you predict and counter the undesirable effects of performance variability?
3. How is performance variability addressed in enterprise data centers?
4. What is a good way to assess if a noisy neighbor problem exists?

5. How can a noisy neighbor problem be prevented?
6. How can understanding a Public Cloud's usage patterns help?
7. What operating systems related strategies can a user consider to optimize Cloud usage?

References

1. https://www.nersc.gov/assets/NERSC-Staff-Publications/2010/CloudCom.pdf
2. Skinner D. (2005). Integrated performance monitoring: A portable profiling infrastructure for parallel application. *Proceedings of ISC2005: International supercomputing conference*, Heidelberg.
3. http://www.industry-academia.org/download/ASPLOS12_Clearing_the_Cloud.pdf
4. http://www.meghafind.com/blog/index.php/2015/12/01/secret-revealed-who-is-faster-aws-or-azure/
5. http://www.meghafind.com/blog/index.php/2015/10/17/why-my-aws-ec2-Cloud-server-suddenly-slows-down/
6. Leitner, P., & Cito, J. (2016). Patterns in the chaos-a study of performance variation and predictability in public IaaS Cloud. *ACM Transactions on Internet Technology (TOIT), 16*(3), 15.
7. Pu, X., et al. (2013). Who is your neighbor: Net i/o performance interference in virtualized Cloud. *IEEE Transactions on Services Computing, 6*(3), 314–329.
8. https://www.zettaset.com/blog/top-challenges-Cloud-security-2018/
9. https://www.idtheftcenter.org/Press-Releases/2017-mid-year-data-breach-report-press-release
10. https://meltdownattack.com/

Chapter 5
Cloud Workload Characterization

5.1 Motivation

In previous chapters, we had elaborated several contexts that justify the use of Cloud services for computational needs. Over time Cloud services have evolved to provide a variety of services. Each category of service poses a different workload challenge. The literature [2–6] is full of such characterization definitions. The definitions emanate from stakeholders that include Cloud service providers, Cloud users, IT and Cloud facility managers, and hardware vendors. In this chapter we will review these definitions and offer a comprehensive workload characterization with appropriate rationale.

Workload characterization employs analytical models [1, 9] and performance metric [15, 16]. Any workload characterization needs to account for resource utilization, job transitions, and such other concerns that require adhering to service-level agreements (SLAs). We will discover in this chapter that different players in the Cloud have different business and technical needs.

Various participants in Cloud Computing, including facility managers, Cloud IT or service providers, Cloud users, consumers, IT managers, and hardware vendors, do not have a common set of definitions of workloads categories. This variation in definitions leads to difficulties in matching customers' requirements with available resources. Several researchers have targeted the problem of meeting SLA (service-level agreement) requirements by optimizing resource allocation for Cloud users by using analytical models [1], high-level performance metrics [2, 3], and characterization [4–6]. Therefore, a key problem is variation in terminologies across various groups and viewpoints. This chapter specifies a set of common definitions between the participants. We will briefly describe the Cloud workload categories to enable their mapping to compute resource requirements. This will help to describe the metrics that can be used to distinguish job transitions between categories. These metrics can also be used to detect resource contention between the workloads and, in conjunction with the categories, minimize contention to improve resource

© Springer Nature Switzerland AG 2020
N. K. Sehgal et al., *Cloud Computing with Security*,
https://doi.org/10.1007/978-3-030-24612-9_5

allocation. Ultimately, the purpose of this chapter is that SLAs, capital purchase decisions, and computer architecture design decisions can be made based upon the workload categories.

Cloud Computing is utilized by a wide variety of applications. These applications have different requirements and characteristics. As inventor Charles F. Kettering said, "A problem well stated is a problem half solved." In order to better state the Cloud Computing workload characterization problem, let us decompose it in four steps, as follows:

1. The first step is to define a comprehensive list of the computer workload categories.
2. The second step is to identify resources required by Cloud Computing.
3. The third step is to map each Cloud Computing workload category to the required computer resources.
4. The fourth and last step is to correlate low-level hardware metrics with the Cloud Computing categories.

The above steps define a problem space, whose solutions include identifying real-time measurements that indicate a change in categories, relating these to SLAs, capital purchase decisions, architecture design decisions, software implementations, and pricing mechanisms.

In order to minimize the capital expenditure of upgrading all servers in a data center, IT managers tend to refresh only 20–25% of platforms each year. This results in a heterogeneous mix of 4–5 generations of platforms in a large data center, assuming that systems original equipment manufacturers (OEM) release a new platform every year. These machines may have differing capabilities, such that the same workload will perform differently depending on the generation of a hardware to which it will be assigned. Furthermore, since workloads in a Cloud are heterogeneous, it creates task-to-machine optimal mapping problems for the Cloud schedulers often resulting in unexpected performance issues.

5.2 Some Background on Workload Characterization

Characterization of computer workloads has been extensively studied with many practical applications. Several existing studies for workload characterization have used targeted benchmarks for resource utilization. Zhang et al. [7] created and used a set of fixed traces to create profiles of workloads based upon machine utilization and wait times. Others have done work on characterizing different types of applications [4, 5, 8–13]. Workload characterization helps to drive the design of a wide range of computing-related systems such as microarchitectures, computing platforms, compute clusters, data centers, and support infrastructures. Benchmark suites such as SPEC CPU 2006 [14] and PARSEC [15] are often used to evaluate improvements to the microarchitecture such as cache designs. These benchmarks typically measure performance of specific computing resources rather than for a particular workload, with the assumption that measuring and optimizing the identified resource is equivalent to optimizing for the workload category.

This section provides references to the underlying needs of certain types of workloads. Our work aims to classify workloads at a higher level and then map different classes to compute resource requirements. Cloud users can use this mapping to better optimize their applications, and Cloud service providers can use it to accommodate their users. Another aspect is the time-based usage patterns for a given workload, such as the following scenarios:

Close to payday twice in a month, bank deposit and withdrawal activities create a usage spike.

Close to an event like elections or sports, lots of traffic on social media, e.g., Twitter.

E-commerce activity close to festivals/holidays, e.g., online gift ordering.

College admission time online traffic, e.g., related to airlines and university Websites.

Jackson et al. [16] evaluate the performance of HPC applications on Amazon's EC2 to understand the trade-offs in migrating HPC workloads to Public Cloud. Their focus is on evaluating the performance of a suite of benchmarks to capture the entire spectrum of application that comprise a workload for a typical HPC center. Through their integrated performance monitoring (IPM) [17] framework, they provide an analysis of the underlying characteristics of the application and quantitatively identify the major performance bottlenecks and resource constraints for the EC2 Cloud. Compared to the slowest HPC machine at NERSC [18], the uncontended latency and bandwidth measurements of the EC2 gigabit Ethernet interconnect are more than 20 times worse. For the overall performance, they conclude that EC2 is 6 times slower than a midrange Linux HPC cluster, and 20 times slower than a modern HPC cluster, with most of the slowdown and performance variability coming from the interconnect. They show that the more an application communicates between nodes, the worse it performs on EC2 compared to a dedicated HPC cluster. Some examples are given below:

CPU intensive activity (like matrix computations using floating-point arrays)

Network traffic intensive activity (like social media, sports event)

Memory look-up activity (like vector data transfer)

Xie and Loh [19] studied dynamic memory behavior of workloads in shared cache environment. They classified workloads based on susceptibility of the activity of their neighbor and the manner of shared cache usage. They grouped workloads into four classes:

1. Applications that do not make much use of the cache (turtle)
2. Applications that are not perturbed by other applications (sheep)
3. Applications that are sensitive and will perform better when they do not have to share the cache (rabbit)
4. Applications that do not benefit from occupying the cache and negatively impact other applications (devil)

Their proposal is not mix incompatible classes, such as mixing a devil with sheep, but may be okay to run devil class of applications with turtle, as the latter doesn't use the cache.

Koh et al. [20] studied hidden contention for resources in a virtualized environment. They ran two VMs at a time on the same physical host and measured the slowdown compared to running the applications alone. They considered a number of benchmark applications and real-world workloads; in addition to measuring performance based on completion time, they reported the following key characteristics related to underlying resource consumption:

- *CPU utilization:* It is an indication of instructions that are being executed at a given moment and how busy is the CPU resource; it is measured as a percentage between 0% and 100%.
- *Cache hits and misses:* It is an indication of memory access patterns. A cache hit means that accesses are localized and required data is found in CPU's local memory, also known as cache, which is very fast. However, if a program requires data not already present in the CPU's local memory, then it is a miss. Data that has to be fetched from a remote memory location takes more time, slowing down the overall execution speed.
- *VM switches per second*: With multiple virtual machines running, also known as multi-tenanted environment. These VMs are switched on a round robin or some priority scheme to allow a fair share of CPU for all VMs. A switch can also be caused by an event such as a cache miss. CPU is not left in wait state while remote memory is being accessed, thus giving compute time to another waiting VM. Too much context switching is not desirable as it causes an overhead in saving and restoring states of CPU registers and flags.
- *I/O blocks read per second*: It refers to the network I/O bandwidth comparison on a per VM basis. VMs share I/O devices, so if one VM is consuming more bandwidth, then less is available for the other VM, slowing it down.
- *Disk reads and writes per second:* Similar to network, storage resources are also shared between different VMs. If a VM performs excessive disk reads or writes per second, then storage access is reduced for other VMs.
- *Total disk read/write time per VM:* Measuring time spent in disk accesses from within a VM indicates the nature of program running in that VM. Other VMs waiting to access storage will experience a slowdown.

Koh et al.'s analysis further shows that applications differ in the amount of performance degradation, depending on their own and other co-scheduled applications' usage patterns. This is expected and is not a new finding. However, Koh et al. clustered applications based on their performance degradation and mapped them to the ten characteristics that they measured. This provides a categorization and means to predict performance degradation in advance.

5.3   Top-Level Cloud Workload Categorization

Cloud Computing users have a variety of needs and wants. Cloud Computing platform (or services) suppliers have resource choices for allocation, planning, purchasing, and pricing. As Khanna and Kumar say, "Systems must handle complexity autonomously while hiding it from users" [26]. This is true and especially difficult in a Cloud Computing environment than any other computing system. The reason is the unpredictable and varied nature of tasks being executed in a Cloud Computing data center. Essential first step to meet this need is a clear, broad, ubiquitous categorization of various workloads. Workload categories can be split in two ways:

1. *Static architecture:* It refers to the implementation of solution architecture such as Big Data storage or a parallel computational setup, e.g., is the application distributed in nature using parallel resources or is it serial in nature? Also, what's the size of database that is needed to support this application? Knowing in advance the type of architecture (parallel or serial) or the size of a database is even more important for layered applications that may be split across different servers, VMs, etc. An incorrect task assignment will result in poor performance for the application. In Cloud, it results for a customer asking for a small or medium size of a virtual machine, e.g., 2 GB memory and 1 CPU core for a task that needs a large size VM.
2. *Dynamic behavior:* This refers to the run-time nature of resource usage or stress that a workload places on the computational resources in a data center, e.g., how much of CPU time, I/O bandwidth, memory, or disk read-write accesses are performed during the application's run-time? In a time-based access pattern, we are interested in both peaks and valleys of usage curves as well as the area under such curves, as shown in Fig. 5.1. A small VM that a customer has requested may be heavily or lightly used depending on the workload that runs in that VM.

Ideally, one wants to know the computational resources required to support an application as well as its loading pattern to ensure that expected performance goals can be met. As an example, a streaming video on demand may be simply reading raw data from the disk and serving video frames through the network, imposing very little CPU loading. It could also be engaged in real-time decompression or encoding of data for a client device, showing a different usage pattern.

Another, orthogonal, way to split workload types is between interactive and batch-mode jobs, e.g., where a real-time response is needed with smaller accesses or computations vs. overall completion efficiency for the large computational tasks. It is important to know the nature of a workload and its users' expectations in advance to do appropriate task placements so any service-level agreements (SLAs) and QoS requirements can be met. There is a subtle difference between the two. SLA refers to a legal agreement regarding the job's completion time and its data security, etc. For example, USA's Health Insurance Portability and Accountability Act (HIPPA) of 1996 requires that any healthcare data must remain on the physical servers located within the USA. Thus, a Cloud company may have worldwide data

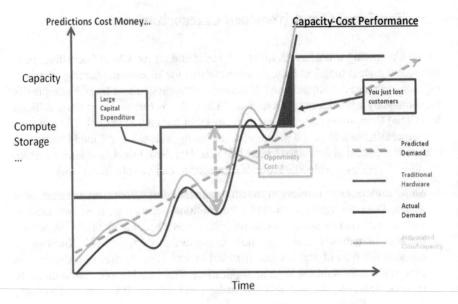

Fig. 5.1 Cloud Computing can be used depending on workload usage variations

centers, but it needs to ensure that any jobs from its US healthcare insurance customers remain within the USA. Other SLAs may refer to the job needing to be done
and results delivered within certain hours or days. QoS, on the other hand, refers to
the run-time performance during the execution of a job. For example, a video being
played on Cloud servers must be jitter-free and thus needs a certain network bandwidth assurance.

5.4 Cloud Workload Categories

This section lists and briefly describes the categories of workloads as differentiated
by the customer and vendor viewpoint. These categories may have overlap, indeed
might be identical, from other viewpoints. Some categories that are included are not
technically Cloud Computing categories. These other categories are included for
multiple reasons. The primary reason is to distinguish them for the Cloud Computing
target of this chapter. A secondary reason is the theory that to an end user, everything appears to be "in the computer or Cloud" whether it runs locally or remotely.
History provides another reason for including non-Cloud workload categories in the
table. In the late 1990s, the concept of ASP (application service provider) was proposed. In this model, the applications would run on servers connected via the
Internet to thin clients, which was called an NC (network computer) [27]. This
model did not last. Therefore, SaaS on the Cloud may evolve; however, the categories described in this chapter will persist. Two key issues with computer workload

categorization that are overcome in this chapter are based upon the perspective used in the definitions. The perspective in this chapter is to define the categories based upon fundamental concepts. For example, some categories are made of many tiny independent tasks as opposed to some single purpose huge tasks. One problem avoided by this approach is a categorization dependent upon the current state-of-the art implementation of computer resources. Our categories are definitely linked to and related to underlying computer resources; however, those resources do not define the categories. This means the categories remain even when the implementation hurdles are removed. Another categorization hurdle is the dependence upon a user's perspective. Our categories incorporate user's perspective; however, the categories still work with the Internet, stand-alone computing, thin clients, cell phone, and automobile GPS systems. The final reason is for completeness. Any list of categories is open to questions about what is left out of the list. This final reason gives an opportunity to answer that question before it is asked.

To reiterate, these categories offer an operational view of Cloud Computing workloads. Although some of the related implementation resources (architectural, hardware, software, and infrastructure) concepts may be listed to help understand the categories, these implementation resource concepts are covered in another section. Although some low-level example hardware metrics, such as cache memory misses, may be cited for clarity in a particular category, those metrics are described in detail in another section of this chapter. The ultimate goal of the research in this area is to relate the ongoing measurement of the low-level metrics to the resource requirements. Secondary goals are to study how those resource requirements change at run-time for the purposes of SLAs, providing Cloud services, hardware capital acquisition decisions, and future computer architectural design decisions.

The categories of computing workloads considered in this section include both Cloud computer workloads and workloads that are not commonly considered Cloud Computing issues. For example, high-end desktop graphics processing is not a Cloud Computing category. The categories in this section include big streaming data, big database calculation, big database access, Big Data storage, many tiny tasks (ants), tightly coupled calculation intensive high-performance computing (HPC), separable calculation intensive HPC, highly interactive single person tasks such as video editing, highly interactive multi-person jobs such as collaboration chats/discussions, single compute intensive jobs, private local tasks, slow communication, real-time local tasks, real-time geographically dispersed tasks, and access control.

The **big streaming data** workload category is characterized by an interactive initiation followed by long periods of huge amounts of data (usually video) sent to an end customer. The long periods of transfer may be interactively interrupted and restarted by the end user. This workload is usually measured by data throughput rather than latency; an example provider in this category is Netflix. The end user will not complain about, or even notice, a 100 millisecond delay at the beginning or in response to an interrupt. However, a 100 millisecond delay between delivered video frames is obvious and irritating. Key underlying hardware resources are the storage server, streaming server, and the network speed to the end user.

The **big database creation and calculation** category is characterized by a large amount of data requiring simple yet massive computation efforts. For example, sorting and analyzing statistics for census data. Another example is the offline keying for search keys of databases. Example suppliers are the US Census Bureau, Yahoo, and Google.

The **big database search and access** workload category is characterized by repeated interactive requests or queries submitted to a very large database. The customer satisfaction for this category is dependent upon the response time or latency. The key resources that limit the latency are the database server software and organization, the disk storage, the network bandwidth from the storage to the database server, and the server load. Examples include Ancestry.com, the Mormon ancestry database, the US Census Bureau, Yahoo search, and Google search.

The **Big Data storage** workload category is characterized by the integrity of large amounts of data which is periodically updated, usually by small increments, which occasionally requires a massive download or update. The end user sees this as an archiving or data backup resource. This workload category in the Cloud is multiuser as opposed to internal solutions that would have the IT department handle archiving. Example suppliers in this category are Rackspace, Softlayer, Livedrive, Zip Cloud, Sugarsync, and MyPC. The success of this category is primarily availability, integrity, and security of the data. The speed (latency, throughput, storage, incremental processing, and complete restoration) is not the highest priority.

The **in-memory database** workload category is characterized by the integrity of large amounts of data that is rapidly accessed. The end user sees this as a real-time database. The limitation of this category is primarily size of the data. The speed (latency, throughput, storage, incremental processing, and complete restoration) is the highest priority. An example workload for this category is real-time business intelligence.

The **many tiny tasks (Ants)** workload category is characterized by several very small tasks running independently. Very small is defined as completely fitting into cache with some room left over for other processes. This might be a lot of cache turtles (as defined by Xie and Loh [19]), not bothering any other process and never being flushed by other processes. These tasks may or may not have some interactive components. This category depends upon the ability to assign multiple processors. Each task is independent, so this can scale easily. Examples of this category are small company monthly payroll, document translation, computer language syntax checkers, simple rendering or animation, format conversion, spelling, and English syntax checkers.

The **tightly coupled intensive calculation (HPC)** workload category is characterized by problems requiring teraflops of computing power. Notice that high-performance computing has multiple meanings. High grid point count matrices of partial differential equations are in this category. The base resource is raw compute power and large memory. However, implementing this with multiple processes adds the resource requirements of network speed and software partitioning. The tight coupling of these equations creates a dependency of the solution at each point on the values at many other points. Classic examples of tightly coupled intensive calculation

HPC workloads are large numerical modeling simulations such as weather-pattern modeling, nuclear reaction simulation, and computational fluid dynamics.

The **separable calculation intensive HPC** workload is characterized by large time-consuming quantities of calculations. This workload category is characterized by the ability to split up the calculations to be performed in parallel or small separable pieces. Examples of separable calculation intensive HPC tasks are computer-aided engineering, molecular modeling, and genome analysis. For example, a section of a gnome slice can be analyzed completely separated from another section. In addition to these two workload categories, the term high-performance computing (HPC) is frequently used to identify a type of computing environment. In such cases, HPC is used to describe the computer systems capability rather than the workload running on the HPC.

The **highly interactive single person tasks** workload category is characterized by a fast task with a single user. The key aspect here is response time latency. The examples for this category include online single-player gaming, online restaurant surveys, and online training courses.

The **highly interactive multi-person jobs** workload category is characterized by the connectivity of jobs, such as collaborating chats/discussions. The resources include the server loading, the bandwidth between servers, and the bandwidth from server to end user. Another example of this category is online multiplayer gaming.

The **single computer intensive jobs** workload category is characterized by high-speed large single-user tasks that have significant user interaction. VLSI integrated circuit (IC) physical design (including timing analysis, simulation, IC layout, and design rule checking) is an example of this category. All computations for a designer are local, but database for the overall chip is on the network or Cloud.

The **private local tasks** workload category is characterized by traditional single-user tasks. Example tasks include word processing, spreadsheets, and data processing. These might be local for convenience or security. In the case of "thin" clients, these could be run in the Cloud.

The **slow communication** workload category is characterized by small amounts of information without a time limit for delivery. The best example of this is email.

The **real-time local tasks** workload category is characterized by hardware measurements fed to a computer system. The key resources here are dedicated CPUs with interrupt-driven communications to the network. An example of this is a refinery or factory production monitoring system. Another example is a home security system. A third example is health monitoring. The nurses need immediate warning when the patient's heart monitor beeps. The hospitals Cloud database needs a continuous storage of each patient's data over the entire stay.

The **location aware** workload category is characterized by utilizing auxiliary location input (such as GPS or telephone area code) data to small amounts of information without a time limit for delivery. The best example of this is following a route on a map while walking or driving. Another example is using the location information as an additional security check for access while travelling.

The **real-time geographically dispersed tasks** workload category is characterized by several dispersed hardware measurements systems feeding data to a network.

The resource key here is dispersed dedicated CPUs with measurement and interrupt-driven communication to the network servers. An obvious example is weather reporting. Another example for this category is power grid management.

The **access control** workload category is characterized by user-initiated requests where the response is to another server authorizing more activity. The local example is the password on your computer. Most systems rely on passwords, and many security experts agree this is a very significant problem [28]. The local example is a bank ATM connected to a network (which may be a Cloud). A Cloud example is online purchasing. The resources required vary widely with the application and level of security desired. The Cloud is inherently open and flexible. A key issue with access control is the conflict with privacy. An example is access control to an email server, where the administrator can observe or alter the accounts of various users, but each user can access his or her account only. The various Cloud providers need to verify the identity of individuals and systems using and providing services. As Bruce Schneier says, "Who controls our data controls our lives," [28] and so it is in the Cloud. Data for access control include private data. Conglomerations of services (i.e., Cloud provider) must present a trusted environment meeting both security and privacy. The access control workload category must balance the trade-off between Cloud provider security and Cloud user privacy.

The **voice or video over IP** workload category is characterized by users initiated requests where the response is through a server to each other. This is not presently a Cloud activity. The resources required are network bandwidth and user local compute power for data compression and decompression. An example is a platform using Chromecast engaging in directing Netflix video traffic.

5.5   Computing Resources

This section lists the various resources that are used to provide Cloud Computing. Specifically, the resources considered will be the ones that have limitations set by SLAs, design decisions, capital purchases, software choices, and implementation choices. The relationship of these resources to lowest level measurements is straightforward in many cases. The relationship of the resources to the higher-level categories is open to conjecture and ripe for extensive research. This chapter takes the first step toward connecting low-level resources to Cloud workload categories.

Persistent storage is a relatively straightforward resource. The user estimates a need, gets an appropriate SLA, and uses until the resource needs an increase. While predicting future needs might be a bit difficult, other tasks are relatively easy. For example, measuring usage, assigning drives, deciding on capital acquisition, and measuring energy consumption and storage system design are eminently solvable.

Compute power/computational capability is measured by CPU time/cycles, number of cores available, number and type of computer nodes available, and the types and capabilities of CPUs assigned.

Network bandwidth has several subcategories, but we are only considering networks involving computers where bandwidth is a resource. For example, privately owned corporate entertainment is accessible only through subscription (monthly fees), requires possession of special hardware and/or software, and can be limited by connection type (DSL vs. dial-up modems). Examples include:

- Games: Playstation Network, Xbox Live, World of Warcraft, Eve
- Movies: Netflix, Amazon Instant Video, Hulu
- Radio: Pandora, Sirius, XM (last two are "satellite radio")

Broadcast transmission receivers such as GPS (global positioning system) or radios. They require a special device added to the computer. For regular radio, it requires an additional receiver to plug into a port. For GPS it also requires clear access to the sky to communicate with satellites and determine location, requires at least four satellites to be visible by the GPS (unless another positioning factor is known such as elevation), and is also limited by interference that creates false images. An example of a computer using GPS on a real-time map-tracking program while one is walking around town.

Data busses within a server such as CPU to memory, cache to main memory, memory to disk, and backplanes on a card.

5.5.1   Data Busses Between Servers

USB (universal serial bus): A full duplex system using a four-pin connector (power, data in, data out, and ground) to connect PCs with other devices at rates up to 5Gb/s (version 3.0).

Network types: A group of computers connected locally, not requiring a long-distance carrier, typically using the Ethernet over twisted pair, maximum data transfer rate is currently limited to 1Gb/s using CAT6 cables.

- **Campus network**: Larger than a LAN and smaller than wide area network (WAN), this network is a collection of LANs in a localized area that doesn't need long haul (leased lines, long distance carrier, etc.) between computers but is usually connected with fiber-optic cables using switches, routers, and multiplexers (MUXs).
- **Backbone network:** Used to connect multiple LANs, campus', or Metropolitan Networks by creating **multiple** paths from one computer to another over a different set of fiber.
- **WAN (wide area network):** Connects different LANs, campus', cities, states, or countries typically using fiber-optic cable along with erbium-doped amplifiers, add-drop MUXs, regenerators, and digital cross connects, current limit is 40Gb/s using OC768 (each fiber).

Cache memory requirements and utilization are important for planning, design, and job assignment to CPUs.

Software capability determines what can be done and how efficiently it can be done. This includes operating systems, programming languages supported, and object code versions available.

Main memory size is a straightforward resource in the sense of measurement, availability, and design. It is also amenable to real-time adjustments.

Cloud providers potentially use all metrics listed here. Efficiency of these metrics can be increased if a workload is known or a practical method of predicting its need is utilized. In some cases bigger is better for computational resources. This may not be always feasible. Disk storage space can be increased by adding as much as needed, but main memory, cache memory, or number of CPU cores requires either a design change or addition of new machines, IP addresses, racks, backplanes, etc. Oftentimes, network bandwidth is capped out and cannot be increased further. This means that network traffic needs to be managed carefully by paying attention to heavy traffic trends and anticipating a change. This coincides with workload characterization and balancing as well as shifts in workload categories.

5.6   Example Workload Categorizations

Table 5.1 relates workload categories to computer system resources. Of course, all computing uses many computing resources. However, some categories of jobs use particular resources to a much greater extent than others. This resource allocation affects all of the different groups. Service providers want to meet customer needs promised in a SLA without overspending on capacity. Customers do not want to purchase more capacity than needed. Computer manufacturers want to enhance hardware or software capabilities to improve performance due to any bottlenecks. The table includes computing jobs that are not Cloud related for two reasons. Primarily the non-Cloud categories are included for completeness. And secondly, in the future, all computing may be in the Cloud.

5.7   Temporal Variability of Workloads

There are two different cases in which the workload category would change. The first case is when the next step or phase of a job is a different category than the current category. Note that this happens in all jobs but only significant if the resource requirements of the two categories are different. Consider the example of a very large HPC computing job, the stages or phases of which are shown in Fig. 5.2. The workload category is considered HPC, and the primary resources are expected to be processors count for computational capability and node-to-node-to-memory data bandwidth. The first phase is the startup and data loading phase. This phase does not require significant resources to change the workload category. As the computation phase begins, the processors and data bandwidth resource availability are the

Table 5.1   Characteristic computing resources for workload categories for the Cloud

Workload category	User view or example providers	Limiting resources	Level of Cloud relevance: "How Cloud heavy is this category?"
Big streaming data	Netflix	Network bandwidth	Heavy
Big database creation and calculation	Google, US Census	Persistent storage, computational capability, caching	Heavy
Big database search and access	US Census, Google, online shopping, online reservations	Persistent storage, network, caching	Heavy
Big data storage	Rackspace, Softlayer, Livedrive, Zip Cloud Sugarsync, MyPC	Persistent storage, caching, bus speed	Heavy
In-memory database	Redis, SAP HANA, Oracle In-Memory DB	Main memory size, caching	Heavy
Many tiny tasks (Ants)	Simple games, word or phrase translators, dictionary	Network, Many processors	Heavy
Tightly coupled calculation intensive HPC	Large numerical modeling	Processor speed, processor to processor communication	Medium
Separable calculation intensive HPC	CCI on Amazon Web services, Cyclone™ (SGI) Large simulations	Processor assignment and computational capability	Heavy
Highly interactive single person	Terminal access, server administration, Web browsing, single-player online gaming	Network (latency)	Some
Highly interactive multi-person jobs	Collaborative online environment, e.g., Google Docs, Facebook, online forums, online multiplayer gaming	Network (latency), Pprocessor assignments (for VMs)	Medium
Single computer intensive jobs	EDA tools (logic simulation, circuit simulation, board layout)	Computational capability	None
Private local tasks	Offline tasks	Persistent storage	None
Slow communication	E-mail, blog	Network, cache (swapping jobs)	Some
Real-time local tasks	Any home security system	Network	None
Location aware computing	Travel guidance	Local input hardware ports	Varies
Real-time geographically dispersed	Remote machinery or vehicle control	Network	Light now but may change in the future

(continued)

Table 5.1 (continued)

Workload category	User view or example providers	Limiting resources	Level of Cloud relevance: "How Cloud heavy is this category?"
Access control	PayPal	Network	Some, light
Voice or video over IP	Skype, SIP, Google Hangout	Network	Varies

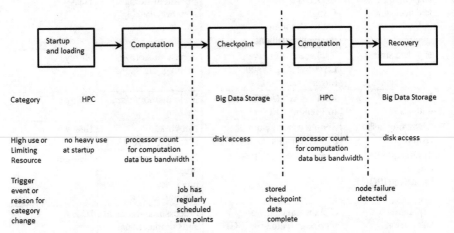

Fig. 5.2 Example of the changing Cloud workload categories for phases/step of an HPC job

limiting resources. Due to the high probability of a single-node failure before the completion of a job, many HPC applications have intermittent checkpoints. As the job enters a checkpoint phase, the resource requirements change. Now disk access becomes a limiting resource, and the category changes to large data storage. After the check pointing is complete, the job returns to the computation phase. Should a node failure occur, the job enters the restore phase, which once again is a transition to a large data storage category with a high disk access requirement.

Another example of a significant category change is paid subscription to online viewing of an event (such as a sports game or a band concert). Specifically, the initial phase is many users with high security login (i.e., payment) followed by a big streaming data phase when the event starts. This example is shown in Fig. 5.3.

The second case is when the job is incorrectly categorized. The incorrect categorization can have an effect on billing, when for example, during the run it is found that a totally different resource is consumed, but the billing agreement set a very high (or low) rate for that resource. An example of this is shown in Fig. 5.4 (example A) where a job was initially characterized as a big database creation and Calculation. However during run-time it was found that the key usage was disk space, not calculations, so it is reclassified as Big Data storage. Another example is shown, also in Fig. 5.4 (example B), where a job was initially categorized as Highly Interactive

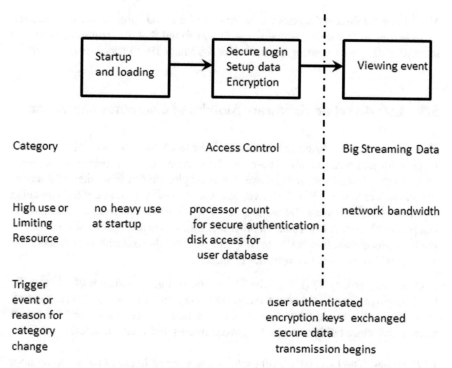

Fig. 5.3 Example of changing Cloud workload categories for phases for paid viewing of an event

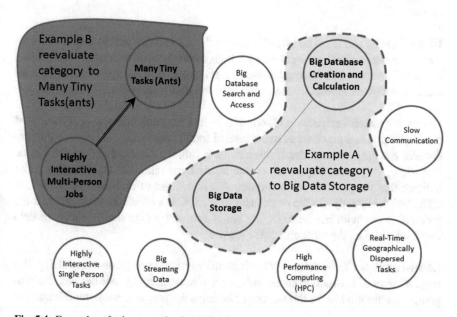

Fig. 5.4 Examples of mis-categorized workloads

Multi-Person Jobs. So the resource to optimize for is I/O communication channels. However, upon measuring actual usage, it was found that the scarce resource was number of simple processing units, such as the Many Tiny Tasks (Ants) requires.

5.8   Low-Level or Hardware Metrics of Computer Utilization

There are several performance counters available on most modern processors for monitoring hardware events. These can be sampled to gain insight on resource usage from the hardware perspective. For example, we can find whether floating-point unit is used at all or, if it is used, and is it so heavily used that it ends up being a performance bottleneck. Below is a list of some metrics relevant to this study. A comprehensive list can be found in processor data sheets and optimization manuals. For Intel processors, Intel's VTune Amplifier XE [29] documentation provides a fair amount of details about performance counters.

Instruction per Ccycle (IPC) The IPC metric is a good indicator of whether the CPU is being fully utilized or not. Although a coarse metric such as CPU usage may show 100% utilization, the CPU may not be running at its fullest potential due to microarchitecture bottlenecks such as cache misses, bus contention, etc.

LLC misses The LLC (last-level cache) misses are an important event to monitor as it gives an idea of how much memory traffic is being generated. The LLC misses counter will help quantify the extent by which the dataset surpasses the size of caches.

L1 data cache misses The level-1 cache is the fastest data provider to the processor and is thus crucial to maintaining a high throughput of instruction completion and a low latency. Monitoring L1 data misses allows us to gauge whether the core working set of an applications is captured in the first-level cache.

Cycles for which Instruction Queue is full The instruction queue is at the end of the fetch stage in a pipeline, or in the case of Intel processors, it is between the pre-decoder and the main decoder. If this counter is full, it indicates that the fetch engine is working very well. Depending on the instructions retired or the IPC, this could indicate that the pipeline is functioning at a high-level of utilization if the IPC is high (which is good). On the other hand, if the IPC is low, it could indicate that the problem is not with the fetch engine but is caused by data dependencies or data cache misses, i.e., the back end of the pipeline.

L1 Instruction Cache misses This is another indicator of front-end efficacy. If a large number of instruction cache misses are observed, it is either because the program sizes are too big for the I-cache or because the program has several branches/

jumps whose targets are not in the cache. In either case, a large number of misses indicate an issue with the front end and the instruction fetch mechanism.

Cycles during which reservation stations are full This would help us characterize the server application. For example, if the floating-point unit (FPU) reservation station is full all the time, it is an indication that the application is probably a scientific application. By observing specific reservation stations, and with some knowledge about the typical usage characteristics of applications, we can identify the applications running on a particular server.

Number of lines fetched from memory As servers have a large workload and the main memory is much larger than for desktop processors, this metric would give us an estimate of the pressure on the main memory. This is another indicator of the memory footprint of the application.

Dynamic, low-overhead low-level monitoring will also help to meet elastic demands in the Cloud. Animoto experienced a demand surge that required the number of servers to grow from 50 to 3500 in 3 days [30]. Although this is an extreme case, low-overhead dynamic monitoring will help in assigning resources to varying demands much more quickly than depending solely on high-level measurements such as application response times. The goal of this is to make the changes seamlessly, without noticeable inconvenience to the end user.

5.9   Dynamic Monitoring and Cloud Resource Allocation

The categories presented in this chapter can be used in conjunction with low-level metric measurements to provide a more robust real-time, dynamic resource allocation. Previous works depend mostly on historical data and metrics such as application response times and high-level resource utilization measurements [2, 3, 31]. Others performed studies using performance counters to guide scheduling algorithms in single-host systems avoid resource contention [20, 32–35]. Future work using the categories will better capture the nuances of workloads and their variability during the run, especially (but not limited to) in a Cloud environment where proper allocation of computing resources to the applications are essential and resource contention between customers' applications can be detrimental to the overall performance of the data center. Research questions that must be answered include:

1. What is the extent of resource contention in various shared cluster settings (e.g., HPC applications running in the Cloud)?
2. Is there enough contention to warrant dynamic real-time monitoring with low-level measurements?

3. How often should these measurements be taken to maximize benefit over their drawbacks (sampling overhead, network bandwidth used for data aggregation, etc.)?
4. How will these measurements help to drive smarter resource allocation, e.g., how do low-level metrics measurements relate to and indicate which resource is the limiting one?

As an example, an HPC cluster augmented with low-level measurements and a robust aggregation framework will be able to answer all of these questions in the domain of HPC clusters. The local measurement of each host can be used to determine the extent of intra- and internode resource contention, and the aggregation of this data can be used to determine the efficacy of doing such measurements. Further study can also be done to determine the optimal frequency of measurement to capture the variability of the workload in time while keeping the overhead to an acceptable level.

As the VM density increases, co-scheduling VMs that are least destructive to each other on the same physical cores is crucial. Low-overhead dynamic monitoring of virtual machines allows prediction of the resources that a VM will require in the next time slice, enabling co-scheduling to minimize resource contention and power consumption. This can identify system-wide effects of an application's behavior, gaining insight on lightweight schemes of classifying applications or VMs at runtime, providing a framework for real-time monitoring, prediction, and optimization of system-wide power consumption.

5.10 Benefits to Cloud Service Providers

Since a Cloud service provider may not be able to anticipate the workloads that customers will be running in the near future, it is not possible to plan ahead on placement of current jobs at hand while leaving sufficient head room for new jobs. Similarly, an option to do live VM migration may not be available in the Cloud due to sheer number of servers and jobs in real time. The best option is to build a profile of jobs run by a customer, and use historic data to anticipate future workloads, and similarly build a time-based usage pattern to predict the near future usage. Another idea is for customers to build a small sample program to mimic the type of applications they will be running in the Cloud. They can then run this sample on a new Cloud machine, guess its capability, and use it to anticipate how their particular job will perform relative to the actual application. Besides security, lack of predictability is the second largest reason why many enterprise customers are shy to adopt Public Cloud for running their mission critical applications. Therefore, a mechanism that prospective users can use to calculate how their application will perform in Cloud will help them to migrate their applications to the Cloud.

As shown below in Fig. 5.5, if Cloud users can anticipate their needs in advance and forecast their demand parameters to the Cloud service providers, their jobs can be matched correctly to the available hardware. It will also identify a shared machine while avoiding heavy-duty users with fluctuating workloads. However, sometimes this is not possible, so service provider must monitor the infrastructure and application performance to take corrective actions, such as moving jobs around or use the past information to do a better placement in the future.

Frequently, a real-life workload is not composed of just one VM, but a combination of two to three VMs, such as to provide a storefront or other interactive service built on business logic and a database in the background. An example would be an airline's Website, where users can do a search for flights, look at alternatives and their prices, select one and apply their frequent flier miles, or use a credit card to complete the purchase, at which time the seat in the flight's database will show as occupied.

This needs a three-layer solution as shown in Fig. 5.6, with presentation layer at the top, followed by middle-layer comprising of business logic such as how much of a flight has been sold and dynamic pricing based on what the competition is offering etc., and lastly a database showing the seats availability, etc. This represents a complex workload with the characteristics of an interactive workload, some computation in the middle and a database read-write access at the bottom layer. Furthermore, this workload's needs will vary in time depending on how many customers are browsing vs. buying at a given time.

Ideally each VM can be placed on a different server optimized for the respective task, and then inter-VM communications will bridge the gaps to give an appearance of a single Web-based elastic task running in the Cloud to the end users. However, any performance problems must be narrowed down and diagnosed in each layer for the whole system to work smoothly. Often, users of a Public Cloud do not know the hardware their VMs will get placed on, thus have to use hit and trial methods to find how many end-users can be supported by a VM.

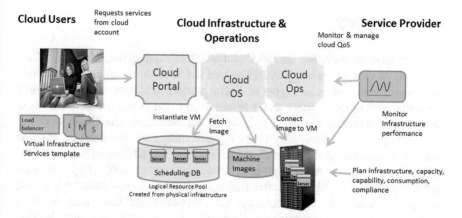

Fig. 5.5 SLA and QoS in IaaS Cloud. Cloud OS can benefit from fine-grained platform resource monitoring and controls to assure predictable performance and efficient and secure operations

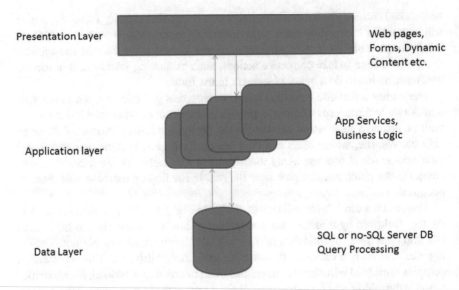

Presentation Layer

Web pages,
Forms, Dynamic
Content etc.

App Services,
Business Logic

Application layer

SQL or no-SQL Server DB
Query Processing

Data Layer

Fig. 5.6 Three-tier Web services architecture

Energy costs are the fastest-rising cost element of a data center. Power consumption is one of the major concerns of these facilities, Cloud, and large IT enterprises. According to senior analyst Eric Woods, "Servers use 60% of their maximum power while doing nothing at all." There are also challenges in managing power at the appropriate times. One solution to monitor and control power consumption of servers is Intel's data center manager (DCM) SDK [36], which enables IT managers to set power consumption policies, and allocate desired power to different servers in a cluster based on their workloads and usage at a given time.

5.11   Summary

In this chapter, we described various Cloud workloads and optimization issues from the points of view of various players involved in Cloud Computing. A comprehensive categorization of various types of diverse workloads is proposed, and nature of stress that each of these places on the resources in a data center is described. These categorizations extend beyond the Cloud for completeness. The Cloud workload categories proposed in this chapter are big streaming data, big database creation and calculation, big database search and access, Big Data storage, in-memory database, many tiny tasks (Ants), high-performance computing (HPC), highly interactive single person, highly interactive multi-person jobs, single computer intensive jobs, private local tasks, slow communication, real-time local tasks, location aware computing, real-time geographically dispersed, access control, and voice or video over IP. We evaluate causes of resource contention in a multi-tenanted data center and conclude by suggesting several remedial measures that both a Cloud service provider and Cloud customers can undertake to minimize their pain points. This chapter identifies the relationship of critical computer resources to various workload categories. Low-level hardware measurements can be used to distinguish job transitions between categories and within phases of categories. This relationship with the categories allows a technical basis for SLAs, capital purchase decisions, and future computer architecture design decisions. A better understanding of these pain points, underlying causes, and suggested remedies will help IT managers to make intelligent decisions about moving their mission critical or enterprise class jobs into Public Cloud.

5.12   Points to Ponder

1. Many applications combine different workloads, e.g., an online map system to support mobile GPS, reading a large file first, building its graph, and then doing I/O.
2. Different applications in a Public Cloud can offer opportunities to load balance and optimize resource usage, as compared to private data centers serving only one type of load, e.g., video playbacks.
3. What is the best way to improve fault-tolerance with multitier architectures?
4. What is the advantage of characterizing your workload in a Cloud?
5. From a Cloud service provider's perspective, how the knowledge of a workload's characterization may help?
6. Can machine learning play a role to improve the efficiency of a Cloud data center?

References

1. Bennani, M. N., & Menasce, D. A. (2005). Resource allocation for autonomic data centers using analytic performance models. *Autonomic Computing, 2005. ICAC 2005. Proceedings of the second international conference on*, pp. 229–240.
2. Appleby, K., Fakhouri, S., Fong, L., Goldszmidt, G., Kalantar, M., Krishnakumar, S., Pazel, D. P., Pershing, J., & Rochwerger, B.. (2001). Oceano-SLA based management of a computing utility. *Integrated network management proceedings 2001 IEEE/IFIP international symposium on*, pp. 855–868.
3. Ardagna, D., Trubian, M., & Zhang, L. (2007). SLA based resource allocation policies in autonomic environments. *Journal of Parallel and Distributed Computing, 67*(3), 259–270.
4. Alarm, S., Barrett, R. F., Kuehn, J. A., Roth, P. C., & Vetter, J. S. (2006). Characterization of scientific workloads on systems with multi-core processors. *Workload Characterization, 2006 IEEE International Symposium on*, pp. 225–236.
5. Ersoz, D., Yousif, M. S., & Das, C. R.. (2007). Characterizing network traffic in a cluster-based, multi-tier data center. *Distributed computing systems, 2007. ICDCS'07. 27th international conference on*, pp. 59.

6. Khan, A., Yan, X., Tao, S., & Anerousis, N. (2012). Workload characterization and prediction in the Cloud: A multiple time series approach. *Network Operations and Management Symposium (NOMS), 2012 IEEE*, pp. 1287–1294.
7. Zhang, Q., Hellerstein, J. L., & Boutaba, R.. (2011). Characterizing task usage shapes in Google's compute clusters. *Proceedings of large-scale distributed systems and middleware (LADIS 2011)*.
8. Arlitt, M. F., & Williamson, C. L. (1997). Internet Web servers: Workload characterization and performance implications. *IEEE/ACM Transactions on Networking (ToN), 5*(5), 631–645.
9. Bodnarchuk, R. & Bunt, R. (1991). A synthetic workload model for a distributed system file server. *ACM SIGMETRICS performance evaluation review*, pp. 50–59.
10. Chesire, M., Wolman, A., Voelker, G., & Levy, H.. (2001). Measurement and analysis of a streaming-media workload. *Proceedings of the 2001 USENIX Symposium on internet technologies and systems*.
11. Maxiaguine, A., Künzli, S., & Thiele, L. (2004). Workload characterization model for tasks with variable execution demand. *Proceedings of the conference on design, automation and test in Europe-Volume 2*, p. 21040.
12. Yu, P. S., Chen, M. S., Heiss, H. U., & Lee, S. (1992). On workload characterization of relational database environments. *Software Engineering, IEEE Transactions on, 18*, 347–355.
13. Calzarossa, M., & Serazzi, G. (1985). A characterization of the variation in time of workload arrival patterns. *Computers, IEEE Transactions on, 100*, 156–162.
14. Standard Performance Evaluation Corporation. (2006). SPEC CPU2006. Available: http://www.spec.org/cpu2006/, 8 Nov 2013.
15. Bienia, C., Kumar, S., Singh, J. P., & Li, K. (2008). The PARSEC benchmark suite: Characterization and architectural implications. Presented at the proceedings of the 17th international conference on parallel architectures and compilation techniques, Toronto.
16. Jackson, K. R., Ramakrishnan, L., Muriki, K., Canon, S., Cholia, S., Shalf, J., Wasserman, H. J., & Wright, N. J.. (2010). Performance analysis of high performance computing applications on the amazon web services cloud. *Cloud computing technology and science (CloudCom), 2010 IEEE second international conference on*, pp. 159–168.
17. Skinner, D.. (2005). Integrated performance monitoring: A portable profiling infrastructure for parallel applications. Proceedings of ISC2005: International supercomputing conference, Heidelberg.
18. National Energy Research Scientific Computing Center. *NERSC*. Available: www.nersc.gov. 8 Nov 2013.
19. Xie, Y. & Loh, G. (2008). Dynamic classification of program memory behaviors in CMPs. *The 2nd workshop on Chip multiprocessor memory systems and interconnects*.
20. Younggyun, K., Knauerhase, R., Brett, P., Bowman. M., Zhihua, W., & Pu, C. (2007). An analysis of performance interference effects in virtual environments. *Performance analysis of systems and software, 2007. ISPASS 2007. IEEE international symposium on*, pp. 200–209.
21. Mell, P., & Grance, T. (2011). The NIST definition of cloud computing (draft). *NIST Special Publication, 800*, 145.
22. Emeakaroha, V. C., Brandic, I., Maurer, M., & Dustdar, S. (2010) Low-level metrics to high-level SLAs-LoM2HiS framework: Bridging the gap between monitored metrics and SLA parameters in Cloud environments. *High performance computing and simulation (HPCS), 2010 international conference on*, pp. 48–54.
23. Carlyle, A. G., Harrell, S. L., & Smith, P. M. (2010). Cost-effective HPC: The community or the cloud?. *Cloud computing technology and science (CloudCom), 2010 IEEE second international conference on*, pp. 169–176.
24. Zhai, Y., Liu, M., Zhai, J., Ma, X., & Chen, W.. (2011). Cloud versus in-house cluster: Evaluating amazon cluster compute instances for running mpi applications. *State of the Practice Reports*, p. 11.

25. Evangelinos, C., & Hill, C. (2008). Cloud computing for parallel scientific HPC applications: Feasibility of running coupled atmosphere-ocean climate models on Amazon's EC2. *Ratio, 2*, 2–34.
26. Khanna, R., & Kumar, M. J. (2011). *A vision for platform autonomy*. Santa Clara: Publisher Intel Press.
27. Chapman, M. R. R. (2006). *Search of stupidity: Over twenty years of high tech marketing disasters*. Berkeley: Apress.
28. Schneier, B. (2009). *Schneier on Security*. Hoboken: Wiley.
29. Intel Corporation. *VTune Amplifier XE*. Available: http://software.intel.com/en-us/intel-vtune-amplifier-xe, 8 Nov 2013.
30. Armbrust, M., Fox, A., Griffith, R., Joseph, A. D., Katz, R., Konwinski, A., Lee, G., Patterson, D., Rabkin, A., & Stoica, I. (2010). A view of cloud computing. *Communications of the ACM, 53*, 50–58.
31. Bacigalupo, D. A., van Hemert, J., Usmani, A., Dillenberger, D. N., Wills, G. B., & Jarvis, S. A. (2010). Resource management of enterprise cloud systems using layered queuing and historical performance models. *Parallel & distributed processing, workshops and Phd forum (IPDPSW), 2010 IEEE international symposium on*, pp. 1–8.
32. Knauerhase, R., Brett, P., Hohlt, B., Li, T., & Hahn, S. (2008). Using OS observations to improve performance in multicore systems. *IEEE Micro, 28*, 54–66.
33. Fedorova, A., Blagodurov, S., & Zhuravlev, S. (2010). Managing contention for shared resources on multicore processors. *Communications of the ACM, 53*, 49–57.
34. Fedorova, A., Seltzer, M., & Smith, M. D. (2007). Improving performance isolation on chip multiprocessors via an operating system scheduler. *Presented at the proceedings of the 16th international conference on parallel architecture and compilation techniques*.
35. Nesbit, K. J., Moreto, M., Cazorla, F. J., Ramirez, A., Valero, M., & Smith, J. E. (2008). Multicore Resource Management. *IEEE Micro, 28*, 6–16.
36. Intel Corporation. *Intel Data Center Manager(TM)*. Available: www.intel.com/DataCenterManager. 8 Nov 2013.

Chapter 6
Cloud Management and Monitoring

6.1 Motivation

Managing a Cloud operation is similar to managing any other shared resource. Imagine checking into a hotel after a long flight, hoping to catch a good night's sleep before next day's business meetings. But suddenly your next-door neighbor decides to watch a loud TV program. Its sound will awaken you for sure, but what's the recourse? Not much, as it turns out other than calling the hotel manager and making a request to noisy neighbor to lower the TV's volume. A similar situation can happen in a Cloud data center with multiple VMs from different customers sharing the same physical server or any set of resources. Your neighbor in this case may be another Cloud user, running a noisy job with too many memory or disk interactions. Obviously, this may cause a slowdown of other jobs running on the same-shared hardware.

6.2 Introduction to Cloud Setup and Basic Tools

In this section, we will introduce some Cloud Computing management terms, with reference to Amazon Web Services (AWS). It is a large Public Cloud owned by Amazon, and following terms are reproduced for readers' convenience. Detailed glossary is available at: http://docs.aws.amazon.com/general/latest/gr/glos-chap.html

1. *Availability zones:*

    I. A distinct location within a region that is insulated from failures in other Availability Zones.
    II. It provides inexpensive, low-latency network connectivity to other Availability Zones in the same region.

© Springer Nature Switzerland AG 2020
N. K. Sehgal et al., *Cloud Computing with Security*,
https://doi.org/10.1007/978-3-030-24612-9_6

    III. A simple example can be a different data center (DC), or a separate floor in the same DC, or an electrically isolated area on the same floor.

2. *Regions*:

    I. A named set of AWS resources in the same geographical area.
    II. A region comprises at least two Availability Zones.

3. *Instance:* A copy of an Amazon Machine Image (AMI) running as a virtual server in the AWS Cloud.
4. *Instance class:* A general instance type grouping using either storage or CPU capacity.
5. *Instance type:* A specification that defines the memory, CPU, storage capacity, and hourly cost for an instance. Some instance types are designed for standard applications, whereas others are designed for CPU-intensive, memory-intensive applications, and so on.
6. *Server performance:* Performance of a server depends on a full set of HW and SW components, as well as any other jobs running on that machine.
7. *Application performance*: An end-user application consumes CPU, memory, or storage resources and reports performance in transactional terms as ops/sec or as latency to complete a task.
8. *Performance comparison*: Server performance is compared by ratio of ops/sec on any two servers.
9. *Elastic load balancing:* Automatically distributes incoming application traffic across multiple Amazon EC2 instances. It enables you to achieve fault tolerance in your applications, seamlessly providing the required amount of load balancing capacity needed to route application traffic.
10. *Load balancers:* Elastic load balancing offers two types of load balancers that both feature high availability, automatic scaling, and robust security. These include the Classic Load Balancer that routes traffic based on either application or network level information and the Application Load Balancer that routes traffic based on advanced application level information that includes the content of the request. The Classic Load Balancer is ideal for simple load balancing of traffic across multiple EC2 instances, while the Application Load Balancer is ideal for applications needing advanced routing capabilities, microservices, and container-based architectures. Application Load Balancer offers ability to route traffic to multiple services or load balance across multiple ports on the same EC2 instance.

6.3 Noisy Neighbors in a Cloud

One of the biggest challenges for an IT administrator in a private or public data center is to ensure a fair usage of resources between different VMs. If a particular VM does excessive I/O or memory accesses, then other VMs running on that same server will experience a slowdown in their access of the same physical resource. This results in a performance variation experienced by a VM user over time, as shown in Fig. 6.1.

It turns out that while memory and CPU cores can be divided by a VMM (Virtual Machine Monitor) for different virtual machines (VMs) on a physical server, there are components such as a memory, network, or disk controller that must be shared by all VMs. If one VM gets burdened with additional workload or malfunction, it will generate excessive I/O traffic saturating the shared controller. Just like a city highway, where an entry or exit ramp-up can become a bottleneck for new vehicles, a new request on the shared controller must wait for its turn. This will result in a slowdown of jobs or Website responses. One way to minimize this is proactive measurements, before your customers notice a slowdown and start complaining. This is accomplished by an agent task for the sole purpose of monitoring a VM's performance over time.

6.4   Cloud Management Requirements

Cloud management tools can be of two types: in-band (IB) or out-of-band (OOB). In-band refers to an agent that typically runs in an OS or VM, collects data, and reports back for monitoring purposes. However, it may interfere with other processes running in that VM, slowing it down or creating additional resource contentions. OOB refers to monitoring tools that typically use a baseboard management controller with own processor and memory system, for observing the main server's health metrics. These do not tend to place additional CPU or memory load on the

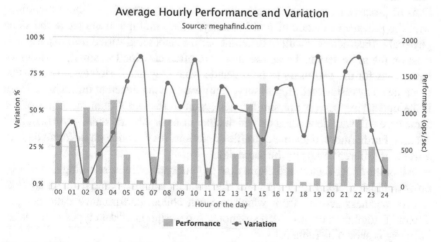

Fig. 6.1 Daily variation of a VM's performance on a server in a public Cloud

server being monitored, but OOB can't report detailed data as an IB agent due to their isolated nature. Rest of our discussion will not focus on OOB methods, which are mainly used to monitor the electrical health of a server by the data center operators, e.g., to check the voltage levels and frequency of a CPU, fan speeds inside the server box to ensure the proper cooling, etc. We will focus more on IB agents, which can report performance-related characteristics of a virtual machine or OS running on a physical server.

One way to limit the load placed by IB agents is to limit their measurement and reporting frequency, e.g., once every 5 minutes. Below is an example of the data types that are observed by an IB monitoring agent.

- CPUUtilization
- DiskReadBytes
- DiskReadOps
- DiskWriteBytes
- DiskWriteOps
- NetworkIn
- NetworkOut

These can be monitored in an alerting mode, e.g., raise an alarm when CPU utilization goes about 80% as CPU has a risk of saturation and slowing down other programs. Once the IT administrator gets such an alert, he or she has an option to move the offending application to another server or kill it based on priority of other applications.

6.5 Essentials of Monitoring

Any Public or Private Cloud data center has a mix of servers, with different generations of processors, memories, networks, and storage equipment. These combined with the prevalent practice of multi-tenancy, where multiple users share the same physical infrastructure, result in different performances and varied user experiences even on the same server. These machines are often classified as small, medium, or large VMs for the end users to help decide what they need. However, two virtual machines may reside on different server hardware giving different throughput. Even if the underlying servers are similar, performance for two virtual machines of the same type will vary depending on which other tasks or jobs are running on their host servers. Furthermore, data center traffic conditions may change on different days of a week or different hours of the same day.

All of the above factors affect run time of your applications, resulting in different costs to finish a task. Monitoring and alerting data can also come from an application or business activity from which one can collect performance data, not just Cloud IT administrator-provided solutions. A description of data types in the latter category is shown in Table 6.1.

Table 6.1 Nature of metrics collected and their units

Metrics name	Description	Units
CPUUtilization	The percentage of allocated compute units	*Percent*
DiskReadOps	Completed read operations from all disks available to the VM	*Count*
DiskWriteOps	Completed write operations to all disks available to the VM	*Count*
DiskReadBytes	Bytes read from all disks available to the VM	*Bytes*
DiskWriteBytes	Bytes written to all disks available to the VM	*Bytes*
NetworkIn	The number of bytes received on all network interfaces by the VM	*Bytes*
NetworkOut	The number of bytes sent out on all network interfaces by the VM	*Bytes*

Table 6.2 Frequency of performance metrics collection in a typical public Cloud

Monitored resources	Frequency
VM instance (basic)	Every 5 minutes
VM instance (detail)	Every 1 minute
Storage volumes	Every 5 minutes
Load balancers	Every 5 minutes
DB instance	Every 1 minute
SQS queues	Every 5 minutes
Network queues	Every 5 minutes

In order to limit the compute load placed by an IB agent on a server, its measurement frequencies are limited, but then short-term spikes may get ignored. An example of measured resources in Amazon Web Services (AWS) is shown in Table 6.2.

More importantly, an IT manager or Cloud user needs to worry about what to do with the data being measured. For example, if a VM is slowing down, then it can be migrated to another server, or if not mission critical can be run at a later time. This may be possible if its output is not desired immediately, as some jobs can be run in the off-peak hours, e.g., late night hours or over the weekend.

6.6   Some Example of Monitoring Tools

The following are three contemporary and popular tools available for Cloud Computing users, first of which is provided for free by AWS:

1. *Cloud Watch:* Amazon CloudWatch [1] is a monitoring service for AWS Cloud resources and the applications running on AWS. Amazon CloudWatch can be used to collect and track metrics, collect and monitor log files, set alarms, and automatically react to changes in AWS resources, such as starting new servers when CPU utilization goes above a certain threshold. Amazon CloudWatch can monitor AWS resources such as Amazon EC2 instances, Amazon DynamoDB tables, and Amazon RDS DB instances, as well as custom metrics generated by

user applications and services, and any log files. One can use Amazon CloudWatch to gain system-wide visibility into resource utilization, application performance, and operational health, as shown in Fig. 6.2. These insights can be used for smoother operation of user applications.

2. *Nagios:* is a free and open-source tool [2] to monitor computer systems, networks, and infrastructure. Nagios also offers alerting services for servers, switches, and user applications. It alerts users when observed quantities cross a threshold, as shown in Fig. 6.3, and alerts when the problem is resolved.

3. *New relic:* is an application performance monitoring (APM) solution [3] that uses agents placed in a VM, say in a Cloud or local server, to monitor how that application is behaving, as shown in Fig. 6.4. It is able to predict a slowdown in services before a user experiences the slow response.

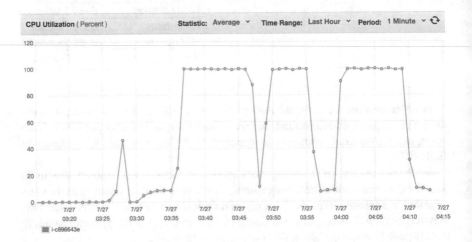

Fig. 6.2 CPU variations as observed in AWS Cloud Watch

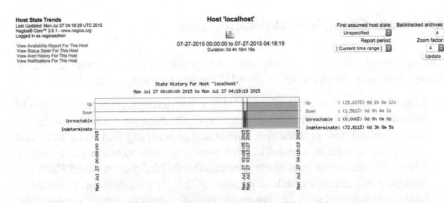

Fig. 6.3 An example of CPU host crossing a threshold and then becoming normal in Nagios

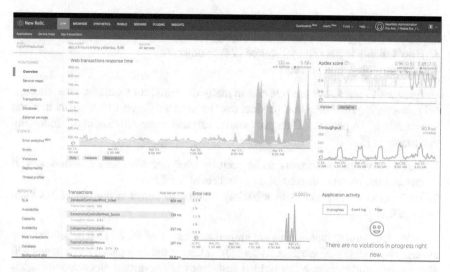

Fig. 6.4 An example of application performance monitoring (APM) in new relic

6.7   Monitoring Use of Images and Their Security

Monitoring requirements in a Cloud can extend beyond the performance of physical infrastructure. An example is the need to track usage of intellectual property stored and available in the public domain.

A significant portion of data in the Public Cloud consists of images and videos. Many of these are proprietary, and copyright owners expect a compensation for usages of their IP (intellectual property). Financial charges for using or copying images (including videos) require a monitoring mechanism. Also, privacy concerns for images require both monitoring and access control. The information (including ownership) related to monitoring and access control is associated with an image's identity.

Ownership and image identity can be separated from the actual images. One solution is to embed information hidden within the image that does not distort the image perceptibly. This is called digital watermarking. Watermark data can be extracted and tracked for usage. The digital watermarking can also be used as a security measure to ascertain ownership. The value of digital image watermarking [4] predates Cloud Computing. However, it presents a solution to the tracking and monetizing of image usage combined with the digital rights management (DRM) problem related to digital images in the Cloud. This is also an example of how an existing proven solution can be used for new class of problems arising in Cloud Computing. One such challenge is to track and record the origin of data objects, also known as data provenance [5]. Needs for privacy and security further complicate managing data provenance in a distributed Cloud environment. Watermarking has been used to address the following questions related to data security in Cloud Computing:

1. What is the original source of data objects?
2. Who is the rightful owner of data objects?
3. How reliable is the original source of data objects?
4. Who modified data objects?

Answers to the above are relevant in many domains; for example, pharmaceutical research requires data and product checks to avoid downstream errors that can irreparably harm patients. Data provenance can use watermarking in the following two manners [5]:

1. Embed a visible watermark to be seen by everyone who sees a data object. It ensures the trustworthiness of data packets in the Cloud.
2. Insert hidden watermarks as a backup facility, to detect if the object has been modified.

First step ensures that origins are known and recorded for all to see. Second step ensures the integrity of the data, as someone altering the data packets or images will also inadvertently change the hidden watermarks. These two steps require additional processing and a slight increase in the upload time of images to the Cloud. However, the added benefit of data reliability overrides these minor costs.

An example of Data Provence application that does not use digital watermarking is by Monterey Bay Aquarium Research Institute (MBARI). It has a large variety and number of data sources related to marine life and ocean environmental conditions. There is no standard format for the incoming streams or files, which are automated or manual. This data is collected from a variety of instruments and platforms. It can come in via Web HTTP, emails, server pulls, or FTP (File Transfer Protocol) interfaces. In order to manage this data, a Shore Side Data System (SDDS) was designed [6]. High-level requirements for this system include:

1. Capture and store data from various sources.
2. Capture metadata about the stored data, e.g., location, instrument, platform, etc.
3. Capture and store data products, e.g., quality controlled data, plots, etc.
4. Provide access to the original raw data.
5. Convert data to the common format for user application tool.
6. Present simple plots for any well-described data.
7. Archive processed data and maintain relationships between datasets.
8. Provide access to data and metadata through an API and Web interface.

SDDS provides a solution to the management of provenance early in the life cycle of observatory data, i.e., between instruments and datasets produced by their deployment. An application of this is a Cloud-Based Vessel Monitoring System [7], which enables protection of marine area by adding a geo-zone or geo-fence. Any vessel entering, crossing, or exiting the predefined boundaries can be identified, reported, and alerted on.

6.8 Follow-ME Cloud

Major consumers of Public Cloud services are the numerous mobile user devices, with their needs of live video calls and always connected social networks, with minimal latency requirements. However, these users are not stationary. A need to provide them with continuous IP (Internet Protocol)-based services while optimizing support from the nearest data center is nontrivial. Ability to smoothly migrate a mobile user from one data center to another, in response to the physical movement of a user's equipment, without any disruption in the service, is called Follow-ME Cloud (FMC) [8]. An example of FMC is shown in Fig. 6.5. Without FMC, a user's device will remain tethered to a data center, i.e., DC1, even though the user may move to Location 2 and Location 3. This may result in need to provide services from a centralized location, placing an undue burden on the intermediary networks. This will increase bandwidth requirements, higher network traffic, larger delays, and extra processing load. End users of a static Cloud location may also experience a reduced quality of service (QoS). A goal of FMC is to avoid such undesirable effects. It also mitigates latency concerns.

One solution is to use a Cloud management middleware [9], which migrates parts of a user's services between data center (DC) sites. These services need to be contained in a set of virtual machines, which migrate from servers in one DC to another. However, it is generally not possible to do a live migration of VMs across facilities unless resource states are replicated in real time. An iterative optimization

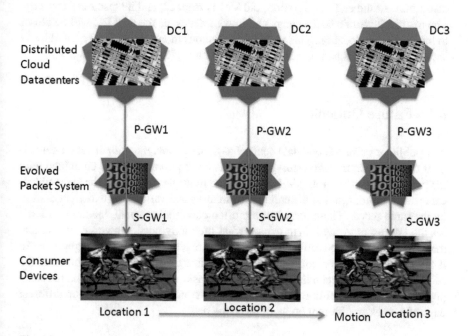

Fig. 6.5 Migration of a user's service with changes in location [8]

algorithm, Volley [10], has been proposed to move application components closer to user's geographical site. Volley migrates user services to new DCs, if the capacity of a DC changes or the user changes physical location. If these DCs are not using the same hypervisor, then additional time is required to convert a VM's memory format. Even if both the source and destination DCs are using the same hypervisor, it will take a finite time to transfer the services and replicate memory states over the network.

To overcome the above challenges, an additional layer of distributed mobile network is proposed [8], as shown in the Fig. 6.5. It provides Serving Gateways (S-GWs) and Packet Data Network Gateways (P-GWs) in the context of a distributed Evolved Packet System (EPS). The proposed architecture results in a federated Cloud spread over multiple regional DCs. These are interconnected yet geographically distributed. Main objective is to ensure an optimal mobile connectivity service. This is achieved by the following three key actions:

1. Replacing data anchoring at the network layer by service anchoring
2. Replacing IP addressing by unique service/data identification
3. Decoupling session and service mobility from DC network layers

When a user moves locations, the middleware layer detects motion by new incoming requests at a new Service Gateway (S-GW) and decides to trigger the VM migration for that user to a new optimal DC location. Until a new VM starts up and replicates services in a new DC location, user continues to be serviced by the initial DC. This may require some overlap of resource usages but minimizes the service disruption. At the end of migration, old VM is stopped, and its resources freed up. The net result will be cost savings and lower latency, as the new DC will be closer to the user's new location. This requires continuous monitoring of server workloads across multiple DCs and optimizations across multiple users' locations.

6.9   Future Outlook

Any public or Private Cloud data center has a mix of servers, with different generations of processor, memory, storage, and network equipment. These combined with the prevalent practice of multi-tenancy, where multiple users share the same physical infrastructure, result in different performances and varied user experiences even on the same server. These machines are often classified as small, medium, or large for the end users to help them decide what they may need. However, two virtual machines may reside on different server hardware giving different throughput. Even if the underlying servers are similar, performance for two virtual machines of the same type will vary depending on which other tasks or jobs are running on the same physical server. Furthermore, data center traffic conditions may change on different days of a week or different hours of the same day.

All of the above factors impact run-times of users' applications and result in different costs to finish a task. A Cloud user may want to maximize their performance or minimize their costs. However, there is no effective tool to predict this before selecting a machine.

We need a new solution to help unveil the physical characteristics of a Cloud machine and see what performance a user can expect in terms of CPU, memory, and storage operations. This will enable a user to model their application by selecting different proportion of these computational elements, or give a performance range, to see how different instance types will perform. Such a solution will also need to take into account applications that are multitiered and use multiple Cloud servers, as it is the composite performance that an end user cares about. An example is a travel agency's Website that upon a query goes in turn to query various airlines and hotels' Web servers, to give a price comparison back to the user. In this scenario, not everything will be in the travel agency's control, but they can place a SLA (service-level agreement) to their data providers or simply timeout and only list the components, which respond in time to avoid losing their customers. In a marketplace, where multiple vendors provide indiscernible services, price and performance are the key to winning.

6.10 Points to Ponder

1. Distance causes network latency, so AWS has set up various regional data centers to be close to its customers and reduce delays. This is one way to mitigate network latency. Are there other methods you can think of?
2. In an AWS region, there are different Availability Zones (AZs) to isolate any electrical fault or unintentional downtime. These could be different buildings or different floors within a data center building. What are other ways to build failure safe strategy?
3. What kind of metrics need to be monitored to ensure the health of a data center?
4. What are the trade-offs of frequent vs. infrequent monitoring of a server's performance?
5. How long the monitoring data should be kept and for what purposes?
6. What's the role of monitoring if a SLA is in place?
7. What are the security concerns of Follow-Me Cloud?

References

1. https://aws.amazon.com/Cloudwatch/
2. https://www.nagios.org
3. https://newrelic.com/application-monitoring

4. Acken, J. M. (1998). How watermarking adds value to digital content. *Communications of the ACM, 41*(7), 75–77.
5. Sawar, M. et al. (2017). Data provenance for Cloud computing using watermark. *International Journal of Advanced Computer Science and Applications*, 8(6). https://thesai.org/Downloads/Volume8No6/Paper_54-Data_Provenance_for_Cloud_Computing.pdf
6. McCann, M., & Gomes, K., *Oceanographic data provenance tracking with the shore side data system*. https://link.springer.com/content/pdf/10.1007%2F978-3-540-89965-5_30.pdf
7. https://www.ast-msl.com/solutions/gap-vms-shore-side-application/
8. Taleb, T., & Ksentini, A. (2013). Follow-ME Cloud: Interworking federated Cloud and distributed mobile networks. *IEEE Networks*, September–October 2013. https://ieeexplore.ieee.org/document/7399400
9. Malet B., & Pietzuch, P. (2010). Resource allocation across multiple Cloud data centres. *Proceedings of ACM MGC*, Bangalore, India.
10. S. Agarwal et al. (2010). Volley: Automated data placement for geo-distributed Cloud services. *Proceedings of the 7th Symposium on Networked system design and implementation*, San Jose.

Chapter 7
Cloud Computing and Information Security

7.1 Information Security Background, Information Context, and Definitions

Information security can be viewed as including three functions, namely, *access control, secure communications*, and *protection of private data*. Alternatively, a common three-topic split is confidentiality, integrity, and availability (the CIA of security policies and objectives). Sometimes, information security is shortened to INFOSEC. Access control includes both the initial entrance by a participant and the reentry of that participant and the access of additional participants. Note that a participant can be an individual or some computer process. The secure communication includes any transfer of information among any of the participants. The protection of private data includes storage devices, processing units, and even cache memory [1, 2].

The first function encountered is *access control*, i.e., who can rightfully access a computer system or data. The access control can be resolved at a hardware level with a special access device such as a dongle connected to the USB port or built-in security keys. Access control is usually addressed at the operating system level with a login step. An example of access control at the application level is the setting of cookies by a browser.

After access control is granted, *secure communication* is the next function, which requires encryption. The most commonly recognized function of a secure system is the encryption algorithm. The most commonly recognized problem in a secure system is the encryption key management. At the hardware level, the communication encryption device can be implemented at the I/O port. At the operating system level, encrypted communications can be implemented in secure driver software. At the application level, the encryption algorithm is implemented in any routine performing secure communication.

Some of the other functions and issues for security systems are hashing (for checking data integrity), identity authentication (for allowing access), electronic

© Springer Nature Switzerland AG 2020
N. K. Sehgal et al., *Cloud Computing with Security*,
https://doi.org/10.1007/978-3-030-24612-9_7

signatures (for preventing revocation of legitimate transactions), information labeling (for tracing location and times for transactions), and monitors (for identifying potential attacks on the system). Each of these functions affects the overall security and performance of a system. The weakest security element for any function at any level limits the overall security and risk. Weakest function is determined by technical issues (such as length of passwords) and also by user acceptance (such as blood samples in an extreme case if so required to authenticate a rare access).

In addition to accepting the security process, a user may have concerns regarding *protection of private data.* Protection of data includes both limiting availability of date to authorized recipients and integrity checks on the data. Specifically, the more information is required to ensure proper access, the more private information is available to the security checking system. The level of security required is not universal. Ease of access is more important for low-security activities, such as reading advertisements. More difficult access is required for medium security such as bank accounts. High security is required for high-value corporate proprietary computations, such as design data for a next-generation product. Very strict and cumbersome access procedures are expected for nuclear weapon applications. These examples provide a clue to security in a Cloud Computing environment with shared resources [3]. Specifically, in the same computing environment, applications are running at a variety of security levels [4]. Security solutions must also consider the trade-offs of security versus performance. Some straightforward increases in the security cause inordinate degradation of performance. As described previously, the security implementations can be done at multiple levels for each of the functions. Because security is a multifunction multilevel problem, high-level security operations need access to low-level security measurements. This is true in monitoring both performance and security.

Three environmental factors directly affect the evolution of information security: computing power available, growing user base, and sharing of information resources. The first factor has been and continues to be the computer power available to both sides of the information security battle. Computing power continues to follow Moore's law with increasing capacity and speeds increasing exponentially with time. Therefore, while the breaking of a security system with brute force may take many years with the present computer technology, in only a few years, the computer capacity may be available to achieve the same break-in within real time. The second environmental factor is the growing number of people needing information security. The world has changed from a relatively modest number of financial, governmental, business, and medical institutions having to secure information to nearly every business and modern human needing a support for information security. The sheer number of different people needing information security has increased the importance of different levels of security. The third environmental change that has a significant impact on the security is the sharing of information resources that is the crux of this chapter.

Specifically, this chapter describes the information security structural changes caused by the spread of Cloud Computing, and more people across the world accessing the Internet not just through PCs and browsers but a myriad of mobile devices and applications. This has dramatically increased the risks and scale of potential damage caused by realization of a security threat on the Internet.

7.2 Evolution of Security Considerations

The two independent factors that have driven the evolution of security considerations on the way to Cloud Computing are the environment (usage models for access and use) and performance (both computation and communication). The computation environment has evolved from a very small select set of specialized users to universal access and therefore a great span of skill sets and abilities. When computers were first created, they were very large, expensive, and difficult to use devices in a physically protected location with physical access limited to the few people capable of using them. The storage of information was based upon devices directly connected to the mainframe and relied upon physical control for information security. Even removable medium, such as magnetic tapes, relied upon physical control for security. Next, an increase in the number of users required access control to be based upon connecting the mainframe computer to users via hardwired terminals separated from the physical computer hardware. The security was based entirely upon access control, which used a user ID and password model. The next generation allowed remote access via telephone lines. This still relied upon user ID and password logins for access control. Note for these two phases, there was no information control (i.e., encryption) of the data transmitted between the user terminal and the mainframe. The next step in evolution is the individual personal computer. This returned to the physical control of the device for access control. However, it added the access control problem of not only a terminal talking to a mainframe but of a computer talking to another computer. The communication was still via dedicated channels (whether a hardwire connection or a direct telephone link). Access control via login or physical access was still the basis for information security. The only limitation to access for authorized users was availability of resources, that is, the number of communication lines or computer time. With the advent of the personal computer, computation was no longer relegated to experts and specialists. With the change come two challenges to security: the huge increase in average people needed easy access and the number of attackers increased greatly. The next step was the Internet. That is a communication protocol that allowed any computational device to communicate with any other computation device with a variety of communication paths. Now the security of the data during transmission needed to be protected in addition to the access control to the devices that were communicating. Also, at this stage removable media had become universal and was in need of information protection. Practically, speaking removable media still relied upon physical control for security.

Now, information communication was routinely, although not universally, protected with encryption. As the evolution of computation leads to Cloud Computing, now everything was available to everyone one. The physical boundaries are gone. When connected to and using the Cloud, the user and even the providers can be using resources anywhere. In fact, that is the goal for Cloud Computing to separate delivering the desired service from the underlying implementation of the capability to deliver the service. This is a huge difference in security, as nothing can be assumed to be secure. In addition to protection of information, now there is the new security problem of protecting access to services. An unauthorized user can attempt to access a resource with enough effort that it interferes with the usage by authorized users. This is called a denial-of-service attack. For example, excessive requests to a Web server can cause a Website to crash. This type of attack, barring physical attacks, did not exist in earlier computer environments. Another security issue is privacy. The information required to access and utilize one service should neither reveal that information to unauthorized parties nor to unauthorized use by parties authorized to have that information. Additional, the environment includes the security risk of attackers falsifying or changing information. This includes changing messages in transit or initiation of false messages. This requires an electron signature or check-sum to detect any data tampering. The current environment requires any Cloud user or provider to consider security in many places. Such security considerations have performance impact and many trade-off points, as depicted in Fig. 7.1. This impact is a measure of drop in the performance due to computation overhead of encryption and decryption. As shown, performance cost of full data and data encryption is very high but needed for highly sensitive data in a Public Cloud. On the other hand, access control and login passwords may be sufficient for single-user devices, such as a personal laptop. Then hash checking and secure handshake, using techniques such as VPN, between a personal device and Cloud connection may be considered sufficiently secured. If there is a desire to balance the security and performance, then sensitive data and code can be partitioned to run in a secure environment in a Public Cloud.

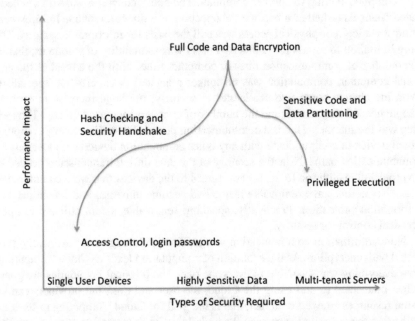

Fig. 7.1 Trade-offs between security required and its performance impact

In parallel to the changing environment affecting the evolution of security considerations, the changes in performance directly affect security considerations. When communication and performance were slow, minimum information security was required because the attackers had simple tools to get access. Simple tools included automated password guessing or brute force encryption hacking. Due to a low level of computational performance, all of these techniques were far too slow to be a serious threat. One method to increase security is to increase the encryption key size. The key size is limited to the performance used to encrypt, as larger keys require more computation. On the other hand, the security of the key size must be large enough to exceed the performance available to the attackers. As computational capacity increases, the keys must get bigger because the attackers have more capacity, and the keys can get bigger because the users have more capacity. Thus, for encryption there is a continuous race between defenders and attackers. The computing environment oscillated between when the communication channel was the performance limiter and when the computational capability was the limiting factor. In the present world of Cloud Computing, the attackers have access to significant computational ability; therefore, security is now a universal serious problem.

7.3   Security Concerns of Cloud Operating Models

Security considerations vary depending on the users' interaction levels at the Cloud [5]. Let us revisit the three operating models for Cloud Computing described below. We shall also discuss key requirements at each of these levels with security boundaries as shown in Fig. 7.2:

1. *Software as a Service (SaaS)* is the highest layer of abstraction focused on end-users of Cloud, to provide them application access, such that multiple users can share the same application binary in their own virtual machines or server instances. Users want to ensure that their data is protected from other users, besides intruders, in the Cloud. Each user's persistent data may need to be encrypted to avoid unauthorized access. An application provider wants to ensure that users have a read-only access to the binary and all patches are updated in a timely manner. This may require a restart of the application instance, if the binary version on disk changes.

2. *Platform as a Service (PaaS)* is focused on application developers, providing them access to elastic servers that can stretch in CPU cores, memory, and storage on need basis. Users want to ensure that platform services, such as various utilities, can be trusted and there is no man-in-the-middle attack, such that user data is compromised. The PaaS requires strong middleware security. PaaS permits integrations of services, so the security perimeters need a greater degree of access control. An example is machine learning and Hadoop clusters that are being offered by several Public Cloud providers. These require user jobs and data to be copied across several machines, so the protocols between software layers and services between various machines on a server cluster need to be secured.

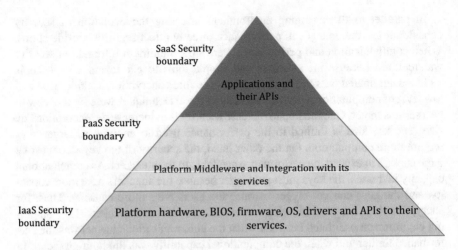

Fig. 7.2 Key requirements and security boundary at each level of Cloud services

3. *Infrastructure as a Service (IaaS)* is the bottommost layer in a Cloud stack, providing direct access to virtualized or containerized hardware. Users want to ensure that hardware level services, such as various drivers and ports, are protected from other processes running on the same physical server. An example is a virtual machine (VM) using the local disk to store data. After that VM's life-cycle is over, another user with a different VM then occupies the same memory and disk space [6]. Now the first user wants to be assured that the storage is zeroed out and the second user can't access any residual data. This can be accomplished by per-user encryption of shared services on a platform, such as storage, networking, and memory, but that will cause run-time penalty every time such a resource is accessed. Another simpler method is to initialize the shared persistent state before the users are switched.

The previous paragraphs described several information security functions and proposed a relationship between the levels and details of implementation in a computer system. The levels and functions for security described throughout this chapter are summarized in Table 7.1.

7.4 Identity Authentication

Traditionally, identity authentication is applied when an individual requests access to a system. For this situation, the three elements or items used for identity authentication are what you have, what you know, and what you are [8]. An example of what you have is a house key. The use of the house key to unlock the door is a sufficient identity authentication for the person to enter through the door. Examples of what you know are passwords, PINs, and your birthday. Examples of what you are

Table 7.1 Information security functions and digital computer system levels

		Access control	Secure communications	Data protection	Monitoring
Software	User application	Some login, usually relies on lower levels [1]	Usually relies on lower levels of implementation	Encrypt or disguise data [5]	Access logs
	Operating system (OS)	Login [1]	In-memory transactions	[1, 5]	Special processes as watchdogs
	Virtual machine layer (VM)	[1, 5]		[1, 5]	[1]
	Hypervisor layer	[1]		[1]	[1]
	Software drivers	From OS	Encryption, security handshake	Encrypt data	
	BIOS/ FW-based system management layer	Privileged execution		Privileged access to certain memory locations	Log files
Hardware	CPU	From OS	Port and bus encryption, secure caches	Separate secure registers and memory	
	Memory cache/main RAM	Encrypted busses, hash checking tables	Data encryption	Partitioning and encryption	Interrupt logs
	Memory disk	Hash, checking tables [5–8]	USB data encryption	Encrypt disk storage, removable devices	Error logs [9, 10]
	I/O	Verify access ID, such as Internet IP address	Encrypt transmissions; trust keyboard, mouse, and audio	Security handshake, coding, encryption	Watchdog processes in hardware and software

refer to biometric information. These are pieces of information that you give to confirm your identity usually in conjunction with an item you have such as a bank card or a check.

The use of passwords is a standard access control method. Passwords have the added value of being accepted by the user community. However, passwords are susceptible to brute force hacking. Therefore, some policies are implemented to increase the size and complexity of passwords. This leads to the practice of users writing down the long complex passwords on a post-it note affixed to the terminal screen or kept in an office drawer. This puts a human limit on the passwords, which leads to the direct prediction of "password exhaustion" [11]. The processing power available can readily hack passwords of a practical size. The conclusion is that the end of password usefulness is at hand. This conclusion is true if a system relies solely on password protection. However, passwords as part of a multipronged identity authentication provide a low-cost, user acceptable increase in security. Passwords can be compared to the locks on car doors. The locks (or passwords) can't protect against the directed high-powered attempts to gain access. However, they do provide adequate security for the lesser or casual attempts. A damaged car lock or a failed password warning provides a signal that a break-in was attempted. This allows intervention before any additional security steps are circumvented.

What you are is a physical characteristic. For example, a picture ID or driver's license is frequently requested to authenticate your identity. A physical characteristic that is measureable is a biometric. Examples of biometrics include height, weight, voice matching, face recognition, iris differentiation, and fingerprints [12]. Measuring the individual characteristic and comparing that measurement to the value of authorized individuals can check biometrics. This is a problem in the Internet world where the biometric measurement device is not under strict control, for example, fingerprints [12, 13] or voice [14, 15]. Cloud Computing introduces a whole new challenge for identity authentication. As seen even in a decade-old survey of IT professionals in Fig. 7.3, security is their #1 concern. This has not changed in the recent years.

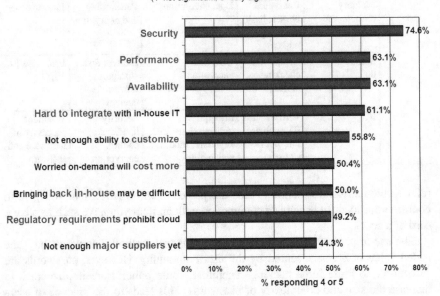

Fig. 7.3 Security is IT professionals' #1 concern in Cloud Computing. (Source: IDC Enterprise Panel, August 2008 n = 244)

An extensive discussion on the definition of Cloud Computing was in the previous chapters. For an identity authentication example, consider that when a program running in Cloud needs to access some data stored in the Cloud, then what you have and what you are criteria are irrelevant. However, the context of access request is relevant and can be used as described in an Amazon Web Services (AWS) Security Best Practices White Paper [9]. It describes an Information Security System Manager (ISSM), which is a collection of information security policies and processes for your organization's assets on AWS. For more information on ISSM, see ISO 27001 at http://www.27000.org/iso-27001.htm. AWS uses a shared responsibility model, which requires AWS and customers to work together toward their security objectives.

AWS provides secure infrastructure and services, while its customers are responsible for securing operating systems, applications, and data. To ensure a secured global infrastructure, AWS configures infrastructure components. It provides services and features that customers can use to enhance security, such as identity and access management (IAM) service, which can be used to manage users and user permissions in a subset of AWS services. To ensure secure services, AWS offers shared responsibility models for different types of service such as:

- Infrastructure services
- Container services
- Abstracted services

The shared responsibility model for infrastructure services, such as Amazon Elastic Compute Cloud (EC2), specifies that AWS manages security of the following assets:

- Facilities
- Physical security of hardware
- Network infrastructure
- Virtualization infrastructure

In this Amazon EC2 example, IaaS customer is responsible for the security of the following assets. Figure 7.4 depicts the building blocks for the shared responsibility model for infrastructure services:

- Amazon Machine Images (AMIs)
- Operating systems
- Applications
- Data in transit
- Data at rest
- Data stores
- Credentials
- Policies and configuration

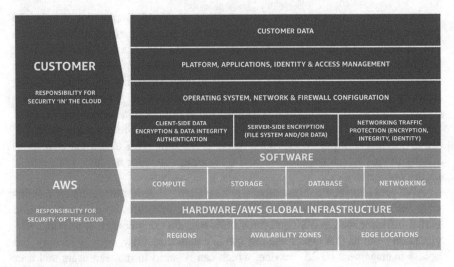

Fig. 7.4 Shared responsibility model for infrastructure services

7.5   Secure Transmissions

Secure transmissions are required when data transfer is required between the islands of security via the Internet. The model assumes a secure source environment and secure destination environment exist. The communication channel between the secure environments is open or unsecure. To protect the information to be transferred, it is encrypted whenever it is in an unsecure environment. The primary approach is to encrypt the information whenever it is "outside" of a secure environment. The primary tool for secure communication is encryption. Many aspects limit the security of an encryption system, such as key management, the strength of the specific encryption algorithm used, the size of the key, and the time and effort required to crack an encrypted message. Much of the security research and security policy decisions are based on the time and effort required to crack an encrypted message. When implementing a secure system, a trade-off decision must be made between the security of the encrypted information (time and effort to crack the cipher) and the efficiency of the system (time and effort to encrypt the message). The trade-off is evaluated within certain boundary conditions, such as the lifetime and value of the information. A message about a transaction occurring tomorrow only needs to stay secured for 2 days, which limits the time and effort that will be available to crack the cipher. An example of this might be a major diplomatic announcement scheduled for the next day. It has a limited time value. In contrast, another message with years of lifetime value offers ample time for very significant effort to break the message. An example of this might be a detailed description of a long-term investment for a new drug discovery and its trial posing a competitive threat to other market players.

7.6 Secure Storage and Computation

The trade-offs between efficiency and security described for transmission also apply to the storage and computation. A common solution for security and integrity checking of networked storage environments is encrypted data file systems [10]. The CFSs (cryptographic file systems) are a significant performance burden. Using a CFS is especially needed when the data storage is farmed out to untrusted storage sub-providers [7]. A big difference is the wide range of storage lifetime. For storage such as copyrighted movies on DVD, there is a longtime value (several months or even years); however, for storage such as main memory, there is a short-time value (perhaps microseconds). The emphasis in the main memory security should be on read-write efficiency. A small loss of time here has a huge impact on the performance of a computer system due to repeated operations.

Hence lighter and faster encryption schemes can be applied to data of ephemeral value, such as being held in a system memory, which will be short lived. For long-term data, companies such as financial institutions have invested in hardware security modules (HSMs). A HSM is a physical computing device acting as a vault to hold and manage digital keys for strong authentication and provides crypto processing services. This can be a plug-in card or an external device, attached directly to a network server. These HSMs are certified as per the internationally accepted standards, such as FIPS (Federal Information Processing Standard) in the USA to provide users with a security assurance.

7.7 The Security Players

As the previous sections demonstrate, the description of a secure system and its participants can get confusing. A common procedure for describing the participants in secure systems, scenarios, and security protocols is to use some nicknamed participants. In various descriptions, key personnel will maintain the same role. For example, person A will always represent the initiator or first participant in the given scenario. Not every scenario has every type of role. In fact, most scenarios have only a few of the roles. For convenience, the parties represented are given nicknames in addition to fixed roles. Several different publications have used this nicknaming technique when describing security systems and protocols [8, 11, 12]. Although these participants do not need to be physical human beings, for this chapter we are adding some virtual entities, as shown in Fig. 7.5. Security attacks at different layers of computer system are shown below in Table 7.2.

Cloud Computing presents a case with different classes of users using the same processes. That is, users requiring very high security are using the same operating system as the users requiring low security and therefore gain the same access. Also, the users requiring high security and the users requiring low security can run the same applications thereby having passed the same access required to run the application software. The implementation of typical real-world multilevel security

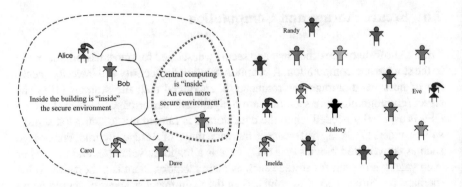

Fig. 7.5 Traditional computing security depends upon physical security

Table 7.2 Security attacks and challenges for digital computer system levels

		Unauthorized data or program changes (malicious by Mallory and accidental by Randy)	Unauthorized observation and copying (intentional eavesdropping by Eve and accidental leaks to Randy)	Denial-of-service attacks (intentional by Imelda and accidental by Randy)
Software	User application	Fake login, or indirect access [13]	Usually relies on lower levels of implementation	Encrypt or disguise data [4]
	Operating system (OS)	Fake login, low-level instruction [1]	In-memory transactions	
	Virtual machine layer (VM)	VM to VM communication [1]	Information leaks [5]	[4]
	Hypervisor layer			
	Software drivers	From OS	Encryption, security handshake	Encrypt data
	BIOS/ FW-based system management layer	Time and date stamps [13]	Secure memory locations	Authentication for execution
Hardware	CPU	Information leaks [15]	Information leaks [15]	
	Memory cache/main RAM		Information leaks [9, 15]	[14]
	Memory disk	Access privileges [9, 13]	Access privileges	

(MLS) is too rigid for most resource sharing activates (such as the Internet) [16] and completely inadequate to meet the Cloud Computing challenge. The players described in this section have the same assignments in both traditional, Internet, and Cloud environments.

7.8   Traditional vs. Internet Security Issues

The traditional approach to information security relied upon physical barriers. In this case, there is a clear physical boundary of what is inside the security perimeter and what is outside. The computing center access is available only to key personnel. The hardware and network connectivity to the main computers are physically identified and can be checked for any taps. Individuals with a higher level of security clearance, such as system administrators, can have special hardware access lines. All of the individual users have a terminal in a controlled environment, such as in their offices or a terminal room.

Humans and devices can both be controlled for monitored access and activity. Devices to control access include card readers and key locks on doors. Human security guards can be posted at key points. Cameras are the primary devices to monitor activity. The most common human support for monitoring activity is the other workers. They can see people they know and what they are doing. People can also look for strangers by observing a person's uniform or badge to determine whether they seem to be doing an appropriate activity or not. The casual monitoring by coworkers is not only effective at catching intruders; it is also a deterrent to causal attempts at security breaches. This physical barrier provides two natural increases in security: limit repeated attempts and time available to perform a breach. In the case of repeated attempts, it becomes obvious to the most casual observer when the same unauthorized person tries to get in multiple times. And it is also obvious when a stranger keeps carrying files or disks out the door. As for the time, it takes minutes to gain access or gather information as the stranger walks around looking at computer screens on workers' desks.

As is shown in Fig. 7.6, the definition of "inside" the security perimeter and "outside" the perimeter is clear. When Alice wishes to get information to Bob, she can just walk over and stay inside the secure perimeter. Eavesdropping by Eve requires a much greater sophistication than the effort to secure the communication. For example, Alice may post documents on her bulletin board facing a clear window open to Eve's apartment across the street, providing her a way to view remotely. Even in the traditional environment information, security leaks occur. Phone conversations, documents, and removable storage media did provide opportunities for information from within the secure environment to get to unauthorized individuals. However, usually enforcing policies based upon the physical boundary was sufficient to provide information security.

The Internet created a new issue – that is, connecting secure islands of information via open channels. The computing center with large computing and storage resources has controlled access and activities. This creates a large island of security where

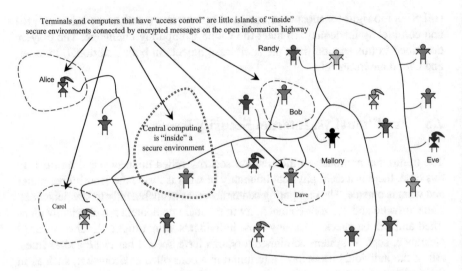

Fig. 7.6 Information security on the Internet

major resources are still controlled and monitored with humans and devices. However, now system operators have access via uncontrolled hardware lines, identical to regular user access lines. Unlike the traditional case, there is no casual monitoring by coworkers. As has been said in the cartoon world, "On Internet nobody knows your real identity." Also, the Internet provides an intruder with unlimited tries to gain access. After an intruder has failed to gain access, the failed attempts cannot be attached to the intruder. Each attempt may appear as a first attempt as IP addresses can be spoofed. Time to attempt repeated access is greatly reduced. Procedures to stop these intruders impact legitimate users. For example, one can stop the high number of repeated access attacks by limiting the number of false attempts by locking an account after some number of false attempts. However, this can prevent a legitimate user from accessing her account. This can lead to another form of security attack called denial of service (DOS). The idea here is not to gain access but to prevent legitimate users' access by hitting the access attempts' limit and thus locking out (or denying service) to legitimate users.

The value of resource sharing via a network for a collaborative work environment is clear and established. One approach to access control to the collaborative environment is based upon people tagging [16]. On the Web, or any shared storage, the physical monitoring of access and activity is greatly reduced. In fact, the case of casual monitoring is completely eliminated. The integrity of shared data can be verified with digital signatures [7]. Integrity checking is needed when access is difficult to control. As shown in Fig. 7.4, the adversaries have access to the open channel. In addition to the open channel, the island of security concept has a time component to the isolation of a resource. A computer connecting to the central resource could be a shared resource such as a computer in a terminal lab. The next user sits down and has access to information that should have been protected but hasn't been erased by the previous user. The concept of attacking persistence in the main memory [17] is to read memory containing passwords or other secret data that is not consciously erased between users.

7.9   Security Using Encryption Keys

As mentioned in Chap. 2, three key elements of Internet security are public key encryption, private key encryption, and secure hashing. We will present details for each of these. The examples chosen (such as RSA, AES, and SHA-3) are commonly used in the current Cloud Computing deployments.

A Symmetric Encryption requires a unique key for each pair of communicators. This increases the number of keys a person must maintain for communicating with lots of different people. For a large number of people to communicate with each other, this method is not scalable, and key management becomes an issue. Asymmetric Encryption uses a different key to encrypt than to decrypt. The name public key refers to an encryption key being made public, i.e., published openly. The RSA algorithm is an example of an Asymmetric Encryption Scheme [17]. Suppose Alice wishes to send a secret message to Bob. Bob generates the keys and publishes his encryption key. Alice uses this public key to encrypt her message and sends it over to an open channel to Bob. Bob uses his secret decryption key to read the message. Bob generates the keys by finding two large different random prime numbers, say p and q, and multiplies them together to create a very large integer n = p∗q. Then a "totient" is created from the large prime numbers and used to create the encryption and decryption keys.

These keys comprise of two numbers: (e, n) is the encryption key set of numbers, and (d, n) is the decryption set of numbers. A text message is encrypted and converted to a block of numbers. First, each character of the text is replaced by its numerical binary representation in the ASCII code. These blocks are sized so that the number representing the block is less than n. Each block from a plaintext number m is raised to the e power modulus of the number n. This plain or clear text number, let us call it m, is then encrypted by Alice to a ciphertext, let us call it c, using the following equation:

$$c = m^e \bmod n$$

The ciphertext c is sent to Bob, who decrypts the message using the secret number d using the following equation:

$$m = c^d \bmod n$$

The three steps of Asymmetric Encryption are shown in Fig. 7.7. It starts with Bob generating a set of keys, a public one to share with senders and a private one for him to decrypt the data. In the next step, Alice uses Bob's public key to encrypt a message. Even if an attacker, such as Eve or Mallory, tries to read or alter the message, they will see only an encrypted text. Lastly, Bob uses his private key to decrypt Alice's message. This is a complex and slow process, as compared to symmetric algorithms. However, it has an advantage over Symmetric Encryption by reducing the key management problem. Asymmetric Encryption is often used to exchange private keys used by Symmetric Encryption.

Symmetric encryption operation is much faster than Asymmetric Encryption. However, for Symmetric Encryption n, the same key is used for both encryption and decryption, so the problem is to secretly get the key to both parties. AES is an example of Symmetric Encryption that is widely used in the Cloud and throughout the Internet [18]. AES encryption is implemented as a series of n rounds, where n depends on the key size. Each round consists of a byte substitution step or layer, a shift row step or layer, and a mix column step or layer followed by a key addition step or layer. To decrypt, the rounds are in the reversed order for encryption. Symmetric Encryption is shown in Fig. 7.8.

Hashing is an information security technique to mark a message to prevent tampering. The information to be hashed is broken into specified size blocks and combined to create a digest number that is much smaller than the input. The raw data and its digest are sent to a recipient, who can rehash the received data, compute a new digest, and compare it to the transmitted digest. If two digests do not match, then the message has been tampered with. It is very difficult to find another input that will result in the same hash value. For example, a long contract is hashed, and an attacker wants to modify the agreed-upon price. Simply changing the price will

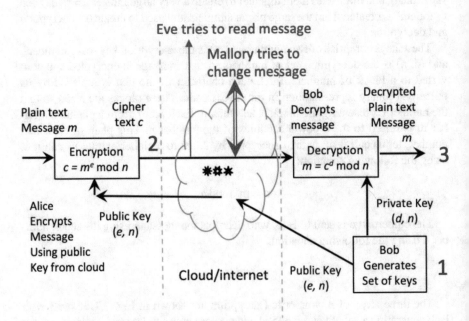

Fig. 7.7 Asymmetric Encryption in the Cloud/Internet

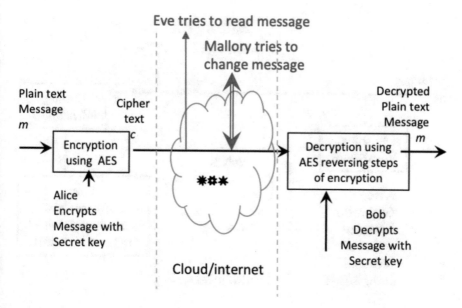

Fig. 7.8 Symmetric Encryption using AES

result in a new hash digest and indicate that the contract has been modified. However, if the attacker can find a few other changes (such as spelling of words) that result in the same hash digest, then the attacker can change the data and not be detected. This digest being equal for both the original data and tampered data is measured as a collision rate. SHA-3 (third generation of simple hashing algorithm) is a standard secure hashing algorithm. The block size is 1600 bits. The blocks are split into r bits of data and c bits of padding, where r + c = 1600. The higher the r is, the faster hashing will run, but a higher c ensures a more secure hash. Hashing on the Internet is shown in Fig. 7.9.

Information security includes the protection of information from reading, alteration, and falsification. Encryption is a key tool for protecting information from unauthorized reading, and hashing is a key tool for preventing alteration or falsification. This section gave example algorithms that are commonly used in Cloud Computing today. Specifically, the Symmetric Encryption example of AES is used for securing large amounts of data for storage or transmission. The asymmetric example of RSA is used for exchanging small amounts of information, most commonly a method to securely exchange keys, which are required for Symmetric Encryption. Finally, SHA-3 is an example of secure hashing that protects information from alteration or falsification [19]. The nature of Cloud Computing exacerbates implementations of these solutions, as we shall discuss in the next section.

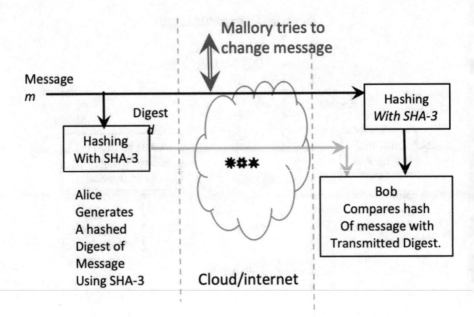

Fig. 7.9 Securing hashing using SHA-3

7.10   Challenges of Using Standard Security Algorithms

System security breaches are more often due to implementation weaknesses rather than flaws in the underlying security algorithm [20]. For example, a message encrypted with a 4096 bit RSA key would take a very long time to be cracked by brute force of trying to decrypt it with every possible key. However, if the private key is kept in a file on the hard drive, one just needs to find it and read it. The standard approach to creating and evaluating the security of a system is to identify the secure boundary and the possible breaches to that boundary. The challenge in Cloud Computing security is the lack of an identifiable boundary.

The first challenge for Cloud Computing security is that many security implementations are hardware dependent [21]. The very nature of Cloud Computing is to virtualize the hardware so that the application can execute on a variety of machines and even utilize different machines for different parts of a large task. Some algorithms will execute much faster on one type of architecture than another. Also, some service providers may have some specialized hardware accelerators. An example of where specialized hardware is sometimes used in the Cloud is for cryptocurrency hash function calculations. One part of security is to authenticate a process to allow access to a data or another process. A hardware implementation support for authentication is the processor ID. So, once an application on a particular process is authenticated, it can be tied to the processor ID for later verification. You may have observed this effect when you login to an established account from a new computer.

The server side determines that you are logging in from a new computer and adds some level of authentication checks. Some sites check by asking user a simple question, "whether or not this is really you." Would an attacker like Mallory ever say "no this isn't me, I am trying to break into Alice's account?" However, hardware specific solutions are not the only challenge for Cloud Computing.

The second challenge for Cloud Computing security is side channel attacks [22–24]. These are described in a later section of this chapter, but this challenge is enhanced in Cloud Computing because resources are assigned and shared among users in such a way as to give access to potentially malicious users on the same hardware.

The third challenge for Cloud Computing in the era of IoT (Internet of Things) and edge computing is that many idle resources are available for overwhelming the service provider. This approach can be used to create a distributed denial-of-service attack regardless of the size and capacity of data center for providing the Cloud services. By massively attempting to access one resource, the service provider will be overwhelmed assigning bandwidth, response, and computation for that one resource. Another denial-of-service attack can be executed completely within the Cloud environment. Here a version of the noisy neighbor problem described in Chap. 13 can be converted from the performance penalty described there to a target denial-of-service attack.

The fourth challenge for Cloud Computing security is backward compatibility versus security breach updates [25]. Information security has been likened to an arms race where every time an attack succeeds, a new defense must be created and deployed. And every time a new defense is implemented, the search for a new attack begins. The attackers have the advantage of picking new parts of the security boundary to attack.

The trade-off between performance, ease of use, and security is not unique to Cloud Computing. This challenge is greater for Cloud Computing due to a wide variety of security requirements implemented within the same Cloud environment. Thus, the Cloud service provider can readily create a charging model for performance versus cost. However, this is not the case for security. The system weakness allowed for a low-cost solution cannot be easily turned on and off for the high-security applications. In summary, the challenge for implementing security in Cloud Computing is the fact that security is implementation dependent, while providing Cloud services is most efficient if it is implementation independent.

7.11   Variations and Special Cases for Security Issues with Cloud Computing

The levels of implementation and functions of a computer system are the same in all three types of Cloud Computing. The difference is in the access and monitoring the usage of each element. In the case of Cloud Computing, there is no guarantee that a particular element be tightly coupled to another element for a particular process. For example, a database access request cannot rely on the fact that the application program on machine A has sole access through the operating system to the I/O port on machine A, as the storage can be shared between several machines. In fact, sharing in Cloud Computing is the very essence of efficiency to increase resource utilization instead of isolated resources. The security of Cloud Computing varies with the models used. Public Cloud model is used as it poses the greatest security challenges.

7.11.1   The Players

Traditional computing environments had a clear delineation between "inside" and "outside." Physically, "inside" might be in Alice's office or "inside" the bank building. With the dawn of networks, and especially the Internet, the networks were partitioned as "inside the firewall" and "outside" which could be anywhere. This is one of the differences between a Public Cloud and a Private Cloud. Secure communication was only needed when the communication went from "inside" to "outside." With Cloud Computing, "inside" is not clearly defined as computers in multiple data centers across different geographies are pooled together to appear as a large virtual pool. Figure 7.10 shows the introduction of Cloud Computing relative to information security. This is true for both Public and Private Cloud because a Private Cloud can use a virtual private network (VPN) that uses the open Internet to connect services only to a restricted category of internal clients. Eve and Malory could have access to the same set of physical resources that public is using. Some

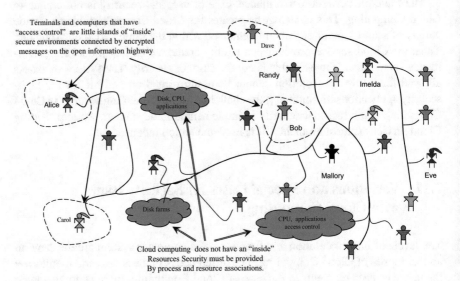

Fig. 7.10 Information security and Cloud Computing on the Internet

people can more easily trust a Private Cloud. Remember that within a Private Cloud, eavesdropping Eve and malicious Malory could be one of your coworkers or someone external.

Also, in a Cloud environment, the unauthorized access could be by some random Randy that inadvertently slipped through the access barrier, as shown in Fig. 7.11. The monitoring and response for security purposes must not only consider the level of secrecy and impact of a breach but also the category of intruders.

7.11.2   Secure Communication

The communication between islands of security through a sea of openness as describe previously is solved by encryption all of the data in the open sea. This is shown in Fig. 7.12.

7.11.3   An Example of Security Scenario for Cloud Computing

As an example of security situation for Cloud Computing, consider a very large medical database. Alice represents the doctor's office administrator that must enter and retrieve patient information for a particular doctor. Bob represents an insurance company administrator that must enter and access patient information for a particular insurer. Carole represents a second medical database administrator with some overlap with the first database. Dave is the doctor at Alice's office and will need to access anything Alice does for reminders, changes, and diagnostic tasks. The patients' pharmacy, hospital, or medical specialists may require additional access to

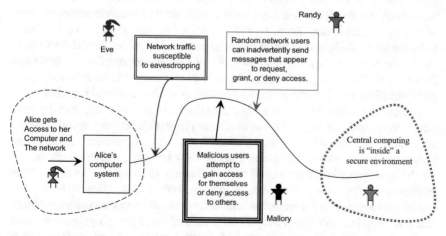

Fig. 7.11 Traditional Internet environment with communication between the islands of security

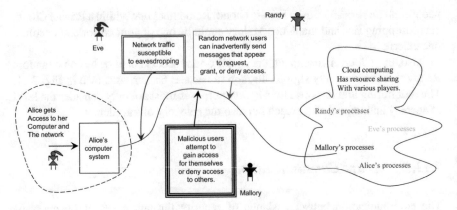

Fig. 7.12 Cloud Computing environment with no central island of security

the database. Consider that Eve is an eavesdropper for a private detective trying to illegally get health information of a patient. Encryption at the connection between two processors within the Cloud does not prevent Eve from using a process monitoring the operating system running the database application for a particular physician. Or, consider that Mallory maliciously accesses information about a patient in the files. He need not access it at the level of the doctor's office; he could access it as a potential insurance agency checking for future clients. Consider Imelda, an imposter who wants to pretend to be someone else. Imelda may have legitimate access as a drugstore person checking on prescriptions. Imelda then uses the little bit of information she has on some patient to perform processes only allowed to Dr. Dave. By pretending to be in an emergency situation, the imposter may gain access to cause a security breach. For privacy concerns, the security breach could be simply gossip about a patient's treatment and condition. Imelda could also interfere with the insurance company by repeated unsuccessful attacks on their database by pretending to be Dr. Dave. Then as an imposter, she may cause a denial-of-service result such that the doctor's office cannot collect timely insurance payments. Even with a trusted arbitrator, Trent, on the job, each database would have a different basis for trust. Mallory can create problems even if just to get the databases to be inconsistent. So for security purposes, Wally the warden monitors transactions. However, if Wally just guards the physical level of encryption, he will miss access problems (false identity authentication) as well as application level breaches.

A security breach can occur at all computational levels or for any security function. A very hard problem is for the warden to distinguish between security attacks and Randy's random mistakes. In the Cloud Computing environment, errors by one person or process (either due to user input or weak security programming) can create the appearance of an attempted security breach. Overreaction to false breach attempts can cripple the performance of any large system. Strengthening the encryption implementation at the hardware level will not solve a weak access implementation at the application level. A simple verification of a user when logging in does not model the verification requirements of various processes, on various CPUs, accessing various storage devices in the Cloud Computing case. Thus a more elaborate threat model is required for Cloud Computing. Victor the verifier must have a multilevel policy that is very fast and efficient. Both Wally the warden and Victor the verifier need access to the virtual security service observer we introduced in The Players section. The virtual security observer Otto must have access to checkers at multiple levels of the computer architecture and across multiple security functions. Notice that Otto does not implement the security solution. This would create the classic problem of "Who is guarding the guardians?" Rather, Otto provides multilevel multifunction checking functions for both security policy implementations as well as performance trade-offs. Otto must have hardware level checking, communication checks, change or variation tracking, storage checksums, operating system messages, and application level tags. Just as car door locks and login passwords provide an indicator mechanism for attempted break-ins, the data from Otto provides Wally with triggers to take further action.

7.12   A Few Key Challenges Related to Cloud Computing and Virtualization

In the scenario of Cloud Computing, people are concerned about the privacy of their data because they do not know where their data is being hosted. In a traditional computing center, anything inside the physical firewall boundaries is considered secure. However, a Cloud does not have clear boundaries because of its distributed features.

Customers need assurance that confidential data from their endpoint devices remains protected in a Cloud-based data center. In a computing system inside a datacenter, as shown in Fig. 7.13, various busses connect components. An adversary could physically tap those busses to gain information. This may be possible since an attacker could replace some modules or insert some components in a computer, for example, during the delivery phase. Hence those busses are not trusted. It is hard for each administrator or privileged personnel from a Cloud service provider to get a universal security clearance, especially when a third party service provider is involved. The providers can use tools such as "QUIRC, a quantitative impact and risk assessment framework for Cloud security," to evaluate the security of their enterprise [26]. Cloud service providers need to gain their customers' confidence in the claims of data privacy and integrity.

Providers using Cloud face additional challenges due to dynamic provisioning of multiple virtual servers on shared physical machines. Such resource sharing is done to achieve the economy of scale. It implies that data from potentially competing sources can reside on the same disk or memory structure. Through an accident or by design, a computer process can violate the virtual boundary to access a competitor's data. Furthermore, a write or data thrashing activity may occur that can go unnoticed. Having secure backups, authentication and data logging for each user's activity inside a Cloud data center can contain such damage.

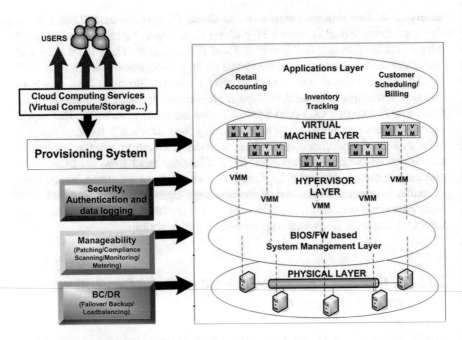

Fig. 7.13 Security inside a Cloud data center

Another specific problem of multiple virtual machines (VMs) is information leaks. These leaks have been used to extract RSA and AES encryption keys [24, 25]. However, it may create another problem of a large volume of log storage, privacy issues of whom else can see a user's accesses.

7.13 Some Suggested Security Practices for Cloud Computing

While Cloud Computing and security practices continue to evolve, many users have already migrated their mission critical applications and high-value data to the Cloud driven by the economic value and convenience factors [27]. This has made both Public and Private Cloud attractive targets for security hackers. Hence, we propose the following practices for the users and practitioners of Cloud Computing to ensure that their assets remain secure:

1. *Continuous Monitoring:* This is needed for any unexpected usage patterns or changes in your cloud resources. You can't protect what you can't see.
2. *Attack Surface Management:* It refers to the set of access points that are exposed to an unauthorized user. An organization needs to limit the devices or methods that can access its mission critical data [28]. Besides the obvious methods of

using encryption, one needs to ensure that the devices that are authorized to access this data themselves are not vulnerable or hacked.

3. *No Residual Footprints:* Looking into bins for any trashed paperwork is an old spying practice. Online equivalent of this is to try reading the leftover bits in the memory or disk, after a target VM stops using these resources in a Cloud. By zeroing out its contents of memory and disk, a VM upon exit can ensure that the next VM will have no residual data to exploit. This operation may cost some extra time and money but is well worth the trouble of avoiding your valuable data falling into wrong hands.

4. *Strong Access Control:* While obvious to any IT manager, many recent attacks came through unexpected entry points. Many companies use Internet-connected heating, ventilation, and air conditioning (HVAC) systems without adequate security, giving hackers a potential gateway to the key corporate systems. An example [29] shows how hackers stole login credentials belonging to a company that provides HVAC services and used that access to gain a foothold on another target company's payment systems. A strong chain is as weak as its weakest link, so analyze your system and its access points, to find its most vulnerable spots.

5. *Damage Controls:* With always evolving sophisticated hacking techniques, no system is 100% hack proof, and it is not a question of if but when a security attack can happen on your Cloud infrastructure or data. Each organization and user needs a plan to minimize the damage in such cases. For an individual, it might be a matter of canceling their credit cards, changing banking passwords, or perhaps closing some online accounts if they are compromised. For an organization, mitigation strategies may be more complex involving an alternative control and command network or quickly shutting down infected servers, etc. [30]

As security intelligence shows [31, 32], available technology is not restricted to firewalls. Your Cloud solutions provider must protect the perimeter in the most effective way possible. As an example, any Cloud email solution should have the following:

- Antivirus
- Anti-spam
- Information leakage control
- The possibility to create specific rules for blocking content, including attachments
- Email traffic monitoring

while any Cloud application solution should have the following capabilities

- Intrusion detection tools
- Application firewall
- New generation firewall
- Attack mitigation tools for DDoS attacks
- Log correlation
- Content delivery network

Above requirements can be including in the SLA (service-level agreement) between a Cloud provider and its users to ensure that both sides understand their roles and tools available.

7.14   Side Channel Security Attacks in the Cloud

Computer security in the Cloud allows unexpected attacks because physical hardware alone does not define the security boundary and is normally not in the control of the Cloud user. Direct attacks attempt to get user's information by attacking the secured data directly. Direct attacks include attempting to decrypt by guessing keys with repeated trials or by attempting to gain access with password guessing. Side channel attacks attempt to gain information by looking at peripheral locations or measuring usage effects rather than the user's data itself [33]. Traditionally, side channel attacks were not significant because they require some level of access to the secured hardware. In Cloud Computing and the IoT, the hardware is everywhere, and in the Cloud the hardware is shared providing access to hackers.

The side channel attacks (SCA) can be grouped into several categories. The first category is cache side channel attacks. In this type of attack, the attacker attempts to access memory locations that are outside its authorization. While separate processes are assigned different segments of main memory, the data must pass through the same cache for both processes. Of course, when data is in the cache, it is assigned a memory tag associated with its own process. However, depending upon the cache coherency algorithm, the cache may not be cleared between process swaps. Also, the bound checking implementation limits what can be learned about the data in the cache. However, with some speculative executions or branch prediction algorithms, the bound checks are suspended to improve run times. It opens up an avenue to carry out successful SCA.

The next category of attack is the timing attack. Here the attacker monitors the target process to measure how long certain operations take. This can be a cache timing tack or a calculation timing attack. For a calculation timing attack, some algorithms for multiplication increase performance by doing a bit by bit add and skip or just skip the add step when the bit is 0. Thus the timing attacker has a sense of how many (and sometimes when) bits are zero in a key. This can significantly reduce the search space for potential keys, making them easier to guess.

A recent example of SCA is Spectre [33], which is a form of timing attack. A malicious VM or a container, in a multi-tenancy Cloud Computing environment, can potentially use this. It uses the branch prediction mechanism of an out-of-order execution in a microprocessor and tricks it to speculatively load cache lines. Then it exploits a side effect of measuring memory access timing difference, to guess the content of a sensitive memory location. The attack is based on the timing difference between a cache hit and miss. This attack can be carried in four steps:

1. Branch prediction logic in modern processors can be trained to hit or miss on memory loads based on the internal workings of a malicious program.

2. Subsequent difference between cache hits and misses can be reliably timed.
3. This becomes the side channel to guess the information of a target location.
4. Results of successive memory reads can be used to read otherwise inaccessible data.

The next category of attack is the power-analysis attack. This is an indirect attack because it is not at all looking at the binary data, although it is using the binary data. For example, the pattern of the power supply current is compared for guessed keys to the normal operation. This happens because electrical charge required for transistors to turn on is higher than to keep them off, which is for binary 1 or 0, respectively. When there is a match in the power usage pattern, then the key has been found. Another category of attack is fault side channel attack. Here the security algorithm or hardware has an intentional error injected while encrypting. The change in the resulting performance gives a hint for the secret information.

There are three categories of side channel attacks that are nor really applicable to Cloud Computing: electromagnetic radiation, photonic (optical), and acoustic. These will not be discussed in detail, but a brief description follows.

The electromagnetic radiation emanating from a chip varies with the operations on the chip. Measuring this radiation gives clues as to what operations are being performed on what data values. What part of a chip is performing an operation gives can be measured by looking from outside the chip package (thermal and photonic or optical side channel attack). This can be used to identify when a secure operation is being performed by what circuitry. Acoustic side channel attacks include measuring ultrasonic emanations from discrete devices (such as capacitors and inductors) and sound patterns from mechanical devises (such as keyboards and disk drives). These three categories of side channel attacks are not presently applicable to the Cloud Computing, unless an insider such as a malicious IT person deploys these. Even then it is difficult, because many users' jobs may be concurrently running on a Cloud server.

A final very important side channel attack is the test access port. Testing standards (IEEE 1148.1 and IEEE 1687) define a common set of requirements for IC and board testing. The goal of design for testability (DFT) is to make as much information available as possible in order to test the device. The goal of secure design is to hide as much information as possible to prevent revealing any secrets. When a chip goes into standard test mode, the date hidden deep inside is accessible. Efforts are made to secure this data while maintaining testability, e.g., by disabling access to some or most DFT ports.

A huge problem for Cloud Computing is the variety of systems employed in a data center. Thus, while one component may be protected against a particular side channel attack, other components will not be protected. This is exacerbated by the legacy hardware problem, which in a widely connected Cloud Computing environments, the IoT includes systems with the latest security precautions and many more old systems lacking sufficient security implementations. One needs to do an end-to-end penetration testing in a Cloud deployment to ensure its security.

7.15   An Introduction to Block Chain for Security

In recent years, digital cryptocurrencies have become popular [34]. Block chain (BC) is the underlying technology for these and many other nonfinancial applications. It is simply a distributed database of records for all transactions of digital events, which have been executed and shared among participating parties. This database or digital ledger is distributed over the network of users, thus making it incorruptible. Block chain technology is composed of three building blocks, namely:

1. *Block:* A list of transactions from a certain time period. Contains all the information processed on the network within past few minutes.
2. *Chain:* Each block is time stamped, placed in a chronological order, and linked to the block before it using cryptographic algorithms. Specifically, the chain is marked with a secure hash to protect against tampering.
3. *Distributed Ledger:* Every node on the network keeps its own copy of the ledger and updates it when someone submits a new transaction.

A majority of participants on the network have to agree for a transaction to be legitimized and then record it in the ledger. This process is called a consensus. Instead of having a centralized consensus, such as a single authority to issue or maintain land records, BC uses a decentralized consensus mechanism, as it is harder for a hacker to manipulate multiple databases simultaneously.

Furthermore, BC database is in a distributed ledger, which multiple parties can use to verify a transaction, but only the owner or an authorized person read the contents of a transaction. An analogy is that your home address may be published on the Internet and thus known to many people, but only you or family members have a key to enter the house.

Some applications of BC include the following:

- *Cryptocurrency:* stores a token of value, i.e., money balance.

  - A value transfer system with no central authority.
  - Encryption prevents any malicious third party tempering. The secure hash function used by many is SHA-256, which is a member of the SHA-2 family of NIST-certified hash functions. An example of SHA-3 is described in Sect. 7.9.
  - Distributed ledger prevents any duplicate transactions.

- *Smart Contracts:* self-managed without a central rule maker.

  - Two (or more) parties agree on the rules/terms.
  - Conditional money release when services fulfilled.
  - Incur penalties if not fulfilled.

- *Trusted Computing:* resource and transactions sharing.

  - Combines BC, decentralized consensus, and smart contracts

- Supports spread of resources in a flat, peer-to-peer manner
- Enables computers to trust one another w/o a central authority

A financial transaction involving cryptocurrency happens in the following steps:

1. Every block in ledger has an input, amount, and output.

   (a) Input: if X wants to send an amount, it needs to show the source from where it came, and this forms a link to a previous entry.
   (b) Amount: how much X wants to spend or send to someone.
   (c) Output: the address of new recipient, e.g., Y.

2. A new transaction is created in form of a block and sent to the network.
3. Every node on the network updates its local ledger.
4. Consensus among the nodes validates a new transaction.
5. New block is thus added to the chain, forming a block chain.

Thus, participants share assets directly with no intermediary. BC enhances trust across a business network; saves time and costs, as there is no central authority like a bank to verify every transaction; and offers higher safety against cyber-crimes and fraud.

7.16   Summary

Computer security issues exacerbate with growth of the Internet as more people and computers join the Web, opening new ways to compromise an ever-increasing amount of information and potential for damages. However, an even bigger chal-lenge to information security has been created with the implementation of Cloud Computing. This chapter gave a brief general description of information security issues and solutions. Some information security challenges that are specific to Cloud Computing have been described. Security solutions must make a trade-off between the amount of security and the level of performance cost. The key thesis of this chap-ter is that security solutions applied to Cloud Computing must span multiple levels and across functions. Our goal is to spur further discussion on the evolving usage models for Cloud Computing and security. Any such discussion needs to address both the real and perceived security issues.

7.17   Points to Ponder

1. Networking gear, such as Ethernet cards, represents the first line of defense in case of a cyberattack; how can these be strengthened (e.g., packet sniffing)? Please see watchdog in appendix.
2. Recent storage crash in an AWS Public Cloud was caused by the scripting error of internal admins; how could it have been prevented?

3. Does having multiple customers in a Public Cloud increase its risk profile, as any failure will hit multiple businesses simultaneously? What operating system strategies you would recommend to prevent business impact?
4. Explain the role of attack surface management to minimize security risks? How would you reduce an attack surface?
5. How may the regular monitoring of a server's usage activity help to detect a security attack?
6. What are the security concerns due to a residual footprint?
7. What type of applications will not benefit from block chain?

References

1. Christodorescu, M., Sailer, R., Schales, D. L., Sgandurra, D., & Zamboni, D. (2009). Cloud security is not (just) virtualization security: A short chapter. *Proceedings of the 2009 ACM workshop on Cloud Computing Security*, Chicago, pp. 97–102.
2. Soundararajan, G., & Amza, C. (2005). Online data migration for autonomic provisioning of databases in dynamic content web servers. *Proceedings of the 2005 conference of the Centre for Advanced Studies on Collaborative research*, Toranto, pp. 268–282.
3. Nicolas, P. Cloud multi-tenancy. Available: http://www.slideshare.net/pnicolas/Cloudmulti-tenancy
4. Bun, F. S. (2009). Introduction to Cloud Computing. Presented at the Grid Asia.
5. Ray, E., & Schultz, E. (2009). Virtualization security. *Proceedings of the 5th annual workshop on Cyber Security and Information Intelligence Research: Cyber Security and Information Intelligence Challenges and Strategies*, Oak Ridge, Tennessee, pp. 1–5.
6. Naor, M., & Rothblum, G. N. (2009). The complexity of online memory checking. *Journal of the ACM, 56*, 1–46.
7. Cachin, C., Keidar, I., & Shraer, A. (2009). Trusting the Cloud. *SIGACT News, 40*, 81–86.
8. Jain, A. K., Lin, H., Pankanti, S., & Bolle, R. (1997). An identity-authentication system using fingerprints. *Proceedings of the IEEE, 85*, 1365–1388.
9. AWS Security Best Practices, August 2016. http://aws.amazon.com/security
10. Juels, A., & Kaliski, Jr., B. S. (2007). PORS: Proofs of Retrievability for Large Files. *Proceedings of the 14th ACM conference on Computer and Communications Security*, Alexandria, pp. 584–597.
11. Clair, L. S., Johansen, L., Butler, K., Enck, W., Pirretti, M., Traynor, P., McDaniel, P., & Jaeger, T. (2007). *Password exhaustion: Predicting the end of password usefulness.* Network and Security Research Center, Department of Computer Science and Engineering, Pennsylvania State University, University Park. Technical Report NAS-TR-0030-2006.
12. Gupta, P., Ravi, S., Raghunathan, A., & Jha, N. K. (2005). Efficient fingerprint-based user authentication for embedded systems. *Proceedings of the 42nd annual Design Automation Conference*, Anaheim, pp. 244–247.
13. Khan, M. K. (2010). Fingerprint biometric based self-authentication and deniable authentication schemes for the electronic world. *IETE Technical Review, 26*, 191–195.
14. Shaver, C., & Acken, J. M. (2010). Effects of equipment variation on speaker recognition error rates. Presented at the IEEE International Conference on Acoustics Speech and Signal Processing, Dallas.
15. Jayanna, H. S., & Prasanna, S. R. M. (2009). Analysis, feature extraction, modeling and testing techniques for speaker recognition. *IETE Technical Review, 26*, 181–190.

16. Acken, J. M., & Nelson, L. E. (2008). Statistical basics for testing and security of digital systems for identity authentication. Presented at the 6th International Conference on Computing, Communications and Control Technologies: CCCT2008, Florida.
17. Rivest, R. L., Shamir, A., & Adleman, L. (1978). A method for obtaining digital signatures and public-key cryptosystems. *ACM Communications, 21*, 120–126.
18. Advanced Encryption Standard (AES) (FIPS PUB 197). Federal Information Processing Standards Publication 197 November 26, 2001.
19. SHA-3 Standard: Permutation-Based Hash and Extendable-Output Functions FIPS PUB 202. https://doi.org/10.6028/NIST.FIPS.202. August 2015.
20. Schneier, B. (1996). *Applied cryptography second edition: Protocols, algorithms, and source code in C.* New York: Wiley.
21. Panko, R. (2003). *Corporate computer and network security.* Prentice Hall, Inc. NJ, USA.
22. Moscibroda, T., & Mutlu, O. (2007). Memory performance attacks: Denial of memory service in multi-core systems. *Proceedings of 16th USENIX Security Symposium on USENIX Security Symposium*, Boston, pp. 1–18.
23. Ristenpart, T., Tromer, E., Shacham, H., & Savage, S. (2009). Hey, you, get off of my Cloud: Exploring information leakage in third-party compute Cloud. *Proceedings of the 16th ACM conference on Computer and Communications Security*, Chicago, pp. 199–212.
24. Osvik, D., Shamir, A., & Tromer, E. (2006). Cache attacks and countermeasures: The case of AES. In D. Pointcheval (Ed.), *Topics in cryptology – CT-RSA 2006* (Vol. 3860, pp. 1–20). Berlin/Heidelberg: Springer.
25. Bishop, M. (2005). *Introduction to computer security.* Boston: Addison-Wesley.
26. Saripalli, P., & Walters, B. (2010). QUIRC: A quantitative impact and risk assessment framework for Cloud security. *2010 IEEE 3rd international conference on Cloud Computing (CLOUD)*, pp. 280–288.
27. Wang, Q., Jin, H., & Li, N. (2009). Usable access control in collaborative environments: Authorization based on people-tagging. *Proceedings of the 14th European conference on Research in Computer Security*, Saint-Malo, France, pp. 268–284.
28. Enck, W., Butler, K., Richardson, T., McDaniel, P., & Smith, A. (2008). Defending against attacks on main memory persistence. *Proceedings of the 2008 Annual Computer Security Applications Conference*, pp. 65–74.
29. http://www.computerworld.com/article/2487452/cybercrime-hacking/target-attack-shows-danger-of-remotely-accessible-hvac-systems.html
30. Al-Rwais, S., & Al-Muhtadi, J. (2010). A context-aware access control model for pervasive environments. *IETE Technical Review, 27*, 371–379.
31. http://www.informationweek.com/Cloud/infrastructure-as-a-service/5-critical-Cloud-security-practices/a/d-id/1318801
32. https://securityintelligence.com/23-best-practices-for-Cloud-security/
33. Kocher, P., Genkin, D., Gruss, D., Haas, W., Hamburg, M., Lipp, M., Mangard, S., Prescher, T., Schwarz, M., & Yarom, Y. (2018). Spectre attacks: Exploiting speculative execution. (PDF).
34. https://www.geeksforgeeks.org/blockchain-technology-introduction/

Chapter 8
Migrating to Cloud

8.1 Cloud Business Models

While some applications, such as social networking, were born in the Cloud and are natural fits, many other traditional products such as CRM (customer relationships management) were adapted by Salesforce.com and now are flourishing in the Cloud. Main attraction for migrating or creating applications in the Cloud is the economy of scale with pay as you need model for usage of IT resources. For customers, it translates to competitive products and cheaper services. There are three types of online business models of interactions between businesses (represented as B) and their customers (represented as C):

1. *B2C:* Business-to-consumers is a classic conversion of traditional brick and mortar businesses, such as bookstores to be online, as demonstrated by Amazon's first foray by putting the physical books for sale online and later offering eBooks via Kindle, etc. In this model, a typical business can reach out to its customers faster and economically using Web services. Many other examples abound, such as booking airlines tickets online instead of calling or standing in a long queue to buy them.
2. *B2B:* Business to business is an example of aggregating many businesses offering similar services, for the ultimate good of consumers. An example is how Walmart manages its supply chain from vendor and inventory among different stores. If an item is selling well at a place, say a winter jacket because it snowed locally, while another store has excess supply, then instead of second store holding a clearance sale, merchandise can be quickly shipped from one place to another. This requires real-time tracking of different goods and services, with automation in decision-making. Another example is Dell ordering parts for a laptop, to assemble and ship it after a customer places the buy order on its Website. Dell's vendors hold inventory of processors, memories, and hard drives in their own warehouses minimizing Dell's costs and risks of price drops, willing to supply the needed parts at a moment's notice. Such an inventory chain is

© Springer Nature Switzerland AG 2020
N. K. Sehgal et al., *Cloud Computing with Security*,
https://doi.org/10.1007/978-3-030-24612-9_8

called just-in-time (JIT), and Cloud Computing makes it even more efficient by linking Private Cloud of Dell with its Vendors' Cloud to make it easy to place orders, track shipments, handle payments and returns, etc.

3. *C2C:* Consumer to consumer is a new model, enabled by Cloud Computing. Before that Internet connected people for exchanging messages or emails, there was no central public place to store or access data, such as eBay. Another example is Facebook, the largest social network till date, which connects over a billion people across all continents to share pictures, messages, and stories. It has helped to create online communities and movements that have driven social and political causes in a manner not possible before.

8.2   A Case Study: B2C

A large business with many outlets and scores of customers spanning across geographical boundaries is the backbone of capitalism. However, managing good inventories in such a scenario is a logistical nightmare, as customer preferences vary across the regions. So winter jackets in a cold place may be sold out, while they may be put on a clearance sale in another area due to unexpected warm weather. This may result in nonuniform pricing, driving customers to do comparison shopping across the stores. Worst case is that this may prompt businesses to have overstock or understock situation resulting in a financial loss. First one to take advantage of this situation on a large scale was Amazon, which started world's biggest virtual bookstore. They didn't need to rent space in every community to sell books and thus could store remotely and mail them at a cheaper price. However, it needed Internet for customers to browse and place orders over the Web and then wait for a few days for the book to show up. Their instant success caused a lot of local bookstores to fail, as they couldn't compete with Amazon's cheaper prices.

Jeffrey P. Bezos started Amazon in 1994 [1]. When many .com companies failed during the year 2000 crash, Amazon managed to survive and is now world's largest online retailer. Amazon has now expanded beyond books to any conceivable thing a customer may want to buy legally, including clothes, food, and electronics merchandise. They often hold limited time deals and offer a Prime membership for $99 per year in the USA, which delivers things free to a customer's home within 2 days. Amazon has attracted other sellers to advertise goods on the Amazon site, sometimes in a used condition at cheaper price, which opened up new revenue stream of commissions for Amazon. An example of e-commerce transaction is shown in Fig. 8.1.

Amazon later expanded to Kindle, an e-book reader, which enables users to download electronic copy of a book instantly and allows carrying many books in a slim form factor of a 7″ tablet. This also enables reading at night without a light, and a special feature allows users to download first few pages of a book for free while they wait for the physical book to arrive by mail. Some other bookstores, namely, Barnes and Nobles, tried to mimic this model but failed even after spending hundreds of millions of dollars.

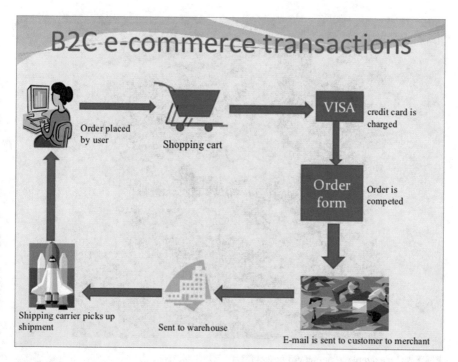

Fig. 8.1 Data and goods flow in a B2C transaction [2]

As Amazon set up an enormous data center to handle incoming customer connections, they entered into the Cloud Computing business by renting unused servers to other businesses and consumers. Their next attempt is to enter online entertainment business by offering movies and some original shows. Their stock price and high price-to-earning (PE) multiples are a testament of how investors feel about this Cloud success story.

8.3   A Case Study: B2B

According to one research firm [3], the total B2B digital commerce in the USA is expected to exceed $1 trillion ($10^{12}$) by 2020, driven by a shift in buying behavior among B2B buyers. This is primarily driven by corporate customers having access to a similar intuitive shopping experience on B2B site as retail customers have on a B2C Website.

There is a 90-year-old Chocolate company that started in Belgium and now an international supplier with a large base in corporate gifting programs. Their business spans multiple other industries such as automotive, legal, finance, retail, technology, and pharmaceutical. In order to reach and leverage its various corporate customers, suppliers, and partners, Godiva decided to use a Commerce Cloud Platform offered

Fig. 8.2   An example of online gift basket [3]

by Salesforce.com. Incidentally, Salesforce itself runs out of Amazon's EC2 (Elastic Compute Cloud) data centers. This Commerce Cloud offers support for critical B2B requirements, such as ordering and pricing lists, restricted login access and permission controls, sales and discounting mechanisms during checkout, etc.

Before adopting B2B Commerce Cloud, Godiva's corporate customers had to enter orders into a spreadsheet and submit it to Godiva's customer service teams, which manually keyed in data. This was time-consuming and error-prone process. Now business customers can directly enter orders and send chocolate gifts in large quantities to various recipients through a single Web order. An example is an auto dealer, who wants to send gifts to each individual car buyer in the previous month, as shown in Fig. 8.2. Their B2B interface is optimized for large split shipments, and volume discounts are automatically applied to large orders. This previously required multiple complex transactions and unsatisfied customers due to human mistakes. Such a large-scale automation is only possible through the advent of Cloud Computing.

8.4   A Case Study: C2C

eBay is world's largest online marketplace [3] with more than 116 million active users that collectively drove more than $75B in sales volume in 2012. Each month, nearly 1/3 of all US Internet users visit eBay, to search for new and used goods. eBay started in America but now has a footprint across all 5 continents and with local sites in over 30 countries.

Pierre Omidyar started an AuctionWeb from his apartment in 1995 to get rid of a few unused things and concept caught on with other sellers and buyers joining in. He would get a small commission from every sale, and in 1997 his company got $6.7 million in funding from the venture firm benchmark Capital. eBay went public in September 1998 at $18 per share, but by early March 1999, the stock was trading at $282 per share [5]. Within the next 3 years, despite a major.com financial crash, eBay had 37 million customers and $129 million in new profits. Such is the power of a successful C2C business, as shown in Fig. 8.3.

Even though Amazon, Google, and Overstock attacked its business model, eBay continues to do well in the online auction market. Their reason for success is the initial focus on antiques and collectibles. Their second growing market segment is car and motor enthusiasts. Customers in the first segment are the antique lovers who are willing to bid very high for a rare item. Since eBay gets a certain percentage of gross sales, it contributes well to their profits while bringing the antique lovers together. The motor lovers story followed a similar trajectory, using the credibility of a leading car collector Kruse, Inc., eBay expanded its categorical collections with sellers and new buyers making eBay Motors as one of the most successful target segments. Every 3 minutes during the fourth quarter of 2016, a car or truck was sold on eBay via a mobile device.

eBay's success is by ensuring that any commercial transaction between a buyer and seller can proceed smoothly and with full integrity. This is assured by holding seller's payment until the buyer pays and receives the merchandize. Then each party ranks and rates the other. This rating is available to all other customers to see, so a bad remark can influence future sales of a seller. People care so much for a good review that they will follow up and often refund or replace an item if the buyer is not satisfied. In cases where this doesn't happen, eBay is willing to step in like a policeman and make things right for the aggrieved party in a dispute. Only credibility

Fig. 8.3 C2C interaction on eBay [4]

worth holding and displaying on an eBay site is one's public rating. Such is the power of Internet in offering a safe trading place, using a combination of private and Public Cloud Computing services.

8.5 Plugging IoT Devices in a Cloud

An outcome of Cloud Computing is to democratize access to expensive servers in a world-class data center, which a small or medium business or a nonprofit organization can't otherwise afford. Taking this effect to an extreme has enabled citizens in developing countries to support a social cause. We will look at the example of Shiv Shankar, a social entrepreneur in Bangalore, who took upon himself to inform the local citizens about their air quality. Recent progress in India comes at a price of higher pollution. Shiv has devised a homegrown solution to measure air quality. He along with a grade 12 student, Varun Ramakrishnan, developed a low-cost IoT (Internet of Things) device for under $100, shown in Fig. 8.4. The team started gathering air-quality data from a dozen different locations, uploading it to Cloud, which enabled them to determine which areas experienced the worst air quality and during which periods of the year. This led them to a location near the Graphite India Junction in Bangalore, where many local residents had complained about soot deposits on their hands and clothes. But until now, there was no scientific data to prove it, and the offending company used unscrupulous methods to get around the local inspectors. Shiv spearheaded a team effort to change that with data.

The team set out to measure particulate matter (PM) 2.5 in air, the safe levels for which the World Health Organization (WHO) specifies at 10 $\mu g/m^3$ annually and 25 $\mu g/m^3$ for 24 hours. They found that PM 2.5 level at a particular location near the offending plant at 3.5 times is higher than other locations, as shown in Fig. 8.5. Armed with this data, the team made headlines [6, 7] in the local papers and TV, which got attention of the media reaching all the way to Supreme Court in India's capital Delhi. The local residents managed to get monetary compensation from the offending company.

Their simple IoT device consists of the following components:

Fig. 8.4 An air-care pollution sensor system developed by Shiv and his student

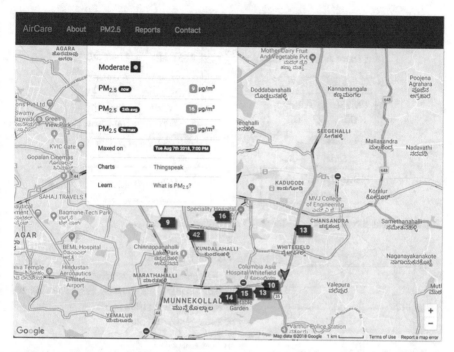

Fig. 8.5 Air-quality sensor data measured and shown on a Bangalore map

1. Raspberry Pi 3 model B with a built-in WiFi and Bluetooth
2. A Plantower PMS3003 air sensor using laser scattering principle. It is used in air purifiers and low-cost air-quality monitors.
3. A 16 GB class 10 micro SD card.
4. A 2 Amp USB charger with backup during any power outages.
5. Estimated nominated power consumption 2.5 W.

On the software side, it uses a Raspbian OS, a Linux variant, with Mongo database for local storage, an air-sensor driver, and a local apache Web server with Python flask for local configuration and management. On the Cloud side [8], the team used an AWS server with Mongo database, an Apache Web server, and a Python flask framework for serving REST APIs. The architecture of this client-server application is shown in Fig. 8.6. Every 5 minutes, the client side application is launched. It turns on the air sensor and writes 240 bytes of measurement data with a time-date stamp in the local database (DB) (Client Mongo DB) and exits. Another client side application launches every 10 minutes. It takes all the accumulated readings to upload to an AWS server. These two tasks are run independently, and local DB size can expand up to 7 GB space. Once the server write is confirmed to the Server Mongo DB with a successful HTTP POST request, then corresponding local entries are deleted. There is a battery backup on the local device serving as an UPS. Security attacks are not considered, as communication is currently only one way from the devices to the servers.

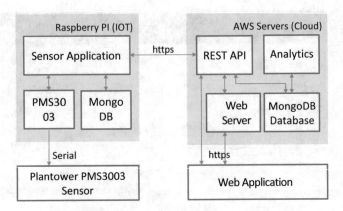

Fig. 8.6 Client-server architecture for Shiv's air-care application

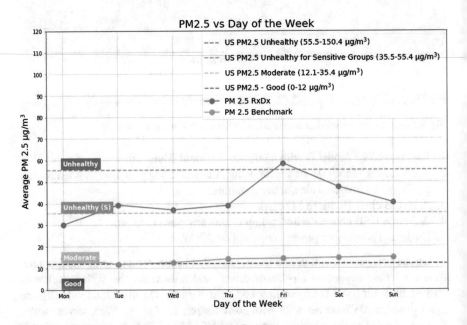

Fig. 8.7 Pollution level changes over a week in July, 2018

On the server side, backend information is stored in a Mongo DB located in the AWS Cloud. Then analytical programs are run every 10 minutes to compute new average and record high values. A backup copy of the database is maintained on the Cloud servers.

Using this data, the team was able to analyze pollution patterns, locations, and times of day when the air quality is worst. A sample weekly pollution pattern is shown in Fig. 8.7. Similar battles of local residents against big corporations have been fought in many developed nations in the past, but their communities had the funding and an established legal system. It is the first time that a grass root effort led by local residents won with the help of Cloud Computing, to gather 24×7 data from 13 different sensors and analyzed it using advanced machine learning techniques, to pinpoint a single local offender.

8.6 Using Multi-party Cloud

If we look at the rising profits of major Public Cloud providers in the USA, one can surmise that their customer base is growing rapidly. In comparison to the total compute consumed by the world's economy, Public Cloud adoption is still a small fraction. There are many reasons for this small growth, one of which is consumers' fear of a vendor lock-in and apprehensions about the failures of Cloud-based services. A possible solution is to enable the multi-party Cloud architecture. It enables customers to distribute their workloads across multiple Cloud environments while mitigating risks associated with individual Cloud environments [9]. This offers the following potential benefits:

1. *Optimized ROI:* Each Cloud is built differently, emphasizing different types of services, offering differing pricing structures, and varying performance. Since a consumer's needs may change over time, having multiple options enable a choice to pick a suitable Cloud at different times.
2. *Better security:* A multi-Cloud infrastructure allows organization to maintain a hybrid environment with local on-site compute. This gives a choice to keep security-sensitive workloads in a Private Cloud or maintain multiple copies in different locations, leading to higher availability. It is assumed that any external data will be secured with strong encryption keys maintained by the data owner.
3. *Low latency:* Any distant location can experience sudden latency due to high network conditions beyond the control of a Public Cloud operator. With multi-Cloud infrastructure, a data center closer to the end users can serve the customer requests with least possible delays.
4. *Autonomy:* Many customers hesitate to put their data and programs in the Cloud fearing a loss of control. This is mitigated by migrating or keeping multiple copies of data and programs in different Cloud.
5. *Less disaster prone:* Most Public Cloud vendors have SLAs (service-level agreements) in place assuring high availability of their compute, storage, and network infrastructure. These agreements often preclude natural disasters, such as an earthquake or lightning strike. Even if such acts of nature are covered, the liability of the vendor is limited to the service fees paid. Having multiple copies of the same data with different vendors assures even higher availability, as the probability of the same disaster striking multiple locations is relatively less.

With the above benefits in mind, a recent study [10] of 727 Cloud technology decision-makers' at large companies found that 86% have a multi-Cloud strategy. Of those, 60% are moving or have already moved their mission-critical applications to the Public Cloud. An example is of pivotal [11], an EMC subsidiary, which provides Platform-as-a-Service (PaaS) solution spanning across several Cloud service providers. Kroger [12], a grocery store chain, uses Pivotal's Cloud Foundry product to launch new applications. These applications help with scanning, bagging, and shipping groceries to customers' homes. Such automation enables over 400,000 employees and managers at Kroger's to track items that are past their due dates and stock them for a better customer experience.

8.7   Software-Based Hardware Security Modules

A hardware security module (HSM) refers to a computing device that securely manages the encryption keys. It provides a strong authentication using crypto processors. HSMs are a part of the mission critical infrastructure in a Cloud to store customer keys. This hardware comes in the form of a plug-in card or an external device that connects directly to a computer or the network server [13]. These devices are certified with standards such as FIPS 140-2 (Federal Information Processing Standard).

HSMs are often clustered for high performance and availability, with their costs running into tens of thousands of dollars. In a Public Cloud, customers typically pay up to 10X for renting a HSM on hourly basis, as compared to the cost of renting a compute server [14, 15]. Thus, there is a strong economic incentive to explore alternate solutions. A startup, Fig 8.8, did just that using Intel's SGX technology [16]. Its cost of implementing and delivering HSM like capability on Xeon servers is significantly lower than using traditional HSMs.

Fortanix offers a Self-Defending Key Management Service (SDKMS), which provides HSM grade security with software-like flexibility. It uses Intel's Software Guard eXtensions (SGX) technology to encrypt and store keys to mitigate against system level attacks. Such a software-based HSM can give multiple orders of magnitude price performance advantage over traditional HSMs, because the keys access will be running at the core CPU speeds, instead of over TCP/IP networks or PCI-E speeds in case of add-on cards. Fortanix's SDKMS provides RESTful APIs, service side clustering, and tamper-proof audit logs along with a centralized management. This creates a run-time encryption platform, which enables applications to process and work with encrypted data. It uses Intel's SGX to create a security enclave that runs signed applications in protected states. One of their early customers is IBM Cloud Data Shield [17], which enables run-time memory encryption for Kubernetes containers without modifying applications, as shown in Fig. 8.8.

The field of using hardware as a root of trust is rapidly evolving; it is much needed to bring up trust in Public Cloud Computing. It will help to overcome security as a perceived barrier for moving mission critical data to the Public Cloud. A recent Forrester's report [18] suggested that 43% of developers surveyed feel that security is a limiter to adoption of containers in Cloud.

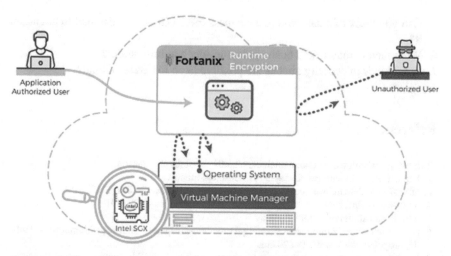

Fig. 8.8 Fortanix Runtime Memory Encryption using Intel SGX [16]

8.8 Summary

Several real-life examples were narrated of companies that have thrived only due to Internet and Cloud Computing in their business models. Cloud Computing has ushered in a new era and proverbial gold rush of opportunities, expanding beyond the above list, such as Uber for ride sharing, BNB for housing, etc., which were not possible previously.

8.9 Points to Ponder

1. Migrating to a Public Cloud may result in IT savings, but are there any potential pitfalls?
2. Business continuity is critical for many mission-critical operations in an enterprise. Can multiple Cloud providers be used to improve BCP (Business Continuity Practices)?
3. If local computing is needed, e.g., when Internet connections are not reliable, then a Hybrid Cloud usage model is desirable with on premise applications and data. In appendix, read about check pointing, backup, disaster recovery, etc.
4. Under the following scenario, would an IT provider prefer a Private or Public Cloud?

    (a) Should a B2C IT service provider prefer a private or Public Cloud?
    (b) Should a B2B IT service provider prefer a private or Public Cloud?
    (c) Should a C2C IT service provider prefer a private or Public Cloud?

5. Can you think of a case where a business tried a Cloud but decided to not adopt it?
6. What factors may lead a business to consider a hybrid model?
7. What kinds of security attacks are possible in an IoT-based Cloud?

References

1. http://www.slideshare.net/annamalairamanath/case-study-on-amazon
2. http://www.slideshare.net/shynajain/b2c-business-models
3. https://www.demandware.com/uploads/resources/CaseStudy_Godiva_ENG.pdf
4. http://www.slideshare.net/SambathMetha/e-bay-presentation-27761391
5. http://teamcaffeine.wikidot.com/ebay
6. https://timesofindia.indiatimes.com/city/bengaluru/air-quality-data-on-whitefield-residents-fingertips/articleshow/66504820.cms
7. https://www.deccanchronicle.com/nation/current-affairs/110818/bengaluru-sensors-to-gauge-whitefield-air.html
8. http://blog.mapshalli.org/index.php/aircare/
9. https://www.Cloudindustryforum.org/content/five-reasons-why-multi-Cloud-infrastructure-future-enterprise-it
10. https://www.techrepublic.com/article/why-86-of-enterprises-employ-a-multi-Cloud-strategy-and-how-it-impacts-business/
11. https://www.zdnet.com/article/pivotals-head-of-products-were-moving-to-a-multi-Cloud-world/
12. https://content.pivotal.io/blog/how-pivotal-is-helping-retailers-improve-the-in-store-experience-through-better-software
13. https://en.wikipedia.org/wiki/Hardware_security_module
14. https://aws.amazon.com/cloudhsm/pricing/
15. https://aws.amazon.com/ec2/pricing/on-demand/
16. https://fortanix.com/
17. https://www.ibm.com/cloud/data-shield
18. https://www.forrester.com/report/Now+Tech+Container+Security+Q4+2018/-/E-RES142078

Chapter 9
Migrating a Complex Industry to Cloud

9.1 Background

As we have seen in the previous chapters, there are three Cloud service models (i.e., IaaS, PaaS, and SaaS), which can benefit both businesses and consumers by migration to Cloud Computing. This has created a race for many businesses to adopt Cloud services. However, that is easier said than done due to privacy and security concerns. While lower IT cost is the main driver, it is not the only consideration for Cloud Computing players and stakeholders. For example, some large financial houses and hospitals are concerned about the security of their data and prefer to maintain their own data centers or create a Private Cloud infrastructure for their employees and customers. One such case is of silicon chip designers. It is interesting because most information in today's world flows through smartphones, laptops, and servers with backend data residing in the Cloud. Hospitals may be worried about exposing patient private medical records in a shared computing infrastructure of Public Cloud, which in turn will violate patient trust and local laws and have adverse effects such as societal or employer reaction, etc. Similarly, large chip design houses are worried about the security of their product design data, lest it falls in the hands of a competitor. A silicon chip design company is comfortable putting its data on a Private Cloud; so internal teams located in different countries can collaborate on complex projects. However, it is reluctant to move the same data to a Public Cloud due to security concerns. Even if economic realities necessitate such a move, the design flows or tools used in a chip design are not available in the Cloud. Following study is an exercise to examine, on what would it take to move such a complex set of tools and design processes to Cloud. The steps used in this exercise can be applied to any set of complex tasks for an industry that is currently not in the Cloud, for determining the possibilities and challenges.

© Springer Nature Switzerland AG 2020
N. K. Sehgal et al., *Cloud Computing with Security*,
https://doi.org/10.1007/978-3-030-24612-9_9

9.2   Introduction to EDA

Electronic design automation (EDA) broadly refers to the software tools, methods, flows, and scripts that are used for very-large-scale integration (VLSI) designs of silicon-based circuits, chips, and systems. The sheer number of transistors (now ranging into several billions) that reside on a single piece of silicon prohibits a hand-crafted design and necessitates automation. The software technologies behind such automation have evolved over the last five decades, starting with university and large research lab tools such as Magic [1] (for basic layout placement), SPICE (for circuit simulation) [2], SUPREM [3] (for IC fabrication process simulation), PISCES [3, 4] (for semiconductor device simulation), SALOGS [5, 6] (for logic simulation), TEGAS [7] (for logic simulation), MOSSIM [8] (for switch-level simulation), SCOAP [9] (for testability analysis), TMEAS [10] (for testability analysis), ESPRESSO [11] (for logic minimization), SICLOPS [12] (for place and route), and Timberwolf [13] (for place and route) to several higher-level logic synthesis solutions in the 1980s. However, these technologies have since moved from academic institutions to commercial companies that specialize in providing EDA tools and necessary support for design teams. The growth rate for EDA companies has been cyclical during the past decade due to various factors and has opportunities for a greater growth [14].

So many industries are evaluating moving to the Cloud that researchers in Dublin have produced a systematic literature review (SLR) of 23 selected studies, published from 2010 to 2013 [15]. The authors produced a database of migrations of legacy software to Cloud, which provides useful information for the EDA companies considering such a move. Also, work has been done to identify performance bottlenecks in Cloud Computing [16]. The authors concentrate on network loading as affected by Cloud workload. Moreno et al. [17] present an analysis of the workload characteristics derived from a production Cloud data center and describe a method for analyzing and simulating Cloud workloads. However, they point out that: "Finally, the workload is always driven by the users, therefore realistic workload models must include user behavior patterns linked to tasks." We endeavor to meet this requirement by relating the general Cloud Computing benefits to the EDA industry-specific workload characteristics. Successful utilization of Cloud Computing by the EDA industry can create an opportunity for greater growth for both the EDA companies and the IC design community. Previous publications have discussed Cloud workload classifications [18] and Cloud security [19]. The purpose of this chapter is to apply that analysis to the EDA industry.

The EDA field is already several decades mature, and one wonders if there has been enough time to solve many key problems in the domain of silicon design automation (DA). The answer is no, as these problems are not static in nature. Many of these are NP-complete or NP-hard in nature, currently only with heuristic solutions. With Moore's law of doubling the number of transistors available for integration every 18 months (current estimates are between 24 and 30 months), any heuristic solution tends to breakdown in a couple of product generations, as the design solution space grows exponentially. This increases the cost of doing designs and is limiting new commercial design-starts, further putting financial pressure on the EDA providers to recover their tool investments [20]. Thus, despite a growing electronic consumer market, the EDA industry enjoys a growth rate slower than for its design customers, at 9.5% over the past 1 year [14]. Paradoxically, a large number of small design efforts can readily use existing tools only if EDA costs come down allowing a broader adoption. This is especially true for new emerging applications, such as Internet of things (IoT), where per-unit silicon cost tends to be measured in cents instead of dollars. Broader access to large pools of inexpensive computing resources and availability of EDA tools to a broad number of current and potential users may result in reducing the cost of new designs. Then more designs can be started by smaller teams and companies [20], especially in the emerging economies. Hence, a clear and present question for the EDA industry is, "How can Cloud Computing enable new growth opportunities?". To address this requires looking at the EDA workloads and mapping them to Cloud workload categories. The Cloud workload mapping in this chapter applies to both Public and Private Cloud. However, EDA tools and flows have evolved over time and will continue to do so. Hence a brief summary of the EDA history, shown in Table 9.1, is essential to understand the rationale behind our proposed workload mapping and to keep it updated with any future tools and flows.

9.3   A Brief History of EDA Tools and Flows

9.3.1   The Nascent Years of the 1970s

Although computer-aided design (CAD) and DA activities started in the early 1970s, for the first decade, almost all development was in large corporate design centers, using proprietary solutions. By the mid- to late 1970s, CAD and DA solutions branched into different engineering domains, such as automotive design, VLSI (very-large-scale integration) circuit design, and printed circuit board designs. The chips of that era were mostly handcrafted, and thus the number of logic gates was

Table 9.1 Evolution of EDA over time

Years	EDA development	Characterization
1970–1979	Hand-crafted, small, and simple designs	Nascent years
1980–1989	Up to one million-transistor designs with schematic entry, layout editing tools	Roaring decade
1990–1999	Layout automation, circuit simulation, logic synthesis, and analysis tools	Growing up
2000–2009	1 billion-transistor designs using IP (intellectual property) blocks, and SoC (system on a chip) products	Maturing
2010–present	On-premise hyper-scale distributed computing with some Private Cloud technologies	Predictable EDA

limited to what a human mind could comprehend in terms of functionality. Circuits were of the simple adder and multiplier types, and point solutions included both analysis (e.g., equation solvers) and synthesis (e.g., simple logic minimization) type of stand-alone tools. During the 1970s, digital electronic CAD and DA tools were mostly for board-level design, with some IC chip tools being developed in-house, such as at Intel.

9.3.2   The Roaring 1980s

In the early 1980s, many companies, of which the top three were Daisy, Mentor, and Valid, created specialized workstations for electronic design. Tools for interactive schematics and layout designs were becoming commonplace, even in the published papers of that era [21], including design synthesis and graphics with batch processing for large analysis problems.

Later, that decade saw specialized high-end workstations created to solve large design problems, such as SUN workstations becoming a hardware platform standard for using CAD/DA/EDA tools. The EDA industry began to focus on IC designs, while the term CAD became more common for mechanical engineering. The chips of that era were able to reach and cross the one million-transistor mark with Intel's N10 design in 1989. Very large analysis (such as design rule checking a whole chip or exhaustive digital simulation for design verification) was still performed on corporate mainframe computers, while incremental or small-scale analysis and simulation were performed on high-performance engineering workstations at each engineer's desk.

9.3.3   Growing Up in the 1990s

This decade witnessed a young EDA industry growing up fast, with many start-ups and subsequent consolidation phases, as EDA tools moved from specialized to general-purpose workstations. The solution vendors and their customers started to talk about flows and frameworks instead of individual point tools. Synthesis and analysis EDA programs were linked in automated script-driven flows with feedback loops. Heuristics-based solutions were used to address the size limitations for mostly NP-complete or NP-hard problems common in the EDA domain. There was an increasing interplay between automatic and human-interactive roles, especially in planning the layout and signal busses, to prune the search space for place and route tools.

The chips in this era routinely reached hundreds of millions of transistors, and tool limitations were overcome with the use of design hierarchies and cell libraries. The latter also turned out to be a major productivity booster [22]. During the mid-1990s, a trend included adding memory, and cache to on-die microprocessors, to speed up the data processing times.

9.3.4 *Maturing into the First Decade of the Twenty-First Century*

With the Internet boom came large data centers and massively parallel user tasks such as database searches. These clearly needed multicore servers, instead of large single-core processor designs. The former became a convenient way to overcome the power and complexity limitations of large out-of-order superscalar single-core designs. This trend benefitted from the concept of hierarchical and cell-based modular designs developed in the previous decade. EDA evolved to increase the inclusion of system-level concepts with the growth of system on a chip (SoC), comprising of intellectual properties (IPs), cores, and custom logic integration supported by the tools.

Using the new EDA solutions, large chip designs, such as Intel's Itanium micro-processors, crossed the 1 billion-transistor mark. This era saw an increased emphasis on low-power designs with the popularity of smartphones, tablet computing, and various new form factors. As individual processors became faster, a number of consumer products evolved to put more functionality in software on standard hardware cores, thereby decreasing the need for many new design-starts in the industry. According to Prof. Michael Porter's competitive five forces theory, in any industry, it is desirable to have many customers instead of a few large customers to minimize the product pricing pressures [23]. This applies to the EDA industry as the uneven split of customers between high-end large designs done by a few companies and the smaller size designs by many companies presents a growth opportunity [20].

9.3.5 *From the 2010s Till Now: EDA Stable*

With the advent of the Internet, higher bandwidth of public and private networks, and consolidation of servers in large data centers came the birth of Cloud Computing. Compute power required to complete a new chip design in acceptable time exceeded the available compute capacity in many design companies, thus creating a natural problem and solution match. However, the lack of security, or the perception of lack of it, prevented movement of proprietary chip designs into Public Cloud [24]. Also, licensing models for EDA tools have not evolved much since the 1990s, requiring user companies to buy yearly contracts for tool licenses, instead of hourly basis pay as you go software rental models popular in the Cloud.

Both of these factors are contributing to the limit on how many simultaneous designs a company can afford to do, thereby also limiting the revenue growth of the EDA industry. Solving this dilemma provides an opportunity for the EDA and chip design industries to grow.

Furthermore, a relentless pressure to reduce the project schedules and costs is causing most design teams to adopt SoC methodologies [25]. This has led many EDA companies to venture into new territory to provide large design components, often sold as IP (intellectual property) blocks, for specific functions such as audio or video processing, USB support, and even CPUs as building blocks for larger and more complex chips. This is one step away from an EDA company doing the complete designs, but doesn't want to directly compete with its customers. These points provide a basis for EDA companies to explore growth opportunities of their services and products in the Cloud [26]. This situation has led to the growth of on-premise hyper-scale distributed computing with some Private Cloud technologies.

9.4   EDA Flow Steps Mapping to Cloud

A typical design flow has dozens of intermediate stages, with up to hundreds of tools generating thousands of design data and log files, requiring millions of CPU hours to complete a design. While actual design flows and tool versions may differ across individual projects, making it hard to share and maintain a consistent design environment, there is a desire to share EDA tool licenses between many design teams to optimize the software cost. An example design flow is shown in Fig. 9.1 and, as an illustration, the mapping of some individual steps to the Cloud workload categories. Ghosh et al. [27] propose a structured approach for Cloud workload analysis and migration aimed at generic Web-based workloads such as ERP and CRM solutions, whereas our approach below goes into the details of silicon design steps to accomplish mapping to Cloud workloads [28].

Instruction set design involves high-level design decisions about the fundamental contract between hardware and software for the target design. Instruction set architecture (ISA) represents the machine-level instructions that a target architecture will natively support. This is an important step, as all optimizations and design decisions follow from this step. For example, if a specific architecture does not need to natively support the division operation, the architects do not need to design division units in the chip and can thus reduce the number of transistors on the chip. In this case, division function can still be supported with software doing repeated subtractions. If the target architecture needs to process large arrays of floating-point elements, performing the same operation on all the elements of an array, the architects may choose to support this directly through the instruction set, by adding vector instructions to the ISA. An example of such an addition is the set of SSE instructions that Intel added to the Pentium 3 processor in 1999. Such decisions are made after careful examination of the performance of most popular applications on the existing architectures and also by simulating the new architecture with proposed instructions. This task is performed during architectural simulation and analysis explained below. The former usually involves scripting a number of runs on the existing machines and automatically gathering the data in a presentable form. Depending on the amount of data being collected, it could involve a moderate number of disk accesses, and depending on the tasks being studied, the overall categorization could be compute-heavy, memory-heavy, or disk-heavy. The bulk of these tasks maps to the high-performance computing (HPC) category. However, for mapping these to the Cloud, one must know the exact architecture and micro-architecture versions of the machines that run these scripts. In the current Cloud Computing model, this matchup might be typically difficult to achieve.

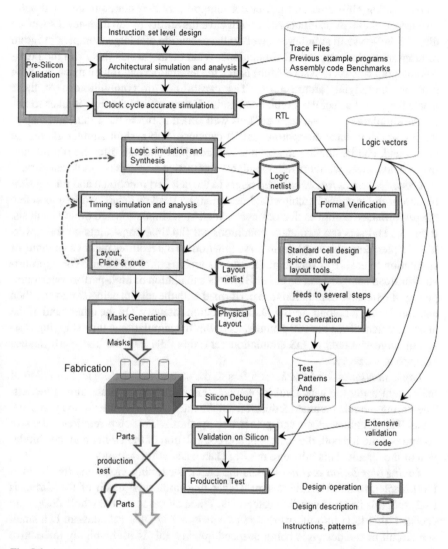

Fig. 9.1 An example chip design flow with several EDA tools

Architectural simulation and analysis is the exploration of micro-architectural design parameters and choices such as the size and organization of CPU caches or how many arithmetic logic units (ALUs) need to operate in parallel to achieve the desired performance and involves setting area and timing goals for a chip. Design architects often work with an instruction set simulator, to see how sample applications will behave with the design configuration changes. Different simulators are used for exploring the design parameters at different levels. For example, full-system simulators can run unmodified operating systems and commercial applications while simulating the hardware structures in less detail, as compared to the component or sub-system simulators that simulate the hardware with a goal of understanding chip area and power consumption. Sub-system simulators usually take traces of events as inputs and do not have the ability to run high-level software directly. Full-system simulation involves the execution of a large, complex program running multiple independent simulations. What distinguishes this from the previous category (instruction set design) is that the simulators can run on machines with different underlying architectures. The capabilities or configurations of these machines do not affect the simulation results. It only affects the time it takes to finish the simulations. These tasks are thus well suited to run in the Cloud and map to the high-performance computing (HPC) category. Sub-system simulators, on the other hand, can be smaller in size and less complex and can often be run interactively. Thus, these would match the highly interactive single person job category.

Clock cycle accurate simulation refers to the detailed functional and timing simulation of the proposed architectural design at the level of detail required to define the performance within each clock cycle. With both simulation and timing analysis, it checks and sets the boundary conditions for the time constraints on the lower-level implementations to perform these functions. This requires massive amounts of simulation to ensure coverage of the various architectural decisions. The appropriate workload category can be either Big Data calculation or high-performance computing. If extensive simulations are performed with the results being stored and then analysis is run, it belongs to Big Data calculation category. On the other hand, if the analysis calculations are run immediately after the simulations, thus skipping storage of large amounts of OS simulation data, this falls into the high-performance computing category.

Logic simulation and synthesis refers to the step of detailed logic simulation of the system using the gate level implementation including accurate timing models. The timing models include feedback from the later step of synthesis and place and route. This includes clock and power tree synthesis. This step requires extensive computer time for both the flattening and compilation of the circuit and the simulation of the circuit. This falls into the Big Data calculation category.

Timing simulation and analysis is the check for timing violations for the low-level implementation of the design. Depending upon how much of the circuit is analyzed, this has two possible categories. When all cases for the whole design are checked, this falls into the workload category of Big Data calculation. If a small sub-circuit of the design is being checked quickly, this is in the highly interactive single person category.

Pre-silicon validation refers to running software OS and associated test content on the model of a chip, before silicon is ready. The idea here is to find any bugs early and avoid the cost of a silicon re-spin. This can be done on an emulation model, derived from RTL (register-transfer level) description or another HLM (high-level model) description or some combination thereof. This involves downloading the RTL model on a large FPGA (field-programmable gate array)-based server boxes and running them, say at a speed of several hundreds of MHz. These server boxes are specialized and tend to be very expensive; hence small design teams can't often afford them. If these boxes are stored in a Private or Public Cloud, then these can be time-shared at an effective cost, thereby reducing the overall product design cycle time and cost. This step maps well to HPC in Cloud.

Formal verification refers to the process of applying mathematical postulates and theorems to the high-level model or its representative code for ensuring that critical logic bugs are detected early, without having to run an exhaustive set of simulation tests. Current generation of chips have many states giving rise to complex combinations of interactions, some of which may result in a deadlock that can be only detected after a long series of events. Such a situation can't be always anticipated or tested with all available test vectors, but may get exposed in the field. Thus, formal verification adds yet another capability to the arsenal of validation team by mathematically ensuring that the flow of events and interactions will be error-free. However, it is also computationally challenged and methods for this are still evolving. Formal verification usually works at an IP level, so it maps well to the single computer intensive jobs category.

Standard cell designs are done on local machines using Spice [2] and hand layout tools. These tasks require real-time editing of a layout figure by mask designers or what-if simulations of circuit timing with interactive device size changes. These steps are performed on engineering workstations or local terminals with graphic displays directly connected to nearby servers. If these servers are moved to a Public Cloud, it is likely to introduce higher latencies due to network delays, which may be intolerable to the human designers. This design task maps well to the highly interactive single person category. This is not to imply that a single person can design a large cell library. Large complex standard cell libraries require large teams and a great amount of compute power. This translates to multiple highly interactive single person task computer workload category.

Layout place and route refers to the process of planning, placing, and connecting design objects. Planning is a top-down process, whereas the placement and routing are done bottoms up, with only the rectangular shape of a lower-level block and its pins visible at the next level up. Transistors are at the lowest level of hierarchy in a standard cell and then cells in higher-level block such as an ALU. Then these higher-level blocks are placed and routed in yet another abstract level function such as to implement a CPU or ISP (image signal processing) unit on the silicon. Such a hierarchical arrangement limits the complexity and number of blocks to be placed and routed to a manageable size in a chip having billions of transistors. Planning is an interactive process, which is mapped to highly interactive single person, whereas place and route tools are batch mode and can be mapped to the category of single computer intensive jobs. Some physical synthesis jobs can take a week or more to complete. A number of such jobs are launched in parallel to run on a server farm and map well to HPC in Cloud.

Mask generation requires some additional processing. With smaller geometries on current silicon designs, there is a need to do some adjustment to account for these dimensions being even smaller than the wavelength of light used for lithography process during chip manufacturing. The geometric steps are collectively called OPC (optical proximity correction) and are very computationally intensive in nature.

Hence, divide and conquer strategy is used to split the final layout in little pieces, which are then sent to a large number of servers for parallel processing and then integrated back into a final layout database. Hence mask generation qualifies as an HPC category with big database creation and calculations.

Silicon debug is performed after a chip is manufactured. This chip is then tested as its power, performance, and even some functional bugs can only be found with actual silicon instead of with EDA models. Also, the chip after a successful power-on can run user applications at full speed, vs. simulation models often running at greatly reduced speeds. Thus the whole HW and SW stack can be tested on an actual chip, exposing some flaws that may require further silicon spins. If better models can be created, then there is a potential for Cloud Computing to add value without waiting for the silicon arrival. However, currently there is no substitute for the post-silicon validation. This is best mapped to a single person interactive task.

The tasks of pre-silicon validation and mask generation clearly fall in the realm of Cloud Computing, if the IP security can be assured. Clearly, security is a boundary condition before any design IP can migrate to a Public Cloud. EDA tasks tend to have elastic computing needs; thus, pay as you go scheme works well. Sometimes bringing in more computing to EDA tasks will help to finish them sooner. This often results in faster run times and helps to reap economic benefits of Cloud Computing.

Production test is the application of a fixed test program to each chip manufactured, in order to identify defective chips and discard them. The creation of the test program requires significant computation for fault simulation and test vector generation. The fault simulation and test generation can be split into several high computation parallel jobs. This falls in the category of Big Data calculation.

Big Data is a relative term. Large databases in the EDA industry are small compared to some in other industries. Of the three categories of Big Data (Big Data storage, Big Data transfer, and Big Data calculation), EDA projects are most dependent upon Big Data transfer. Specific EDA challenges involve getting large design database from the customer into the Cloud and then getting the modified Big Data back to the customer.

Some workload categories are best performed on a local computing resource such as a desktop workstation. Other workload categories benefit from the availability of large amounts of computing capacity, whether calculation or storage. Because different EDA tools utilized at different steps of a design flow map to different workload categories, the decision on what resources should be allocated is different during the different steps of design process. Therefore a decision to utilize Cloud Computing for the EDA design flow is more appropriately made at the individual steps rather than for the whole process. An interesting example is data mirroring for a multi-site design team. For some steps in the design flow, data mirroring maps to the Big Data storage category. For other steps in the design flow, data mirroring maps to the highly interactive multi-person jobs. The allocation of resources is exactly what Cloud Computing solves and is needed for a multi-category process such as the EDA design flow. In summary, a one-size-fits-all answer to the EDA design tool migration to the Cloud is replaced by a tool-by-tool or step-by-step decision based on the needs and requirements. The mapping of the workload categories provides a basis for the broad adoption issue considerations described in the next sections.

9.5 Considerations for Cloud Computing Adoption

Beyond the workload category mapping as described in a previous chapter in Sect. 4.6, there are some broad considerations that the EDA industry must address to enable a migration to the Cloud. The considerations for EDA moving to the Cloud include tool licensing, information security, Cloud provider desire, lessons from past failed attempts to use the Cloud, tool delivery mechanism, and customer demand. HPC capabilities for meeting performance computing demand of EDA are a must.

First, consider the licensing of EDA tools, which often prohibits sharing a tool between different geographical sites of the same company. An EDA tool at run time must renew its token from a central server at regular time intervals. A Cloud by definition has distributed computing; so a centralized token implementation is a hindrance to sharing the same EDA tool across different servers or data centers within a logical Cloud. It is common for a design flow to include EDA tools from multiple EDA ISVs. Fixing the licensing scheme for Cloud is beyond individual EDA vendor's effort and will need the EDA industry to work together. The current EDA licensing model can be in harmony with the Cloud Computing paradigm of pay as you use when the EDA industry players address the appropriate licensing mechanism [29].

A second consideration is that information security in the Cloud must be solved to the satisfaction of all stakeholders in the EDA industry [30]. Information security can be viewed as including three functions: access control, secure communications, and protection of private data. Access control includes both the initial entrance by a participant and the reentry of that participant or the access of additional participants. The secure communication includes any transfer of information among any of the participants. The protection of private data includes storage devices, processing units, and even cache memory. These and more information security issues especially related to the Cloud are described in [19]. An example security problem is denial-of-service attacks. Ficco and Rak [31] point out the need for Cloud providers to monitor, detect, and provide host-based countermeasures. From the user viewpoint, security is an important part of the problem. For example, Ghosh et al. propose a framework to allow a customer to evaluate Cloud providers using several factors, including different security issues [27]. The EDA industry does not in and of itself pose new information security issues [32]. However, the Cloud suppliers and the EDA vendors must work together with the EDA user community to find a solution that meets the needs of all stakeholders.

A third consideration for EDA vendors is whether the Cloud suppliers even want EDA tools in the Cloud [33]. Both Amazon and Rackspace have publicly stated a desire to attract HPC workloads in their data center, to showcase their prowess and increase revenue. The list of Cloud service providers (CSPs) that may be potentially interested in EDA is fairly long [34] and spreads across the world. Their services range from SaaS (Software as a Service), PaaS (Platform as a Service), and IaaS (Infrastructure as a Service), including hosting customers' servers in CSP-owned data centers. Of this list, the most relevant is IaaS because a typical EDA company and its users will come with their own software and data and need physical or virtual servers to host them. Figure 9.2 shows how extra compute resources can reduce time-to-market.

A fourth consideration is that moving to Cloud has been attempted in recent years. Several large EDA companies have already experimented with offering Cloud services using their own data centers, with mixed responses from customers. Synopsys is an example with a verification-on-demand solution using VCS, which provides massively scalable verification via Amazon Web services. According to Synopsys Website, VCS [35] is the first EDA product to become available on a Public Cloud. While this does not demonstrate that EDA in the Cloud is a success or failure, it does show that the work has started.

A fifth consideration is how the EDA tools are offered to the end users. As it often happens in the software industry, some companies are aggregators of others' tools. This model may work well for EDA in the Cloud. Transim Technology

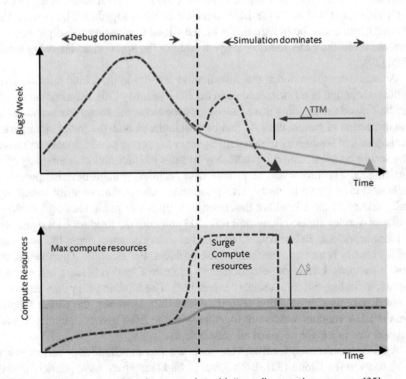

Fig. 9.2 Design teams can reduce time-to-market with "surge" computing resources [35]

Corporation hosts several Cloud-based engineering applications [36]. Advantest is the company that launched the Cloud Testing Service (CTS) aimed at design verification engineers, fabless company, and any institutions with limited resources [37]. The service is a pay-per-use approach applied to IC prototype testing, debugging, and troubleshooting.

A sixth consideration is the EDA customers' demands. Cadence and a start-up Nimbic (purchased by Mentor Graphics) found limited customer demand for moving to the Cloud [38]. However, Synopsys has recognized the demand for EDA tools in the Cloud [35]. A model which may satisfy EDA customers' demand is a hybrid one, with some of the EDA tools being in the Cloud and others in the captive data centers. As Fig. 9.3 shows, 60% of participants in a recent DAC (Design Automation Conference) survey predicted that EDA licensing will move to the Cloud model in the near future.

A seventh consideration is that the current EDA tools and flows were not designed with Cloud Computing in mind. Therefore, to fully utilized Cloud Computing, these may need to be redesigned. Potential benefits of redesigning the EDA tools for Cloud may include an order of magnitude improvement. This clearly requires further study and experiments on adopting EDA algorithms for Cloud Computing [39].

This chapter has concentrated on the case study of the EDA industry adoption of Cloud Computing. An excellent analysis of other issues for the adoption of Cloud Computing by small- and medium-sized enterprises (SME) described that the larger barriers to adoption are nontechnical [40]. First and foremost it is worth noting that they found that most companies see the benefits of adopting Cloud Computing. The top seven reasons that companies gave for not adopting nor intending to adopt Cloud Computing solutions were security (including data loss), loss of control of data and infrastructure to manage data, lack of ability to measure cost-benefit analysis, availability and quality of service, data lock-in (inability to change providers), and data privacy and associated legal requirements. Their chapter emphasizes that in some cases, such as security, adoption of Cloud Computing has actually improved the situation. This is particularly important for security because it is listed as a top concern in the EDA industry. Cloud Computing adoption decisions should be made on a case-by-case basis for different categories of computing. We illustrated this with a case study of the EDA industry.

Fig. 9.3 EDA tool licensing model [TB: Time Based, Cloud, Royalty (on IC production), or Other (mostly combination)] [30]

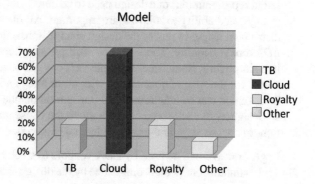

We have described broad considerations that must be addressed in order to take advantage of opportunities that Cloud Computing provide to the EDA industry. These considerations include licensing, security, Cloud providers, past EDA attempts to move to the Cloud, delivery mechanism, and customer demand. The approach to meeting these considerations may vary with different steps (hence workload categories) in the design process or with different vendors, customers, and Cloud providers. As we go to the press, a new report on EDA in the Cloud has emerged [41] stating, *"this represents a fundamental shift on many fronts. On the design side, chipmakers barely paid any attention to cloud offerings until last year, despite the best efforts by Synopsys, Cadence and Mentor, a Siemens Business, to promote cloud capabilities. But several things have occurred to make this more promising. First, TSMC has endorsed design within the cloud last October when it announced its Open Innovation Platform for RTL to GDSII would be available on Amazon Web Services (AWS), Microsoft Azure, Cadence and Synopsys clouds."*

9.6 Applying Block Chain to Security

As we studied the basics of block chain (BC) in Chap. 7, it offers higher levels of security with a distributed ledger and transparency of transactions. However, it is not a panacea for all applications that requires a higher level of security. In the context of migrating EDA to a Public Cloud, or even using it in a shared Private Cloud, the following opportunities are available to apply BC techniques to EDA flow:

1. *Cell-based vs. custom designs:* In a custom design, each circuit is carefully hand-crafted to meet the timing, area, and power constraints. These parameters are traded-off against the design productivity, which is met by using a cell-based design. In a cell library, changes are often required to fix bugs or to optimize the underlying design constructs. It is therefore vital to keep a trail of all changes made in a shared cell library, and this is where block chain will be useful. BC can serve as the root of trust for updating cell databases across multiple design teams.
2. *Incremental requirement changes:* During a design cycle, customer requirements may change, causing changes in the RTL, circuit, and layouts. However, these different representations of a design need to be consistent. In addition, a history of all changes and ability to rollback are important. All these needs can be met with a distributed ledger to record the design change transactions and resulting log files.
3. *EDA tool versioning:* There may be multiple suppliers for different design tools in a single flow. Sometimes multiple versions of the same tool need to be maintained for the sake of consistency with other tools and design versions. All such information can be maintained in a transparent manner with a distributed ledger, so designers at each stage can reproduce the cause and effect of a design iteration, to retrace steps and move the team design process forward.

In general, BC can be used by EDA vendors to track tool license usage and billing and maintain contractual compliance by the design teams in the Cloud. Similarly, cell library providers and third-party designer can contribute to a shared design database via smart contracts maintained in the BC distributed ledger.

As an example, below is a scenario for Cloud-based hierarchical design verification solution:

1. Let's say a design uses other third-party cells and IP libraries.
2. Each cell provider maintains own ledger, showing who is using its cell and which sub-cells are in its own hierarchy.
3. Only the top-level design knows interconnects between all of its cells and thus owns its confidential design IP.
4. Running a hierarchical verification puts out the inputs given to each cell on the BC network, and in turn that cell owner responds to the node that sent the inputs with its output.

   (a) Sender doesn't need to know where its cells are residing.
   (b) A cell doesn't know how it is being used.
   (c) Only the top-level design owner knows all inputs and outputs.

No need to have a single or flattened database.

Any incremental cell changes are easy to roll up.

The above-proposed flow enables multiple cell library providers and different design teams to interact together in a transparent manner.

9.7 Summary

The EDA industry software, tool methodologies, flows, and scripts for VLSI design have evolved from individual programs on mainframe computers, through collections of tools on engineering workstations, to complete suites of tools with associated methodologies on networks of computers. Electronic design automation has had a significant impact on circuit and hardware design, which in turn has contributed to the vast progress in the field of computing. Design automation is one of the reasons why computer chips with upward of billions of transistors can be designed. One can assert that the server farms that form the back end of the Cloud would not have been around without the EDA industry. Thus, it is interesting to see whether Cloud Computing can, in turn, facilitate future growth of the EDA industry.

While Cloud Computing is often thought of as a monolithic entity, it comes in many flavors and is comprised of several subsystems and services. Similarly, EDA tools are sometimes thought of as part of a system built with each tool having similar attributes. Based upon a previous categorization of Cloud Computing workloads, this Chapter maps the sub-tasks of an example silicon design flow to the types of workloads. The mapping of workloads is applicable to both Private and Public Cloud Computing. This mapping can serve as an example for EDA companies and hardware design firms as they look to explore the Cloud for hardware design tasks. Our method can potentially open new doors and customer bases for enabling EDA growth. This chapter also provides examples of some early adopters, the issues they faced, and new emerging challenges, whether real or perceived [42]. Additionally, some considerations were mentioned, such as licensing and delivery mechanisms that go beyond the mapping of tasks to workloads. The major contribution of this chapter is a proposed method for mapping EDA tools to Cloud Computing categories to facilitate the decision of which EDA tools are candidates for moving to the Cloud. Such a step is needed to migrate any established industry to adopt Cloud Computing.

9.8 Points to Ponder

1. For mission-critical application, such as EDA or health data, what precautions need to be taken? EDA companies want to recoup R&D investments faster by selling private licenses, while their customers are hesitant to put confidential design data in a Public Cloud. Given this situation, would it ever make sense to migrate EDA workloads to Public Cloud?
2. How do IP concerns influence the decision to move to Cloud? What precautionary measures need to be considered?
3. Are there any other enterprise or Private Cloud customers who faced similar dilemmas, but economics enabled their migration to Public Cloud?
4. What can Public Cloud service providers do to accelerate migration of EDA workloads?
5. What benefits can an EDA solution provider get by migrating to a Public Cloud?
6. What benefits can EDA customers get by migrating to a Public Cloud?
7. What are the challenges in block chain adoption for EDA industry?

References

1. Ousterhout, J. K., Hamachi, G. T., Mayo, R. N., Scott, W. S., & Taylor, G. S. (1984). Magic: A VLSI Layout System. *21st Design Automation Conference*, pp. 152–159.
2. Nagel, L. W. (1975). *SPICE2: A computer program to simulate semiconductor circuits.* Berkeley: ERL-M520, Electronics Research Laboratory, University of California.
3. Beebe, S., Rotella, F., Sahul, Z., Yergeau, D., McKenna, G., So, L., Yu, Z., Wu, K. C., Kan, E., McVittie, J., & Dutton, R. W. (1994, December). Next generation Stanford TCAD---PISCES 2ET and SUPREM OO7. *IEDM 1994 Proceedings*, San Francisco, CA, pp. 213–216.
4. Pinto, M. R., Rafferty, C. S., & Dutton, R. W. (1984, September). PISCES-II - Poisson and continuity equation solver. Stanford Electronics Laboratory Technical Report, Stanford University.
5. Acken, J. M., & Stauffer, J. D. (1979, March). Part 1: Logic circuit simulation. IEEE Circuits and Systems Magazine.
6. Acken, J. M., & Stauffer, J. D. (1979, June). Part 2: Logic circuit simulation. IEEE Circuits and Systems Magazine.

7. Szygenda, S. A. (1972, June). TEGAS-Anatomy of a general purpose test generation and simulation at the gate and functional level. *Proceedings of 9th Design Automation Conference*, pp.116–127.
8. Bryant, R. E. (1981, July). MOSSIM: A switch level simulator for MOS LSI. *Proceedings of 18th Design Automation Conference*, pp. 354–361.
9. Goldstein, L. H. (1979). Controllability/observability analysis of digital circuits. *IEEE Transactions on Circuits and Systems, 26*(2), 685.
10. Grason, J. (1979). TMEAS, a testability measurement program. *Proceedings of 16th Design Automation Conference*, pp. 156–161.
11. Hachtel, G. D., Hemanchandra, L., Newton, R., & Sangiovanni- Vincentelli, A. (1982, April). A comparison of logic minimization strategies using ESPRESSO: An APL program package for partitioned logic minimization. *Proceedings of the International Symposium on Circuits and Systems*, Rome, pp. 42–48.
12. Preas, B., & Gwyn, C. W. (1978). Methods for hierarchical automatic layout of custom LSI circuit masks. *DAC*, 206–212.
13. Sechen, C., & Sangiovanni, A. (1986). Timberwolf 3.2: A new standard cell placement and global routing package. 23rd DAC, pp. 432–439.
14. http://esd-alliance.org/wp-content/uploads/PDFs/2017/MSS_Q2_2017_Newsletter.pdf
15. Jamshidi, P., Ahmad, A., & Pahl, C. (2013). Cloud migration research: A systematic review. *IEEE Transactions on Cloud Computing, 1*(2), 142–157.
16. Guan, H., Ma, R., & Li, J. (April-June 2014). Workload-aware credit scheduler for improving network I/O performance in virtualization environment. *IEEE Transactions on Cloud Computing, 2*(2).
17. Moreno, I. S., Garraghan, P., Townend, P., & Xu, J. (April-June 2014). Analysis, modeling and simulation of workload patterns in a large-scale utility cloud. *IEEE Transactions on Cloud Computing, 2*(2).
18. Mulia, W. D., Sehgal, N., Sohoni, S., Acken, J. M., Lucas Stanberry, C., & Fritz, D. J. (2013). Cloud workload characterization. *IETE Technical Review, 30*(5), 382.
19. Sehgal, N. K., Sohoni, S., Xiong, Y., Fritz, D., Mulia, W., & Acken, J. M. (2011). A cross section of the issues and research activities related to both information security and Cloud Computing. *IETE Technical Review, 28*(4), 279–291.
20. http://www.eetimes.com/document.asp?doc_id=1326801
21. Joyner, W. EDA: The First 25 years. Presented at the 50th Design Automation Conference, Jun. 2013. [Online]. Available: www.dac.com
22. Sehgal, N., Chen, C. Y. R., & Acken, J. M. (1994). An object-oriented cell library manager. *Proceedings of ICCAD*, pp. 750–753.
23. Porter, M. E. (2008). The five competitive forces that shape strategy. *Harvard Business Review*. http://hbr.org/2008/01/the-five-competitive-forces-that-shape-strategy
24. Kuehlmann, A., Camposano, R., Colgan, J., Chilton, J., George, S., Griffith, R., et al. (2010). Does IC design have a future in the Cloud? *Design Automation Conference (DAC)*, 47th ACM/ IEEE, pp. 412–414. IEEE.
25. Rutenbar, R. (2013, June). EDA: The Second 25 years. Presented at the 50th Design Automation Conference. www.dac.com
26. Schneier, B. (2015, March). The hidden battles to collect your data and control your world. https://www.schneier.com/books/data_and_goliath/
27. Ghosh, N., Ghosh, S. K., & Das, S. K. (2015). SelCSP: A Framework to Facilitate Selection of Cloud Service Providers. *IEEE Transactions on Cloud Computing, 3*(1).
28. NIST Special Publication 800-145.
29. Ralph, J. Challenges in Cloud Computing for EDA. Chip Design Magazine. http://chipdesign-mag.com/display.php?articleId=4492
30. Stok, L. (2013, June). EDA: The Next 25 years. Presented at the 50th Design Automation Conference. [Online]. Available: www.dac.com
31. Ficco, M., & Rak, M. (2015). Stealthy denial of service strategy in Cloud Computing. *IEEE Transactions on Cloud Computing, 3*(1).

32. Brayton, R., & Cong, J. (2010). NSF workshop on EDA: Past, present and future. *IEEE Design and Test of Computers, 27*, 68–74.
33. EDAC Forecast Meeting, San Jose, Mar. 14, 2013. http://www.edac.org/events/2013-EDA-Consortium-Annual-CEO-Forecastand-Industry-Vision/video
34. http://talkinCloud.com/tc100
35. http://www.synopsys.com/Company/Publications/SynopsysInsight/Pages/Art6-Cloud-IssQ2-11.aspx
36. "Designing in the Cloud", Electronic Specifier, 7th November 2012. http://www.electronic-specifier.com/design-automation/designing-in-the-Cloud-arrow-dr-uwe-knorr-transim-technology-es-design-magazine. June 2015.
37. Elliott, M. (2015, June). Productonica: Cloud clears way to IC testing. Electronic Specifier, 2013, November https://www.electronicspecifier.com/test-and-measurement/advantestcx1000d-productronica-cloud-clears-way-to-ic-testing
38. https://www.mentor.com/company/news/mentor-acquires-nimbic
39. Schwaderer, C. (2014, May). EDA and the Cloud. Embedded Computing Design. http://embedded-computing.com/articles/eda-the-Cloud/
40. Trigueros-Preciado, S., Perez-Gonzalez, D., & Solana-Gonzalez, P. (2013). Cloud Computing in industrial SMEs: Identification of the barriers to its adoption and effects of its application. *Electronic Markets, 23*(2), 105–114.
41. https://semiengineering.com/more-data-moving-to-cloud/
42. http://www.techradar.com/us/news/Internet/Cloud-services/security-is-a-red-herring-in-the-Cloud-1163167

Chapter 10
Cost and Billing Practices in Cloud

10.1 Cloud as a Service (CaaS): The Billing Imperatives

The usage Cloud is through a service provider, wherein the two primary stakeholders are the consumers and provider of the service. However, unlike utilities such as electricity, water, postal, or city services, the Cloud Computing services offer opportunities for many interesting and innovative provisions, which we shall discuss later in this chapter. In this section, we compare and contrast some facets of billing in traditional utility services vs. Cloud Computing.

10.1.1 Cloud Business Practices

Consumer services are often competitive and market-driven. Therefore, a service provider must operate in the customer first mode of operations to survive and succeed. In the traditional terms, a utility should be easy to access and provide customer delight. In Cloud Computing context, this translates to:

1. Easy to use interface
2. Attract customers to repeat experience with greater vigor

The second item above is important to make it attractive for higher volumes and greater capacity utilization of Cloud resources. Note this is not different from traditional utility services, which often offer cost-saving options for volume and frequency of business.

A Forrester study [5] observed that patron patience lasts no more than 8 seconds with not-so-friendly Web interface. In traditional utility services, there is not much transactional traffic. So, what this translates to in Cloud Computing context is that the service provider must provide a straightforward customer interface for multimodal transaction traffic. For example, multimodal transactional traffic should

© Springer Nature Switzerland AG 2020
N. K. Sehgal et al., *Cloud Computing with Security*,
https://doi.org/10.1007/978-3-030-24612-9_10

not result in password fatigue or delays with repeated exchanges for access requests. A single sign-on for a session and strong authentication is desirable.

A good traditional utility practice is to provide notifications to patrons. This is very easy in Cloud Computing context. It also is an opportunity for prompt cross-selling options. Depending upon the nature of usage and patterns of operations, the Cloud Computing service can offer the most beneficial pricing. Later, in the chapter we will describe how Amazon adapts it's billing to the services by employing auto scaling and other practices.

An important consideration that goes with utility services companies is observance of some standards. Cloud Computing billing practices do have service-level agreements, as described in a later section of this chapter, but there are no known common standards as of now. In particular, there are undefined areas on customer assistance, default or minimal obligations and options, emergency or breakdown coverage, termination of services, and transfer of accounts following mergers and breakups. This is still an open area and evolving.

10.2 Load Balancing

Before any discussion of Cloud costs and customer billing, it is important to understand the concepts of load balancing. It refers to efficiently managing incoming tasks across a group of servers, to ensure an optimal loading of the machines. Too many jobs running on a server can slow it down. It may happen even if only one part of the server hardware is overloaded, such as memory accesses, while CPU is waiting. Thus it is important to distribute the jobs evenly but critical to ensure that job types are optimally mixed. A task can be CPU-bound, memory-bound, or storage- or I/O-bound. It is easy to determine after a job has run, depending on what kinds of resources were consumed. However, in a Cloud, the operator typically doesn't look into a customer virtual machine (VM) for privacy reasons. In this case, an almost blind placement is done, and if the later monitoring reveals that a server is running slow, then some jobs can be migrated to a different machine. This may result in an interruption of the service, which Cloud customers do not like. So, as we see, that load balancing is a nontrivial task.

One can think of a load balancer as an airport officer directing incoming customers to relatively less crowded counters. This keeps the lines moving and clears passenger traffic faster. A load balancer is located at the gateway of a data center and routes customer requests across all servers. It has the objectives to maximize speed and capacity utilization. If a server goes down, or is removed for maintenance, the load balancer will redirect traffic to the remaining online servers. Similarly, when a new server is added to the rack, the load balancer will start sending requests to it. Thus, a load balancer has the following functions:

- Evenly distribute customer requests across multiple servers.

- Provide flexibility to add or subtract servers as the demand dictates.
- Ensure high reliability by sending new requests only to servers that are available.
- Monitor performance and detect a slowdown in existing servers.
- Depending on IT policy, either migrate jobs for balancing or kill offending jobs.

In order to assist with the above, several algorithms are available:

1. *Round Robin*: If all servers and jobs are alike, then simply distribute incoming tasks across a group of servers sequentially.
2. *Least Connections*: A new request is sent to the machine with least users or customer connections. This works well in an eBay-like environment, with interactive user session. There may be an agent running on each server to communicate back the number of active users.
3. *IP Hash*: A customer profile is maintained, based on the past jobs run, to forecast the type of future tasks and compute resources that a customer will demand. Then the IP address of a client is used to determine which server will be better suited. This is an excellent application for machine learning. One can learn from the pattern of usage and determine an optimal mapping for each type of user. In this scheme, servers are pooled based on their hardware resources, e.g., machines with graphics processors may be used for game players or crypto currency miners.

In a later section, we will review Amazon's Elastic Load Balancer.

10.3   Pay as You Go

The entire Cloud Computing usage model is based on a pay as you go or pay for what you consume concept. This enables Cloud customers to not buy IT hardware or software products, but rent them on a time and/or volume basis, e.g., how many CPU cores are needed for how long, with how much memory and persistence storage space, etc. The persistence storage also comes in various forms, such as disks directly attached to the compute instance for faster access, but at a higher cost, or stored across the network in a long-term storage repository at a cheaper cost and higher latency, etc.

At the end of a month, or agreed-upon usage period, a Cloud service provider (CSP) will send the bill to its users. If upfront menu choices are not made wisely, then the user may be in for a surprise, but little can be done retroactively to reduce costs.

To create innovative and interesting billing propositions, major Cloud providers such as Amazon EC2 offer additional choices such as On-Demand vs. Spot Instances, ability to do auto scaling, Elastic Load Balancers, etc. The following sections will explain each of these concepts in detail. Although these details are pertinent to Amazon EC2, other major CSPs offer similar capabilities with different names. Let us consider a Cloud customer who is hosting an end-user facing Website. This customer may experience variable user load during day and night hours. The rationale for offering a choice to customers is to ensure that their needs can be met and they fill up the installed Cloud capacity.

10.4   Amazon EC2 Motivations and Setup

Amazon offers their compute capacity in availability zones (AZs), which are distinct locations within a region that are engineered to be isolated from failures in other availability zones. As explained in Chap. 6, an availability zone is a distinct location within a region that is insulated from failures in other availability zones. It also provides inexpensive, low-latency network connectivity to other availability zones in the same region. A simple example can be a different data center (DC), or a separate floor in the same DC, or an electrically isolated area on the same floor. Whereas a region is a named set of AWS resources located in a particular geographical area, such as Northern California, each region must have at least two availability zones.

Motivation for multiple regions is to provide worldwide customers a server location closer to their usage points, so that signal transmission latencies are reduced. Otherwise, each click from a customer in India will potentially need to travel to a data center in the USA, and then sending the response back will require several hundreds of milliseconds, slowing down every Web transaction. Furthermore, different customers have different usage patterns and financial abilities to pay for Cloud Services. Hence, someone who is an occasional user of Cloud may want to pay by the hour with no long-term commitments or upfront fees. Amazon's On-Demand Instances is the best choice for such a consumer. Another customer with a large fluctuating demand may want to use Amazon's Spot Instance bidding. An example of such a user is an accounting or tax preparer site. Yet another customer with a long-term horizon and predictable demand, such as an engineering design firm with multiple contracts, would be interested in Amazon's Reserved Instances to save money and get guaranteed capacity. Last, but not the least, is the category of customers who must operate on isolated and dedicated servers to ensure maximal security and performance. This includes healthcare providers, financial sector, certain government and police services, etc. Amazon's Dedicated Hosts and Instances are the best way to go for such customers.

There are no minimum fees to use Amazon EC2 [1]. Currently, there are four ways to pay for what you consume in their Cloud Computing based on the following usage patterns:

1. On-Demand Instances
2. Spot Instances
3. Reserved Instances
4. Dedicated Hosts and Dedicated Instances

The following section will present a detailed description of how each of these works and their usage patterns. However, before that let us take a look at Amazon's Partner Network (APN), which supports AWS customers with technical, business, sales, and marketing resources. We will only review the first category as relevant to the Cloud Computing context. APN Technology Partners provide software solutions that are either hosted on or integrated with the AWS platform. Technology Partners include independent software vendors (ISVs), SaaS, PaaS, developer tools, and management and security vendors. An example is AWS Proof of Concept (PoC) Program, designed to facilitate AWS customer projects executed via eligible APN Partners. To accelerate a customer's AWS adoption, PoCs provide funding with free AWS usage or co-funded professional services. The goal is to accelerate customer onboarding of enterprise workloads to AWS and to develop customer references and/or case studies that APN Partners can use to demonstrate their AWS expertise.

10.4.1   Amazon On-Demand Instances

With On-Demand Instances, customers pay for compute capacity by the hour with no long-term commitments or upfront payments. They can increase or decrease the compute capacity depending on the demands of their applications and only pay the specified hourly rate for the instances used. On-Demand Instances are recommended for:

- Users that prefer the low cost and flexibility of Amazon EC2 without any upfront payment or long-term commitment
- Applications with short-term, spiky, or unpredictable workloads that cannot be interrupted
- Applications being developed or tested on Amazon EC2 for the first time

There are no guarantees that users will always be able to launch specific instance types in a timely manner, though AWS best case efforts to meet their needs.

10.4.2   Amazon Spot Instances

Amazon EC2 Spot Instances allow customers to bid on spare Amazon EC2 computing capacity for up to 90% off the on-demand price. Spot Instances are recommended for:

- Applications that have flexible start and end times
- Applications that are only feasible at very low compute prices
- Users with urgent computing needs for large amounts of additional capacity

Spot price fluctuates based on the supply and demand of the available compute capacity. It's the leftover capacity of AWS to be used on as available basis. There is no difference in the performance compared to On-Demand Instances, and it is usually cheaper than the On-demand Instances as there is no guarantee of the availability. The user can choose a start time and end time for the instances or can make a persistent request (no end time specified) for this service. This service is preferable for computing needs that are not tied to any deadlines, whose computing needs are large and the interruption of service is acceptable.

Users can bid for Spot Instances. Once your bid exceeds the current spot price (which fluctuates in real time based on the demand and supply), the instance is launched. The instance can go away anytime the spot price becomes greater than your bid price. You can decide your optimal bid price based on the spot price history of the last 90 days available on AWS console or through the EC2 APIs.

10.4.3   Amazon Reserved Instances

Reserved Instances provide customers with a significant discount (up to 75%) compared to the On-Demand Instance pricing. In addition, when Reserved Instances are assigned to a specific availability zone, they provide a capacity reservation, giving customers an ability to launch instances whenever needed.

For applications that have steady-state or predictable usage, Reserved Instances can provide significant savings compared to using On-Demand Instances. Reserved Instances are recommended for:

- Applications with steady-state usage
- Applications that may require reserved capacity
- Customers that can commit to using EC2 over a 1- or 3-year term to reduce their total computing costs

AWS introduced a single type of Reserved Instance with three payment options: one upfront payment for the entire Reserved Instance term (1 or 3 years), pay for a portion of the Reserved Instance upfront with installments for the rest over the course of the 1- or 3-year term, or nothing upfront except the commitment to pay over the course of a year.

10.4.4   Amazon Dedicated Instances and Dedicated Hosts

A Dedicated Instance is a virtual machine assigned to a customer and hosted on a specific host. A Dedicated Host is a physical EC2 server dedicated for that customer's use. No other VM will share resources with it, and resources like CPU/memory will not be shared with anyone else. Dedicated Hosts can help to reduce costs by allowing customers to use your existing server-bound software licenses, including Windows Server, SQL Server, and SUSE Linux Enterprise Server (subject to the licensing terms), and can help to meet compliance requirements:

- Can be purchased on demand (hourly).
- Can be purchased as a reservation for up to 70% off the on-demand price.
- Can provide one or more EC2 instances on Dedicated Hosts

10.5 Motivation and Methods for Right Sizing Customer VMs

Since each user application has a certain need of CPU, memory, and storage resources, it is important to provision these in advance to avoid adverse run-time performance effects. However, giving an application more resources may not necessarily make it run faster, but will certainly cost more in a Cloud. To start with, every time a new VM is launched in the Cloud, it has a new IP address. This may not be desirable for a public facing Website as it needs a DNS (Domain Name Server) entry mapping its name to a fixed IP address. This problem is addressed by Amazon's use of *Elastic IP*. Such a fixed IP can map to one or more servers with an *Elastic Load Balancer* (ELB) in between. It makes sure that the users' traffic is evenly distributed. Having two or more servers running in parallel also ensures that the Website will be resilient, even when a server suffers an outage or slowdown. Another challenge is that usage of a public facing Website may vary over time. An example is a digital newspaper hosting site, which will have more users in the morning and evening, given that its readers may be working at their regular jobs during daytime and probably a very few during late nights as most readers would be sleeping. In order to accommodate a variable workload, the newspaper site wants to duplicate its content on multiple servers and bring them online as the users grow, such that each will get an acceptable response time. Any Website with a slow response during heavy traffic time stands to lose its customers and suffer a business loss. This problem can be easily solved with *auto scaling*, which monitors the CPU usage of existing servers. When the utilization grows, it will start new servers to evenly distribute the load.

Thus it is important to understand an application's requirement and plan for it in advance. In this context, the following AWS concepts are introduced:

10.5.1 Elastic IP

When an instance is launched, AWS assigns it a public IP address. Every time the instance is stopped, this IP address is freed up. However, an application's users may expect a permanent IP address for their future sessions. AWS' solution is the use of Elastic IPs (EIPs).

An Elastic IP address (EIP) is a static IP address designed for dynamic Cloud Computing. With an EIP [2], one can mask the failure of an instance or software by rapidly remapping the IP address to another instance. An EIP is associated with customer's AWS account, not a particular instance, and it remains associated with the account until the customer chooses to release it.

10.5.2   Elastic Load Balancing

Elastic Load Balancing [3] automatically distributes incoming application traffic across multiple Amazon EC2 instances. It enables achieving fault tolerance in user applications, seamlessly providing required amount of load balancing capacity needed to route application traffic.

Elastic Load Balancing (ELB) ensures that only healthy Amazon EC2 instances will receive users' traffic by detecting unhealthy instances. ELB reroutes traffic to the remaining healthy instances, as shown in Fig. 10.1. If all of a customers' EC2 instances in one Availability Zone A are unhealthy, but additional EC2 instances were setup in multiple Availability Zones, E will route traffic to the healthy EC2 instances in those other zones, such as Availability Zone B.

Elastic Load Balancing offers two types of load balancers that feature high availability, automatic scaling, and robust security. These include the Classic Load Balancer that routes traffic based on either application or network-level information and an Application Load Balancer that routes traffic based on advanced application-level information, such as Web page load delays. The Classic Load Balancer is ideal for simple load balancing of traffic across multiple EC2 instances, while the Application Load Balancer is ideal for applications needing advanced routing capabilities, microservices, and container-based architectures. Application Load Balancer offers ability to route traffic to multiple services or load balance across multiple ports on the same EC2 instance. Amazon also offers EC2 Container Service (ECS) that supports Docker containers and allows running applications on a

Fig. 10.1 An example of a server going bad in Zone A and traffic getting routed to Zone B

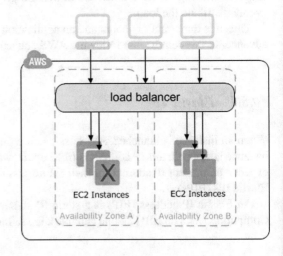

managed cluster of server instances. Amazon ECS eliminates the need to install, operate, and scale user's own cluster management infrastructure. With simple API calls, one can launch and stop Docker-enabled applications and query the complete state of a cluster. APIs also access features such as security groups, Elastic Load Balancing, EBS volumes, and IAM roles. Amazon enables scheduling the placement of containers across the clusters based on resource needs and availability. Users can integrate their own or third-party schedulers to meet business or application specific needs.

10.5.3 Auto Scaling

In this section, we will be using four concepts that are uniformly used in the IT community and can be defined as follows:

1. *Scale up* refers to increasing compute capacity of a server, by increasing either its density or performance, e.g., a CPU with more cores or more memory, resulting in higher throughput on a single threaded application. More memory can be of higher capacity or faster speed components.
2. *Scale out* refers to adding multiple units of computing, e.g., more servers for a distributed application to share the load.
3. *Scale in* is the process of reducing or compacting the computing capacity for efficient reasons, e.g., when load is light then moving virtual machines from many physical servers to run on a fewer servers, resulting in reduced operating costs.
4. *Auto scaling* is the ability of a system to scale in or out depending on the workload.

Auto scaling in EC2 [4] helps to maintain application availability, by scaling Amazon Server capacity to go up or down automatically, according to the conditions you define. You can use auto scaling to help ensure that you are running a desired number of Amazon EC2 instances. Auto scaling can also automatically increase the number of Amazon EC2 instances during demand spikes to maintain performance and decrease capacity during lull periods to reduce customer costs, as shown in Fig. 10.2.

When the load is low, only a few servers are needed. However, as more customers use a Website or an application hosted in Cloud, their experience may deteriorate as existing server slows down. In this case, new servers are started and brought online to share the increased load. This can happen automatically, by checking the CPU utilization of current servers. If it increases beyond a threshold, then new servers are started. Similarly, when demand falls below a threshold, then additional servers are shut down to save money, and the remaining customers are moved to fewer servers. It is called customer consolidation. Scale up or down often requires a stop and restart of application on a new server, unless a live migration of virtual machine is available. Scale in or out is suitable for distributed applications that are

Fig. 10.2 An example of
auto scaling

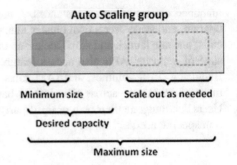

typically multi-threaded and have many tasks that can benefit from additional machines at the run time.

Auto scaling is well suited to applications that have stable demand patterns or that experience hourly, daily, or weekly variability in usage. Auto scaling enables following the demand curve for customer applications closely, reducing the need to manually provision Amazon EC2 capacity in advance. For example, one can set a condition to add new Amazon EC2 instances in increments to the auto-scaling group when the average utilization of Amazon EC2 is high; and similarly, one can set a condition to remove instances in the same increments when CPU utilization is low.

10.6   Cost Minimization

If customer demand is fluctuating, and application is scalable, then instead of getting a large computer, it may be better to run multiple copies of an application. These can be hosted on identical EC2 instances, such that auto scaling manages launch and termination of these EC2 instances. As shown in Fig. 10.3, one can use Amazon CloudWatch (CW) to send alarms to trigger scaling activities and Elastic Load Balancing to help distribute traffic to customer instances within auto-scaling groups. CloudWatch monitors CPU utilization of servers on which user applications are running. If and when the utilization goes above a certain threshold, say 85%, then an alarm is generated, which in turn can alert the application owner and based on a predetermined scaling policy bring up new server instances to share the application load. When done correctly, CPU utilization of previous servers as well as latency experienced by new users will both go down. Thus auto scaling enables optimal utilization of Cloud resources.

If a customer has predictable load changes, then one can set a schedule through auto scaling to plan scaling activities, as shown in the life cycle of Cloud application in Fig. 10.4. At the start, a new auto-scaling group and scale-out events are defined. Then one or a few set of instances is launched, and whenever load increases new instances are added to the group. This can be done with a CloudWatch trigger as defined previously or based on a fixed schedule such as people waking up on Monday morning to check their emails, which will need more email servers.

Fig. 10.3 Using Cloud
utilization alarm to trigger
auto scaling

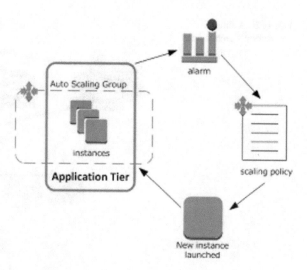

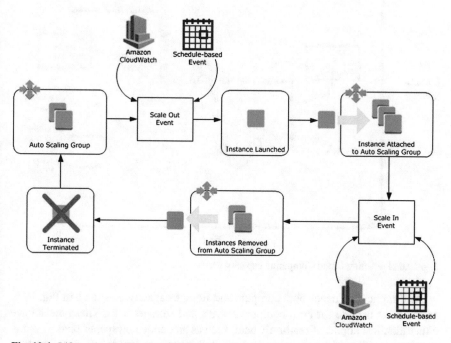

Fig. 10.4 Life cycle of a Cloud application

Similarly, in Step 5, a *scale-in* event is defined, which can also be based on
CloudWatch monitoring CPU utilization going below a threshold, say less than
50%, or a scheduled event such as the start of a weekend when fewer people log in
and check their work emails.

Fig. 10.5 A simple
application's architecture

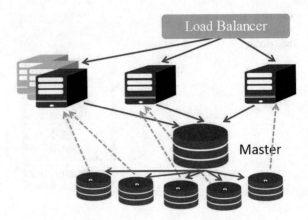

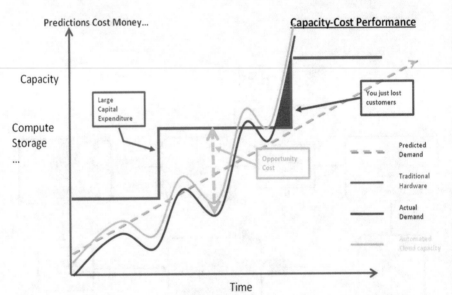

Fig. 10.6 Predicting Cloud Computing capacity needs

Any application needs both compute and storage capacity, as shown in Fig. 10.5, so scale out is done for both compute servers and storage disks. These disks have either identical copies of read-only data, such as an online newspaper read by many users, or partitioned information such as different users looking at different segments of a large database, e.g., airline flights or an online auction site, which can be independently partitioned.

The key is to understand an application's need of Cloud Computing resources and then predict and provision them in advance, to avoid losing customers or paying for unused capacity, as shown in Fig. 10.6. The dotted inclined line shows predicted compute demand growing over time. Blue line shows a step function, as large capital expenditures occur to install more server racks, networking switches, storage devices, power supplies, air conditioning etc. Once the dotted line reaches the limit of blue step, the next step of capital expenditure must be incurred. However, actual compute demand as shown by the solid red line fluctuates over time. It should always stay within the installed capacity envelope as defined by the blue step function. Their difference, as shown by the green arrow, represents unused installed capacity, which is the opportunity cost as that money could've been used elsewhere in the business. In a case where the red actual demand exceeds the blue envelope, there is no compute capacity available to run the actual tasks, so it represents unhappy or lost customers. Of course, with Cloud Computing, the blue line can expand elastically up or down, to follow the red curve, thus minimizing the green opportunity cost and no lost customers.

10.7   Capacity Forecasting

Every business organization wants to grow, but growth brings its own problems as quality of service may suffer causing slowdowns for the existing customers. In order to ensure that business operations run smoothly, IT department needs to understand the current bottlenecks, build future growth models, and predict users' demand in advance. This problem was exacerbated before Cloud, as it took weeks if not months to order, procure, and deploy new servers. In an elastic Cloud, new servers can be brought online in a matter of minutes.

An example is a movie serving Website, which could experience growth in a particular region so the servers need to be located in the same or nearby region. This will ensure minimal latency for the new and existing customers. Also, it needs to understand the bottleneck, as new customers may demand more of the existing movies, or new films. Serving more customers with the same set of movies will need duplication of storage servers and network resources. Adding more movies to the repository will require new compute servers to encode the new movies, and then new storage devices to host them, while maintaining the older movies on the existing servers. Thus it is important for a business to understand its users, their needs, and growth patterns. Furthermore, using spreadsheets, various scenarios need to be worked out on what to host and where, so the cost is understood and budgeted in advance. The following example will make this clear:

An application needs four EC2 instances and S3 storage, where the application running on EC2 requires 8 GB setup minimum. If the cost of each Server instance is $0.10/hour, and storage cost is $1.00/GB/month, what's the yearly budget for this organization?

$$\text{Server Instance cost / year} = 4 * \$0.10 * 24 * 365 = \$3504 / \text{year}$$

$$\text{Storage cost / year} = 8 * \$1 * 12 = \$96 / \text{year}$$

$$\text{Total outlay} = \text{Instance Cost} + \text{Storage Cost} = \$3600 / \text{year}$$

Let's consider a business of online newspapers, magazines, or on-demand video services, where many users can access this read-only data through a Web application. If this business expects to grow by 50% per year, their storage cost may remain constant due to the *read-only* nature of operations; then server instance cost/year will grow by 50% to support 50% more users, bringing the new total to $5352/year.

Now let's say another set of 50% users demand new magazines, which needs new computers and storage, by the same proportion, requiring 50% more instances and storage. This will require new expenses to be $8082/year, which is more than 2X of the first year's expense. More users is generally good news, only if accompanied by a commensurate growth in the compute infrastructure. Else result will be similar to doubling the number of cars on existing roads, causing traffic jams slowing down all users.

10.8   Optimizations Across Cloud

Since multiple Cloud service providers (CSPs) exist, it makes sense to be able to compare their offerings on a common set of metrics and then choose what best meets a user's criteria. Furthermore, it may be desirable to have multiple accounts and make an application portable across CSPs, to avoid a lock-in. This is easier said than done, as each Cloud provider offers different sets of tools and different flavors of solutions, with an aim to entice and lock in their customers. Currently, there are three large Public Cloud providers in the USA.

Provider	Offering name	Description
Amazon	AWS	A comprehensive, evolving Cloud Computing platform provided by Amazon.com. Web services are sometimes called Cloud services or remote computing services. The first AWS offering was launched in 2006 to provide online services for Websites and client-side applications
Microsoft	Azure	Microsoft Azure began as a Windows Server-based Cloud Platform but has since grown to become a very flexible Cloud Platform that enables developers to use any language, framework, or tool to build, deploy, and manage applications. According to Microsoft, Azure features and services are exposed using open REST protocols. The Azure client libraries, which are available for multiple programming languages, are released under an open-source license and hosted on GitHub
Google	Google Cloud	A Cloud Computing platform by Google that offers hosting on the same supporting infrastructure that Google uses internally for end-user products like Google Search and YouTube. Cloud Platform provides developer products to build a range of programs from simple Websites to complex applications

Providers' costs are changing constantly, e.g., AWS has changed its prices 50 times since it launched in 2006. Although other benefits are associated with Cloud infrastructure, the low upfront costs and inexpensive payments for servers attract a large segment of customers and are always mentioned as a major incentive for Cloud adoption. It is important to understand provider pricing to make informed decisions on IT spending optimization. When considering Cloud infrastructure's total cost, businesses need to be asking the following questions:

1. What should the size of VMs be?
2. How many VMs of each size are needed?
3. How long does one want to commit to a Cloud provider?
4. How many peripheral services, such as storage, security, scaling, etc., are needed?
5. What discounts are available?

Pricing models vary across AWS and Azure and other providers, so above factors are a mere start to compare different CSP offerings. In the coming years, focus may shift from Cloud Computing to specific Cloud services, such as for HPC and gaming. Specialized Cloud will begin to emerge providing services like those highlighted in Fig. 10.7.

Specialized Cloud type	Infrastructure	Application platform	Application	Vertical-specific Cloud
Services provided	Storage Computer Networking Management	Middleware Design Development Deployment Testing Operations	Vertical Applications Healthcare Retail Finance Manufacturing Enterprise Consumer	Healthcare Retail Finance Manufacturing Others

Specialized Cloud Type:	Infrastructure	Application Platform	Application	Vertical Specific Cloud

Services Provided:	• Storage
	• Computer
	• Networking
	• Management
		• Middleware
		• Design
		• Development
		• Deployment
		• Testing
		• Operations
			• Vertical Applications
			  o Healthcare
			  o Retail
			  o Finance
			  o Manufacturing
			• Enterprise
			• Consumer
				• Healthcare
				• Retail
				• Finance
				• Manufacturing
				• Others..

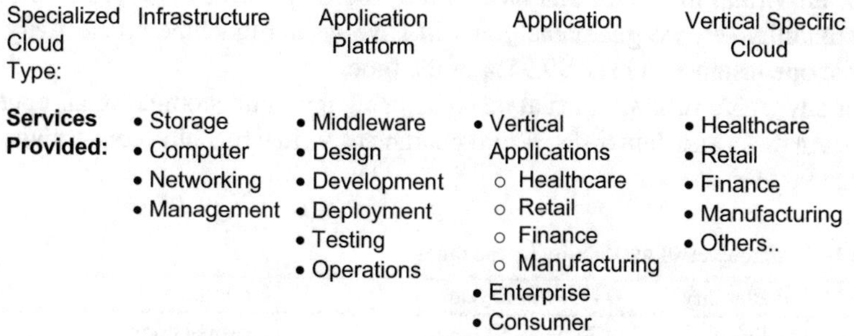

Fig. 10.7 Evolution of Cloud service providers

10.9 Types of Cloud Service-Level Agreements

A service-level agreements (SLAs) is a contract-negotiated and agreed between a customer and a service provider. The service provider is required to execute incoming requests from a customer within the negotiated quality of service requirements for a given price. Due to variable load on a server in Cloud, dynamically provisioning of computing resources to meet an SLA, and allow for optimum resource utilization, is nontrivial. An example is the guarantee of network bandwidth or data read-write speeds for an application. These tasks are generally conducted on shared resources, such as a network interface card or a memory controller on a server, with multiple virtual machines or applications accessing data over the same set of physical wires. Result is a conflict of demand for resources and their availability in a shared multi-tenanted server, causing unexpected wait times. The only way to guarantee a certain bandwidth is to reserve it in advance and leave it unused until an application requests it. However, that will cause unused capacity and higher cost for the Cloud operator. Hence they often over-allocate the same resource to multiple customers and avoid specific SLAs with precise guarantee of quantified performance.

Most Cloud SLAs specify only uptimes [7, 8], stating that their compute servers, storage devices, and networks will be available for a certain percentage of the time. Service is measured by "9s" of availability, such that 90% has one 9, 99% has two 9s, 99.9% has three 9s, etc. A good high-availability (HA) package with substandard hardware has three 9s, whereas an enterprise class of hardware with a stable Linux kernel offers five or more 9s. Below is an example of typical availability levels in Table 10.1.

Microsoft Azure SLA offers commitments for uptime and credits for any downtime. An example of virtual machine connectivity, which encompasses server and network uptimes, is given below, followed by credits offered for any downtime in Table 10.2:

- Monthly Uptime = (Minutes in the Month - Downtime)/Minutes in the Month
- For all virtual machines that have two or more instances deployed in the same availability set, we guarantee you will have virtual machine connectivity to at least one instance at least 99.95% of the time.
- For any single instance virtual machine using premium storage for all operating system disks and data disks, Azure guarantees virtual machine connectivity of at least 99.9%.

Table 10.1 Examples of availability by the nines

9s	Availability	Downtime/year	Examples
1	90.0%	36 days, 12 hours	Personal computers
2	99.0%	87 hours, 36 minutes	Entry-level business
3	99.9%	8 hours, 45.6 minutes	ISPs, Mainstream business
4	99.99%	52 minutes, 33.6 seconds	Data centers
5	99.999%	5 minutes, 15.4 seconds	Banking, medical
6	99.9999%	31.5 seconds	Military defense

Table 10.2 Example of MS Azure Credits in event of downtimes

Monthly Uptime Percentage	Service credits
<99.95%	10%
<99%	25%
<95%	100%

Table 10.3 Example of Google Cloud Credits in event of downtimes

Monthly Uptime Percentage	Percentage of monthly credits
99% to <99.95%	10%
95% to <99%	25%
<95%	50%

Similarly, Google Cloud Platform offers a stringent SLA and credits back for any downtime as shown in Table 10.3:

- "Downtime period" means a period of 60 consecutive seconds of downtime. Intermittent downtime for a period of less than 60 consecutive seconds will not be counted toward any downtime periods.
- "Monthly Uptime Percentage" means total number of minutes in a month, minus the number of minutes of downtime suffered from all downtime periods in a month, divided by the total number of minutes in a month.
- "Maximum financial credit" refers to the aggregate number of financial credits issued by Google to the customer for all downtime periods in a single billing month which will not exceed 50% of the amount due from customer for the covered service for the applicable month. Financial credits will be in the form of a monetary credit applied to future use of the covered service and will be applied within 60 days after the financial credit was requested.

Besides availability, SLAs should specify downtime, mean time to repair, and mean time to respond. Scheduled maintenance of updating systems doesn't generally count as downtime. Cloud service providers offer multilevel SLAs, e.g., Rackspace offers managed and intensive levels of support [6] as described below:

- 100% network uptime guarantee
- 1-hour hardware replacement
- URL, port availability, and hardware monitoring
- Managed firewall and VPN access
- OS patching
- Bandwidth and backup performance utilization

The following services are offered only to intensive-level customers for a higher cost:

- Server virus scanning
- Guaranteed response times for support requests
- Advanced system performance and device monitoring
- Custom configuration trend performance

In addition, the following optional services are available for an additional cost:

- DDOS (Distributed Denial of Service) attack mitigation
- Managed security services
- Encrypted backups

However, Rackspace customers may not get predicted, measurable, and quantified performance levels. Similarly, Amazon offers dedicated servers, but customers have to be content with whatever performance occurs even on those machines. This is one of the key limitations currently observed in the Public Cloud.

10.10   Summary

The entire Cloud Computing usage model is based on a pay as you go or pay for what users consume. This enables Cloud customers to not buy IT hardware or software products, but rent them on a need basis, e.g., how many CPU cores are needed and for how long, with memory and persistence storage space, etc. The persistence storage also comes in various forms, such as disks directly attached to the compute instance for faster access, but at a higher cost, or stored across the network in a long-term storage repository at a cheaper cost and higher latency, etc. It is up to a user to understand their needs, pick a Cloud service provider, and wisely choose additional services or capabilities to do auto scaling and load balancing and plan for future growth.

10.11   Points to Ponder

1. NIST has well-defined Cloud Standards; compare how Amazon, Google, and Microsoft Public Cloud offerings comply with NIST standards.
2. Amazon is innovating many additional services to attract and keep its EC2 customers. Give some examples.
3. OpenStack has been using open-source code to build interoperable Cloud solutions, but its adoption has been slow. Why?
4. A start-up wants to save money, and its tasks require 4 GB memory. Should it rent a small-size VM (virtual machine) that costs less, but supports only 2 GB, or a medium size that costs double?
5. What's the value of a constant IP address for a Website hosted in the Cloud?
6. Under what circumstance will you prefer to use a load balancer vs. an auto scaler in a Cloud?
7. What criterion you would use to compare and contrast different Public Cloud service providers?

References

1. https://aws.amazon.com/ec2/pricing/
2. http://docs.aws.amazon.com/AmazonVPC/latest/UserGuide/vpc-ip-addressing.html
3. https://aws.amazon.com/elasticloadbalancing/
4. https://aws.amazon.com/autoscaling/
5. Webinar "Beyond the Bill" by Forrester Research. 2011.
6. https://www.rackspace.com/en-us/managed-hosting/service-levels
7. https://azure.microsoft.com/en-us/support/legal/sla/
8. https://Cloud.google.com/pubsub/sla

Chapter 11
Additional Security Considerations for Cloud

11.1 Introduction

Edge computing represents a combination of distributed computing connected to the centralized servers. Historically, centralized versus distributed models have alternated as computing and communication capabilities have grown, while the limiting factor has alternated between computational capability and communication capacity. The present environment of Cloud and edge computing is a complex mixture of computing capability, communication capacity, and security considerations. In this chapter, we will focus on the security aspects of edge computing. Any such investigation must include multiple subtopics, e.g., protecting information content from observation and alteration, protection of operational capability from unauthorized access, protection of normal operation in the presence of malicious overloaded requests, etc. Solution components need to consider prevention from and response to any security threats [1]. Examples of prevention include encryption to protect content from observation and alteration, access checking protocols to prevent unauthorized accesses, tracking mechanisms to identify attempted attacks, and blocking messages except from trusted devices.

Today's information technology environment contains a wide variety of computing resources and a multiplicity of communication channels between the various computing resources. Economics drove creation of large data centers, and Cloud Computing was born to utilize this emerging enormous computing power. As capability of inexpensive computing continued ahead of the communications capabilities, computational power moved back to the end nodes of a system. The age of IoT (Internet of Things) arrived a decade ago as demonstrated by the fact that more things were connected to the Internet than people in the world [2]. The "things" connected to Internet include sensors, controllers, and intelligent devices [3]. These devices have limited power to create security problems, but they have even more limited ability to provide security solutions. To date the biggest security breaches in the IoT world have been instructions sent to the IoT devices, which then launched

© Springer Nature Switzerland AG 2020
N. K. Sehgal et al., *Cloud Computing with Security*,
https://doi.org/10.1007/978-3-030-24612-9_11

massive denial-of-service attacks on central servers. The top three examples are
Mirai, Hajime, and Persirai codes [4]. Mirai botnet code infected poorly protected
Internet devices, such as home surveillance cameras, which were still using their
factory default username and password. It used them to launch a massive DDOS
attack. Hajime was an IoT worm that built a huge peer-to-peer (P2P) botnet, but its
real purpose remains unknown. Persirai is yet another botnet that infected 120,000
IoT cameras to carry out DDoS attacks.

To visualize a wide variety of elements and security requirements in the IoT
domain, consider Fig. 11.1. The standard Internet communication security approach
(including virtual private networks, i.e., VPN) is to establish a link between Alice
and Bob using access control to identify the authorized individuals and then to use
encryption for information exchange between the "islands" of security containing
Alice and Bob. Alternatively, Dave may want to do a transaction with his bank.
Dave's transaction requires a higher level of security than Dave's normal activities.
This need should be scrutinized, authenticated, and validated. In parallel, Carol may
want to turn on her light bulbs at home since she will be arriving after dark. While
this does not require a high level of security, Carol certainly does not want some
random person turning her lights on and off. Since the potential damage due to this
action is less than from an unauthorized bank transaction, it also needs some authentication, but the security bar is lower for turning on lights in a home vs. accessing a
bank account. Security is all about raising the barrier, but the extent of that depends
on the potential damage by an authorized action. A higher barrier costs more in
terms of implementation costs and lowers user convenience. Other examples of low

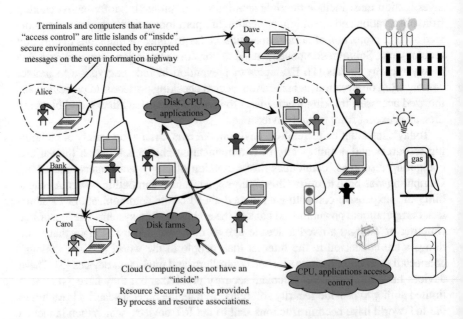

Fig. 11.1 The variety of elements connected in the IoT world demonstrates security challenges,
especially with a wide range of security requirements

levels of security are the household appliances, such as a toaster or a refrigerator. The high levels of security examples include opening a home garage, accessing banks, or operating factories.

11.2   Web Threat Models

Client-side browsers may contain vulnerabilities that may enable remote code execution by Websites. A Google study in 2007, "The Ghost in Browser" [5], found Trojan software on 300,000 Web pages and adware on 18,000 Web pages. The URLs (uniform resource locators) or Web addresses for these sites were examined, to find issues of following types:

1. *HTML*: Attackers can identify Web applications with vulnerabilities and insert small pieces of HTML code in Web pages. This HTML code is then used as a vehicle to launch large collections of exploits against users who visit the infected pages [6].
2. *Rendering content*: Each browser has an execution model, with windows or frames to load content and render it. Rendering refers to processing HTML and scripts to display Web pages, which may involve images, sub-frames, etc. By having a transparent frame overlaid on a legitimate and expected frame, attacker is able to get the click data as well as information such as login passwords, credit card numbers, etc.
3. *Remote scripting*: The goal here is to exchange data between a client-side app running in a browser and a server-side app without reloading pages. Examples include a Java applet, ActiveX controls, or Flash. A page can maintain bi-directional communication with browser, till a user closes the page or quits the Web app. By injecting scripting commands on behalf of an active user, the attacker can cause the server side to behave in an unexpected manner, for example, deleting all files in the user's account.
4. *Cookies*: These are used to store state on a user's machine. Think of these as simple and small text files that are used by a Web browser to maintain persistent context across Web sessions. Cookies can contain information such as user authentication, personalization, usage tracking, etc. A browser may store typically 20 cookies/site, with 3 Kb of information per cookie. Cookies provide confidentiality against network attacker, e.g., a browser will only provide cookies over HTTPs, but not integrity protection, e.g., a network attacker can rewrite secure cookies and alter their content.
5. *Frames and frame busting*: A Web page may contain frames from different sources, which can be rigid or a floating in-line iFrame, as shown in the Fig. 11.2. Their purpose is to designate screen areas to display content from another source. The browser is supposed to provide isolation based on frames. A browser's security mechanism is based on such isolation, as each frame of a page has its own origin or URL. However, if frame policies are not set properly, then one

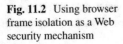

Fig. 11.2 Using browser
frame isolation as a Web
security mechanism

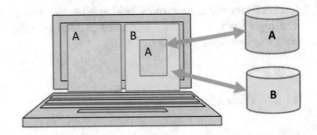

frame can interfere with the data or cookies of another frame within its own
hierarchy. An example of this is clickjacking [6], which uses features of HTML
and Javascript to trick the victim to perform undesirable actions, such as clicking a button that appears to perform a different operation. This is a "client-side"
security issue that affects a variety of browsers and platforms, using page over-
lays. Once the victim is surfing on the fictitious Web page, he thinks that he is
interacting with the visible user interface, but effectively he is performing
actions on a hidden page. The attacker can deceive users into performing actions
they never intended to perform through an "ad hoc" positioning of the elements
in the Web page, such as typing their passwords or credit card numbers in a
realistic appearing page overlaid on a hacker's page. Frame busting can prevent
this. The aim of this technique is to prevent a site from functioning when it is
loaded inside a frame.

11.3 Open Web Application Security Project

In view of the threats presented in the previous sections, a not-for-profit foundation
called Open Web Application Security Project (OWASP) [7] was established in
April, 2004, to conceive, develop, and maintain applications that can be trusted. All
of the OWASP tools, documents, forums, and chapters are free and open to anyone
interested in improving application security. An example of end-to-end threat model
and mitigations espoused by OWASP is shown in Fig. 11.3.

Web application attackers can potentially use many different paths through
applications to harm businesses or organizations. Each of these paths represents a
risk that may, or may not, be serious enough to warrant attention. Sometimes, these
paths are trivial to find and exploit, and sometimes they are extremely difficult.
Similarly, the harm that is caused may range from nothing all the way to shutting
down a business. To determine the risk to an organization, one should evaluate the
likelihood associated with each threat agent, attack vector, and security weakness
and combine it with an estimate of the technical and business impact to the organization. Together, these factors determine the overall risk. OWASP has identified [8]
and listed top 10 risks in 2013, as enumerated below:

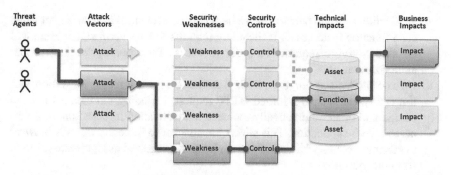

Fig. 11.3 An illustration of application security risks

1. *Injection*: Injection flaws, such as SQL, OS, and LDAP injection, occur when untrusted data is sent to an interpreter as part of a command or query. The attacker's hostile data can trick the interpreter into executing unintended commands or accessing data without proper authorization.
2. *Broken Authentication and Session Management*: Application functions related to authentication and session management are not implemented with adequate protections. This allows attackers to compromise passwords, keys, and session tokens or to exploit other implementation flaws, to assume other users' identities. Another way this can happen is through a playback attack, e.g., by storing and replaying a user's login session.
3. *Cross-Site Scripting (XSS)*: XSS flaws occur whenever an application takes untrusted data and sends it to a Web browser without proper validation. XSS allows attackers to execute scripts in the victim's browser, which can hijack user sessions, deface Websites, or redirect the user to malicious sites.
4. *Insecure Direct Object References*: A direct object reference occurs when a developer exposes a reference to an internal implementation object, such as a file, directory, or database key. Without an access control check or other protection, attackers can manipulate these references to access unauthorized data.
5. *Security Misconfiguration*: It is important for an organization to have a good security policy, which needs to be adhered to in its security implementation. It needs to be defined and deployed for the application, frameworks, application server, Web server, database server, and platform. Secure settings should be defined, implemented, and maintained, as defaults are often insecure. Additionally, software versions should be kept up to date with latest patches.
6. *Sensitive Data Exposure*: Many Web applications do not properly protect sensitive data, such as credit card information, tax IDs, and authentication credentials. Attackers may steal or modify any weakly protected data to conduct credit card fraud, identity theft, or other crimes. Sensitive data deserves extra protection such as encryption at rest or in transit, as well as special precautions when exchanged with the browser.
7. *Missing Function Level Access Control*: Most Web applications verify function level access rights before making that functionality visible in the UI. However,

applications need to perform the same access control checks on the server when each function is accessed. If incoming requests are not verified, attackers will be able to forge requests in order to access functionality without proper authorization.

8. *Cross-Site Request Forgery (CSRF)*: A CSRF attack forces a logged-on victim's browser to send a forged HTTP request, including the victim's session cookie and other automatically included authentication information, to a vulnerable Web application. This allows the attacker to force the victim's browser to generate requests for vulnerable application, disguised as legitimate requests from the victim.

9. *Using Components with Known Vulnerabilities*: Components, such as libraries, frameworks, and other software modules, almost always run with full access privileges. If a vulnerable component is exploited, such an attack can facilitate serious data loss or server takeover. Applications using components with known vulnerabilities may undermine application defenses and enable a range of possible attacks and impacts. A recent example of such an attack is WannaCry ransomware [9], which exploited vulnerability in Windows Server Message Block (SMB) v1 protocol. WannaCry used a combination of the RSA and AES algorithms to encrypt files and then demanded money in Bitcoins to decrypt an infected user's files. Microsoft addressed this vulnerability with Security Bulletin [10].

10. *Un-validated Redirects and Forwards*: Web applications frequently redirect and forward users to other pages and Websites and use untrusted data to determine the destination pages. Without proper validation, attackers can redirect victims to phishing or malware sites or use forwards to access unauthorized pages.

11.4   Emergence of Edge Computing

Another consideration in edge computing is multiple connection paths for each device. Each element on the edge can connect using a choice of paths or even multiple paths between the same endpoints. Specifically, any computing element on the edge can connect via the Internet, telephone lines, cell phone connections, wireless local area service networks (WiFi), or local wireless point-to-point connects such as Bluetooth, NFC (near-field communication), etc. See Fig. 11.4 for multiple paths from Alice to Bob, to a local server hub, to the Internet, or to the house alarm system. Edge computing continues to mature and encompass more of our world. Standards are being created such as Waggle [11], which is an open sensor platform for edge computing that has been introduced to reduce some of the foreseen compatibility problems. Edge computing security issues encompass end-to-end devices and the networks in between.

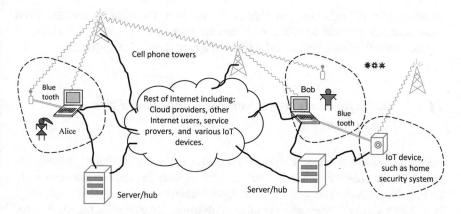

Fig. 11.4 Communication connectivity from the edge

11.4.1 Status of Edge Computing Security and Recent Breaches

The security issues for edge computing often overlap with existing security problems. Access control using identity authentication is especially difficult in the IoT environment. Edge computing greatly increases the number of devices that need authentication. The pairwise authentication problem increases faster than exponentially, as it follows the N curve where N is the number of pairs. Added to the authentication problem is the problem of corrective action when unauthorized access is detected.

One of the largest attacks that the Internet ever experienced was recently launched using unsecure routers, digital video recorders (DVRs), and online surveillance cameras [12]. A collection of devices called botnet (an army of infected devices) was used to launch a distributed denial-of-service (DDoS) attack on KrebsOnSecurity. com, the Website of a security journalist who had previously exposed cybercriminals. This attack generated >660 Gbps of traffic, making it the largest attack on record in terms of data volume. In another case, a pair of researchers showed that they could remotely hijack a Jeep's digital systems over the Internet. It led to a recall of 1.4 million vehicles [13], which required a costly fix after it was shown that a moving Jeep's steering wheel could be turned, which caused unintended acceleration and brakes disabled remotely. Many homes have Internet-enabled devices including thermostats, garage door openers, smart TVs, etc. Such devices may contain vulnerabilities, enabling hackers to compromise a home, including changing the heating or cooling settings, opening garage doors, and using TVs to connect with PCs on the home networks for stealing personal data [14].

Threat tracking and tracing are difficult for the IoT environment, but there are only a few channels through which an attack may travel. With edge computing, definition and enforcement of the virtual protection boundary are difficult. Therefore, monitoring and responding to threats is the key. Fortunately, the increased computational ability of the elements at the edge also offers the potential for increasing the sophistication of the security monitoring and corrective responses.

11.4.2   Security Modeling Specifically Targeting Edge Computing

Perimeter defense has long been insufficient for IoT security. Fixed protocols for boundaries of security with individual devices' security implementations are likely to fail, because devices can have multiple channels of communications across boundaries. Each of these can be configured dynamically bypassing the fixed protocols. In addition, a fixed universal security policy is inadequate. However, components throughout the edge computing environment must be adaptive in the sense that each device builds an individual trust model with the other devices to which it connects. This model must include monitoring to determine the level of trust applied to each individual connection between devices. The source device's trust (which sets the specific security policies and actions) increases based upon a history of successful connections and transactions with the responding devices. The source device's trust decreases based upon measured or detected failures for connections and transactions with the responding device. The decreased trust invokes increased security measures as will be described in a later section. Therefore, each device must learn who to trust and what level of trust to extend to other individuals and devices.

Each device may be part of the community of edge devices and Cloud services. This community is similar to online communities of individuals. Hamilton et al. describe the trust in an online community as a function of loyalty to the community [15]. Each edge device evaluates its trusted partners based upon preference, commitment, consistency vs surprise, and decisions or actions to be taken. The preference and commitment is established by the quantity and time spread of past communications. The measure of trust from one edge device to other entities is either increased by exchanges consistent with past exchanges or decreased by any surprisingly different exchanges. Thus, consistency increases trust, and inconsistency decreases trust. The level of trust (based upon the past) and the immediate request drives a decision or action on the part of either the edge device or the Cloud service component. A key to the success is the ability of each entity to learn and improve the measurement of trust.

The status of trust of A toward B can be trust, doubt (or mistrust), neutral (as in no previous contact), or cautious (indicating a mix of doubt and trust). The state of trust will change status based upon the experience A has with B. Figure 11.5 shows the state of trust for A with respect to B. The black arrow shows the initial condition of neutral, which means A neither trusts nor doubts B. Many positive experiences are possible, such as correct authentication, successful transfer of data, legitimate

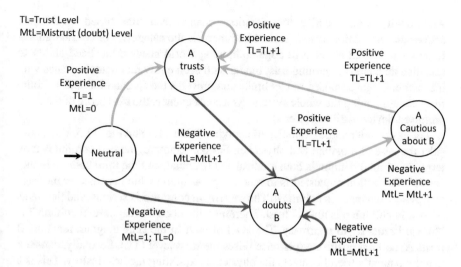

Fig. 11.5 Element A's state of trust level of element B

response, familiar identification, repeat of past success, etc. Many negative experiences are also possible, such as incorrect authentication, failure to transfer data, unexpected response, excessive attempts at contact, transfers of corrupted data, inappropriate requests for status of requestor, etc. Note that the trust of A for B and the trust of B for A are independent of each other. Starting with no experience, A neither trusts nor doubts B, which initializes the trust level to zero (TL = 0) and the mistrust level to zero (MtL = 0). After a positive experience with B, the TL is set to 1, and A now is in a state of trusting B. As future positive experience occurs, the state of A stays in the state of trusting B with the trust level increasing with each positive experience (i.e., TL = TL + 1). However, with a negative experience, the state of A transitions to the state of doubting B and the mistrust level is increased (i.e., MtL = MtL + 1). Upper values of these counters can be bound or suitably classified in different numerical ranges, such as a high, medium, or low category. Also, note that from the state of neutral, a negative experience transitions to the state of A doubting B with the MtL set to 1. As the negative experiences continue, the status of A stays in the state of doubting B and the mistrust level increases with each experience (MtL = MtL + 1). With a positive experience, the state of A transitions from the state of doubts to the state of cautious and the trust level in increased by 1. While in the cautions state, a positive experience transitions the state of A to trusting B, but a negative experience transitions the state of A to doubting (or mistrusting) B. The decision on what action to take in response to a request is a function of the particular relationship that nature of the request and the trust level vs the mistrust level.

The previous discussion proposes that information security is far more complex in the current computing environment. Not only does each participant (device, element, or person) require different security considerations, but each relationship between each pair of participants requires different security considerations.

Additionally these security considerations change over time based on the past actions and new information. Table 1 summarizes the categories of considerations. It shows that each element in edge computing world needs a localized ability to establish an adaptive learning trust model with each entity that communicates with the element. Our proposed model limits and prevents the spread of a device failure from contaminating the whole system. As a consequence, the trust score of the compromised device shall be lowered.

Let us consider some examples of applying Table 11.1 and Fig. 11.5. First, consider the case of a patient and physician. For our example, the first column is long term, the second column is both medical and financial, and the third column is successful. The action requested is to renew a prescription which is "data or message request or exchange" in column 4. The severity in column 5 is serious, and the status of trust in column 6 is mutual trust. Therefore, the doctor's response in column 7 is "forward request" to pharmacy. The level of trust in the state diagram remains, B trusts A, and the positive experience raises the trust level (TL). Secondly, consider that the patient's friend contacts the physician requesting medical history. This is a new, first contact, and column 2 is medical; history is neutral; action request is data request; and severity is serious, but the status is neutral. Now for medical requests, the response is multiple in both responding to the requester that this is protected information and alerting the patient that the request was made. The level of trust in the state diagram moves to mistrust because this was an unexpected and not previously authorized request resulting in negative experience. This will be modified with the patient's response to the notification from the physician.

Finally, consider interactions between two devices, for example, a connected car and a Cloud Computing resource. Specifically, the car's computer contacts the automotive maintenance center to schedule a regular maintenance. From Table 1, column 1, we see this is a medium-term relationship with few contacts. From column 2 we see it is both scheduling and financial. From column 3 we have successful. Therefore, from state Diagram 3, we have a positive trust level for between both the car and the maintenance shop. The action request column is for data message exchange of data, time, and financial commitment. From column 5, the severity is casual as it is not urgent or serious. As mentioned before, in column 6 we have mutual trust based upon the history and the state diagram. The action is to respond. Now consider that the car maintenance shop attempts to contact the car and drive it. The first column is still a medium-term relationship with few contacts. However, in column 2 the purpose of the relationship does not match the action request from column 5. Because columns 3 and 6 point to some level of trust, the severity of the action from column 5 leads to a response of "alert" and "forward request" but not perform action.

The previous discussions concentrate on the trust level between two entities. However, in reality there are multiple entities involved in some trust relationships. As an example, some security protocols include a third party security certification. In addition, there are some security situations where a third party monitors or records transactions. These considerations are yet to be explored.

Table 11.1 Categories of security considerations for connection from A to B

Length and frequency of relationship	Purpose of relationship	History	Action request	Severity and urgency	Trust status	Response
Length *One*	*Multiple*	*One*	*Multiple*	*One*	**A to B** *One*	*Multiple*
New	Casual	Neutral	Data or message delivery	Emergency	Trust	Ignore
Short term	Medical	Successful	Data or message request or exchange	Critical	Cautious	Store
Medium term	Legal	Failure	Monetary transfer	Casual	Doubt	Respond
Long term	Financial	Mixed successes	Physical action	Serious	Neutral	Forward request
	Schedule or calendar		Verification	Unclear		Alert
Frequency *One*	Employment	Past success, recent failure	Open connection		**B to A** *One*	Perform action
First contact	Political	Past failure, recent success	Attestation		Trust	
Few contacts	Religious	Relationship change			Cautious	
Many contacts	Ownership/property				Doubt	
	None/just information				Neutral	
	National security					

The application of deep learning for speech recognition is advancing [16], and it could be applied for speaker recognition for authentication and other security evaluations. The concept is to push some of the security decisions to the edge computing devices. The additional compute power at the edge is already being applied for decision-making using machine learning [17, 18]. The future of security with edge computing and the Cloud is a mix of central protocols in the Cloud and decision-making at the edge based upon machine learning, monitoring, and analyzing communication activity [19]. A machine learning environment may allow the identification and defense against unexpected and unpredictable security challenges. However, ML is a double-edged sword as hackers with access to training data can corrupt the learning process or alter their attack code to specifically bypass a predetermined security model [20]. There is a no silver bullet to ensure the security for all devices participating in edge computing, so a community-based adaptive trust model may present an optimal solution.

11.5 Security Solutions for IoT Devices

Security solution constraints involve technical factors such as limited internal processing, memory resources, and power consumption demands. Vendors often reduce the unit cost of devices by minimizing parts used and product design costs. It is more expensive to design interoperability features into a product and test for compliance with a standards specification. A non-interoperable device may lack in standards and the documented best practices. It may also limit the potential use of the IoT device. Absence of these standards can result in deviant behaviors due to proprietary nature of designs and interfaces.

It is recognized that traditional trusted compute boundary (TCB) expands with edge computing to include domains that are physically outside the control of remote device or central data center owners. The best they can do is to monitor/track a threat, identify an attacker, launch a recovery, and prevent false positives. These steps are outlined below:

1. *Monitor/track a threat*: This is possible by establishing a normal usage pattern for the IoT device. As an example, a home security camera uploads data whenever motion is detected. If the regular pattern for a home is no more than a couple of dozen data uploads during a day, then hundreds of data loads to the central server within a few minutes may indicate that the device has been compromised. It could be an attempt to cause a DOS attack.
2. *Identifying attackers*: Once a threat is detected, the attacker needs to be identified. Identity of the attacker could be its IP address that is repeatedly pinging a central server to launch denial-of-service attack.
3. *Attack recovery*: Recovery can start with blocking the offending IP address. However, an attacker may corrupt the critical data before the attacker's presence

is detected. Instituting frequent checkpoints, to facilitate rollbacks to known good state, can be a mitigation option.

4. *Accidental failures confused with security attacks*: Any detection method suffers from the risks of false positives, not unlike a spam filter categorizing valid emails as junk. An example of this is a stock market trading computer that detects unusual activity, which may be genuine yet may flag it as an attack. Similar situations can happen with security alarms due to spurious sensor activity. A learning system that becomes smarter over time could help reduce false positives.

5. *Data integrity protection*: We previously described a system level attack involving multiple devices writing to a central server database. This can be protected by assigning a virtual partition or container to the data coming from each distinct source and checking the address range of each access to prevent data integrity of other users on the same server.

Internet Engineering Task Force (IETF) has identified the problem of interoperability, as many suppliers build "walled gardens" that limit users to interoperate with a curated subset of component providers, applications, and services.

Interoperability solutions between IoT devices and backend systems can exist at different layers of the architecture and at different levels within protocol stack between the devices. The key is standardization and adoption of protocols, which should specify when and where it is optimal to use standards. More work is needed to ensure interoperability within the cost constraints for edge computing to make it pervasive.

There are other regulatory and policy issues at play, such as jurisdictional boundary adjudication for data collected in one place and stored in a different location. In such cases which laws apply to data retention and leaks? This is especially important for personal data, such as financial information or health records. One of the ways privacy is protected for consumer data in the Cloud is by controlling its access to authorized users only. As we shall see in the next section, several automated methods are available to identify the rightful owner of personal data located on the shared servers. Such techniques are being used by SaaS providers in Private or Public Cloud [21].

11.6 Metrics for Access Control Using Speaker Identification

When designing and evaluating a speaker identification system for access control, there is a trade-off involved. It is between successfully allowing authorized access, successfully rejecting unauthorized access, falsely rejecting authorized access, and falsely allowing unauthorized access. Consider that M is the number of matches, where a match is defined as the biometric measurement matches the person or expected data. For example, a recorded voice is compared to a stored voice. If the comparison method indicates that the measured and stored values are from the same source, it is considered a match, where NM is the number of non-matches. For example, a recorded voice is compared to a stored voice, and if the comparison method indicates that the measured and stored values are from the same source, it is considered a non-match. Matching or non-matching is used to determine authorization for access. If the match occurs for a person, it is called a true match. Sometimes a match occurs for a person that is not authorized to have access. This is a false match. TM is the number of true matches, and FM is the number of false matches. Usually when a non-match occurs, the person rejected is not authorized to access, but sometimes an authorized person gets a non-match. TNM is the number of true non-matches, and FNM is the number of false non-matches. Therefore, the correct rate (CR) is the sum of the true matches (TM), and the true non-matches (TNM) are divided by the total number of trials, as shown in the following equation:

$$CR = (TM + TNM) / \# \text{trials} \tag{1}$$

The total number of samples is also equal to the total number of access requests, which is the sum of the authorized access requests (A) plus the unauthorized access requests (UA). The number of matches (M) is the sum of the correct or true matches (TM) and the number of incorrect or false matches (FM), as shown in the following equation:

$$M = TM + FM \tag{2}$$

The number of non-matches (NM) is the sum of the true non-matches (TNM) and the false non-matches (FNM), as shown in the following equation:

$$NM = TNM + FNM. \tag{3}$$

So, a simple measure of the security of the system is the rate of correct or true results out of the total. However, this measure obscures the relative importance of mistakenly granting access to that of mistakenly rejecting access. The likelihood of mistakenly granting access is the false match rate (FMR), which is the portion of matched data that is for the wrong person. The likelihood of mistakenly denying access is the false non-match rate (FNMR), which is the portion of non-matches that represent denial of access to an authorized person. These are metrics for the accuracy and effectiveness of an identity authentication system. For a very secure facility, one wants a very low FMR even at the cost of a high FNMR. That is, keep out those bad guys even at a high cost. For relatively open environments, one can sacrifice a high FMR, which is allowing unauthorized access, in order to achieve a low FNMR, which is to avoid turning away anyone who deserves access. Table 11.2 shows the various possible results of identity authentication requests.

A biometric provides a measurement that can be used for identity authentication. One can choose from several different biometrics. For example, biometrics that could be used for identity authentication are face dimension, hand dimension, height, weight, finger prints, iris scan, DNA, and voice print. The characteristics of biometrics that can be used to evaluate and compare different biometrics are cost, time, universality, distinctiveness, permanence, collectability, acceptability, cir-

Table 11.2 Authentication match and non-match cases

	Authorize person should get access	Imposter should not get access
Measurement matches Access granted	True match (TM)	False match (FM)
Measurement does not match Access not granted	False non-match (FNM)	True non-match (TNM)

cumvention, accuracy, repeatability, storage requirements, and availability of technology [22]. Cost parameter includes money, time, equipment, and expertise for implementation of the system and collection of the measurements. The time characteristic is specific to the measurement collection and analysis time. That is the time from when Alice requests identity authentication until the access is granted or denied. Universality is a measure of how many of the sample population are able to meet the requirements of the system. For example, everyone has DNA, but not everyone has hair. So a DNA test is universally applicable, but hair color is not applicable to people without hair. Distinctiveness is a measure of how unique or different the measurements for an individual will be from other individuals. Fingerprints are very distinct, whereas weight is not. Permanence is a measure of the change in the biometric with the passage of time. For some people weight can change dramatically with time. However fingerprints only have small changes with time, such as cuts and scrapes. Collectability is the characteristic indicating how easy it is to get samples for the biometric. Blood is tough to get. Height is very easy to get. Acceptability is a subjective measure and represents the willingness of a person to submit to the biometric measurement. An example of very low acceptability is requiring a blood test just to enter a convenience store. On the other hand, we readily submit to height measurements for carnival rides at the state fair. Circumvention is the ease or cost to trick or falsify the measurement. Measuring weight is easy to falsify by carrying lead in one's pockets. A falsified eye scan is a bit more difficult. The accuracy of a biometric is the probability that an individual will be properly authenticated. Specifically, it includes the probability of properly authenticating the identity or access for authorized individuals and properly rejecting the identity or access for unauthorized individuals. Repeatability is the variance of the biometric measurement over repeated trials. The data storage requirement is evaluated both for the individual measurement and the total database of each individual measurement. Although in practice, the size of the total database is rarely a significant factor in evaluating a biometric. The local storage at the acquisition point is a very significant issue for analysis, storage, comparison, and transmission. The availability of the technology is a make or break decision, as well as, a quantitative measure. A biometric is not an option for immediate deployment if it requires a technology that does not currently exists. However, even if the technology exists, the ready availability of the technology is a factor. For example, many computers and recording devices have the ability to capture a voice or a picture, but not many people have ready access to DNA or fingerprint collection devices. The selection of a biometric based upon these characteristics clearly involves many trade-offs. These components have been applied to choosing a biometric for an Internet-based environment [21]. The most obvious evaluation can be made based upon the characteristic of availability of technology. As history has shown, new technologies pop up, and some limited technologies become very widespread. Based upon these metrics, speaker recognition is a very good choice for biometric authentication used for access control.

Authentication using voice is called speaker authentication, which is a subset of speaker recognition. A match is defined as two values within a given threshold of each other. The threshold for a match is varied based upon the desired match rate, non-match rate, and level of security. One interesting observation is that as the number of matches increases, the true match rate decreases. This represents an increasing likelihood that an unauthorized person is granted access.

Distributed systems, especially the Internet, have a variety of equipment. The use of standard or readily available equipment is essential for any ubiquitous solution. For the three elements of identity authentication, each element has a different equipment requirement. In the case of knowledge, the question of what you know can easily be answered with a keypad or keyboard. In fact, this simple equipment requirement has driven most of the security implementations. For the element of possession, the question of what you have is most commonly handled by a credit card reader. However, most Internet-connected computers do not have credit card readers. Of course, typing in your credit card number converts the element from possession of the credit card to knowledge of the credit card number. To keep with the possession element concept, some systems require insertion of a specific CD. However, as USB ports have become very common, the trend is to use a USB plugin as a "hardware key." Finally, for the element of biometrics, the answer to the question of what you are is most readily answered by your voice! That is because, among the biometrics, the most commonly available measuring device is a microphone.

Speaker recognition is a leading contender for the biometric element in a distributed environment, so the effect of equipment variations is critical. The range of frequencies for the human voice will influence the equipment used for identity authentication. The human voice typically ranges from 20 Hz to 4000 Hz and can go as high as 11,000 Hz. The microphone or equipment used may have a dynamic range less than the human voice.

Speaker recognition, the biometric of voice, is a means to fulfill the biometric signature of identity authentication [23, 24]. Speaker recognition measures characteristics of a person's voice to authenticate their identity. In distributed environments, system developers have little or no control over the equipment utilized. Multi-site company networks or bank systems with account access via the Internet, phone, and the bank proper are examples of these distributed system. For speaker recognition systems in a distributed environment, such as the Internet, microphones are certain to vary. Speaker recognition is a dynamic or behavioral biometric as opposed to a physiological or static biometric. A static system measures purely what you are, such as a retina scan. A dynamic biometric measures your actions, such as facial expressions, signatures, behavioral patterns, or voice generation. Speaker recognition, a dynamic voice biometric, can be used as a means of authentication as well as identification. However, identification is a broader problem facing many challenges [25]. Authentication is a closed set problem. Identity authentication asks, "Is the person who the person claims to be?" whereas identification is an open set problem that asks, "Who is the person?" Systems are divided into one of two groups, text dependent or text independent. A text-dependent system is one in which the phrase(s) that one speaks to enroll (teach the system who you are) is the same phrase(s) used when requesting authentication for system access. Simple pattern-matching algorithms can be used, with some systems, to verify the proper person is saying the proper phrase. This is an example of two-factor authentication in one operation. Text-independent systems use fundamental voice data buried in voice signals to do speaker recognition. Because text-independent systems analyze basic voice information and not how a particular user says a particular phrase, it does not matter what phrase is spoken. There are several voice attributes that can be analyzed to verify identity. They can be divided into two basic groups, high-level and low-level information. High-level information could include accent, common word, or phrase usage or pronunciation. The low-level information breaks voice signals into small segments and analyzes the basic structure of one's voice, i.e., tone, frequency, etc. The transducer effect has some consequence on these attributes. Table 11.2 shows the four possible outcomes of a speaker recognition system. How transducers affect the false match rate (FMR is for acceptance) and false non-match rate (FNMR is for rejection) rates of the speaker recognition system limits the strength of using speaker recognition. The strength of the access control is a serious consideration for system security. For both the enrollment and testing stages, the individual speaks into a microphone, and the signal is converted to a digital signal. The first step in the process, collecting and digitizing voice samples, is one of the variants in the process. Identity authentication can be attempted at numerous terminals, i.e., cell phone, landline, laptop, home and work PCs, etc. These terminals, using different types and models of transducers, have varying effects on the conversion of a voice signal to an electric signal. Evaluation of a system with various tests and repeating the authentication process while varying the threshold produces a detection error trade-off (DET) curve. The simple threshold algorithm sets a threshold (LR) for matching and measures how many authentication attempts were improperly accepted or improperly rejected. The point where the false accept (FA = FMR) rate and false reject (FR = FNMR) rates meet is called the equal error rate (EER).

A more secure system may set a threshold where the false accept (FA = FMR) rate is very low, while the false reject (FR = FNMR) rate is higher. A system where user convenience is a primary concern may rather have a higher FA rate while keeping the FR rate low [22]. The results from one example system are shown in Fig. 11.6. For this system, when the LR threshold is 0.065, EER = FA = FR = 0.45. This threshold setting is very relaxed, as many authorized and unauthorized individuals are accepted. For a more secure system when the match LR threshold = 0.08, then the false accept rate drops to 10%, but the false reject rate jumps to 75%. The price of keeping more imposters out is to keep more authentic people out. For a

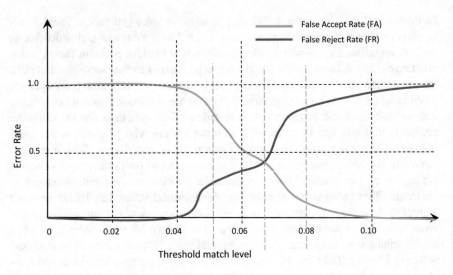

Fig. 11.6 Example detection error trade-off (DET) curve

lower security requirement, a LR threshold of 0.05 yields a false accept rate of 80% with a false reject rate of 30%.

For a high security environment, this would not be used as the only authentication element. However to demonstrate and evaluate the question of equipment variation effects, this simple algorithm is adequate. A more secure system utilizes multifactor authentication.

The authentication metrics are applied differently to each of three elements or items used for identity authentication. For the element possession, it is the evaluation of the validity of the item presented. For the element of knowledge, it is a comparison of the submitted information with the expected information. And for the biometric element, it is a comparison of the measured feature to the distribution of expected values. For example, consider an Internet terminal access point. The knowledge element is a password. A longer password is harder to guess or crack. "Cracking" a password can be done with software that repeatedly guesses at a password and keeps trying until access is granted. A long password takes a long time to guess or crack. This keeps the FMR low. However, even valid users can forget or mistype long passwords. This raises the FNM (false non-match rate or rejection) to unacceptable levels. For the possession element, a USB plugin module can be used. This creates a strong guard against false accepts (i.e., a low FMR) if it is unique. However, in an institution one might need multiple identical access plugins (such as multiple keys to the lab door) to avoid excessive FNMR. For the biometric element, the speaker recognition approach is used. Consider that these elements are statistically independent. For example, my password does not correlate with my voice. The concept worth noting is that the FNMR is the sum of the FNMR for each of the elements. This is because any one of the elements flagging a non-match is a non-match for the authentication request. However, the FMR is the product of the separate elemental FMRs. This is because the false match requires getting all of the elements correct. Therefore a slight increase in one element TMR is a significant increase for the overall system. The overall system parameters should be set after some policy decisions are made. For example, first decide which element is the primary protection element. Set the security for FMR for this primary element. Then set the lower priority elements with laxer thresholds. With a priority on the password, a very long password can be required. When passwords exceed a person's ability to remember it, the person takes shortcuts. Consider that no amount of additional characters will increase the access security when a person writes their long password on a post-it note next to their terminal. However, even a very lax FMR of 10% would significantly increase the access security in this case. This shows that even with the weaknesses of the biometric of speaker recognition, by adding it as a third element with a lax threshold, the overall system security is greatly enhanced without significant cost. A secure system includes a step of authenticating the identity of the individual requesting access. Adding a biometric component improves the security of the identity authentication of the system. Speaker identification is a biometric used for access control. The level of security is set as a threshold for matching that involves a trade-off between reducing successful imposter intrusions (a low false match rate) and reducing rejection of authorized individuals (false non-match rate). Just like fingerprints, one's voiceprint is a natural marker that provides security [21].

11.7   Real-Time Control of Cyber-Physical Systems

A cyber-physical system (CPS) is a physical system with an embedded computer control. It contains various physical devices being controlled by system software and measurement devices. Examples of cyber-physical systems include IoT-based control systems, smart grid, autonomous automobile systems, medical monitoring, process control systems, robotics systems, and automatic pilot avionics [26]. Cyber-physical systems utilize connections to the Cloud for data storage and Big Data analytics as a basis and updates for the local cyber-physical control system. The local cyber-physical system must provide the real-time control, as short response times are critical for most applications. However, with the proliferation of computer control systems and associated measurement devices, collection of large amounts of data requires Cloud Computing capability for the storage and Big Data analysis.

Consider a factory with many stages for production, some of which are completely manual, others with automated robots, and some stages that have a mix of manual operations and computer-controlled operations to perform. These stages can be used for fabrication, assembly, and monitoring. There may be additional monitoring stations throughout the factory.

At each stage, the control of fabrication and collection of monitoring data must be secured from both intrusion and unauthorized observation incidents. A Cloud provided policy directive for a particular machine might specify the order of mechanical operations, such as put the nut on the bolt and then tighten it to a particular tension level, measure the applied torque and then check whether to tighten it more, etc. An intrusion attack could modify the tension measurements, so the nut never gets fully tightened, and a defective product is sent to the next stage. The fabrication security includes the protection of instructions from being tampered when sent from the Cloud to a local server and then to the cyber-physical control at the machine. It also provides for local authentication of the source for these instructions, verification of the instructions, and what steps to take in the event of a detected security problem. The observation security includes checking the validity of the collected data, protection of content of the data in the transmission back to the Cloud, authentication of identity of the sender of the data, and appropriate actions when any errors are detected.

Besides the fabrication or machine-specific monitoring, there is monitoring throughout the factory which includes temperature, fire detection, and intrusion detection. The temperature and smoke monitoring systems are the key elements to fire detection. As these become more integrated with the overall Cloud Computing and local cyber-physical control system, they also become targets for remote integrity attacks that support an attacker.

In addition to information protection, security considerations are a serious safety concern at the mixed operations stations. Typically here the monitoring must be conducted in the presence of human operators. Many machines are of sizes and functions such that faulty operations can pose a threat of bodily harm to the nearby humans. An example is of a welding machine in a car assembly line. Here the security and safety concerns overlap. For example, a machine directive might be that whenever a dangerous part of the machine gets too close to a human, it stops. The security issue is twofold. One is reduced productivity; the malicious attacker can change the distance measurements such that the machine is shutting down so frequently that very little gets done at the station. Secondly, it results in diminished levels of safety, such as a denial-of-service attack of sending messages to the machine so fast that it doesn't respond to shut down before harming the humans.

Basically, cyber-physical systems must meet several separate objectives, as listed below:

1. Security of data and instructions from the Cloud.
2. Protection of data gathered and information sent back to the Cloud.
3. Protect the cyber-physical system from information flow interfering with its real-time control responsibilities.
4. Protecting the integrity of monitoring data.
5. Finally, although arguably the most important, is the security of the system to ensure human safety.

Designing and deploying a cyber-physical production system can be done using a 5 C-architecture (connection, conversion, cyber, cognition, and configuration), as shown in the Fig. 11.7 [27].

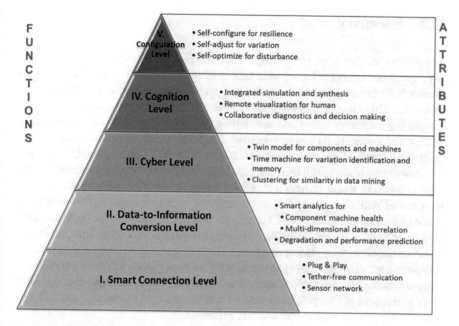

Fig. 11.7 A pyramid representation of 5 C-architecture for the design and deployment of a cyber-physical system [27]

In the "connection" level, devices are designed to self-connect and self-sensing for their correct behaviors. In the "conversion" level, data from self-connected devices and sensors is used to measure the features of critical issues. This refers to self-aware capabilities, i.e., machines can use self-aware information to predict potential issues. In the "cyber" level, each machine can create its own "twin" image by using the instrumented features. These help to characterize the machine health pattern based on a "time machine" methodology. The twin model enables a physical operation to be coupled with a virtual operation by means of an intelligent reasoning agent. The "digital" twin of the real machine operates on a Cloud platform and simulates the health conditions in real time, hence the concept of "time machine." This uses the integrated knowledge from both data-driven analytical algorithms and available physical knowledge.

The established "twin" in the cyber space can perform self-compare for peer-to-peer performance reviews for further analysis. In the "cognition" level, outcomes of self-assessment and evaluation are presented to the users based on an "infographics." Its intent is to show the content and context of any potential issues. In the "configuration" level, a machine or production system is reconfigured based on the priority and risk criteria to achieve resilient performance.

11.8   Summary

The present state of edge computing is an environment of vastly different computing capabilities connected via a wide variety of communication paths. This situation creates both great operational capability opportunities and unimaginable security problems. This chapter emphasized that the traditional approaches to security of identifying a security threat and developing the technology and policies to defend against that threat are no longer adequate. The wide variety of security levels, computational capabilities, and communication channels require individualized and responsive approaches with learning capability. We propose that each element in the edge computing world utilizes a localized ability to establish an adaptive learning trust model with each entity that communicates with that element. We review the scope of various IoT (Internet of Things) devices in the field that are bidirectionally connected to data centers (in-house or Cloud) via various networks. Then we looked at the nature of security issues, and mechanisms to quantify risk associated with the complete hardware and software stack, with an example of a typical surveillance camera system. We proposed a method to calculate system security and suggested ways to improve it. Our proposed method can be extended to evaluate any IoT system and improve its end-to-end security profile.

11.9   Points to Ponder

1. How can Web browser cookies be both helpful and harmful at the same time?
2. What is security misconfiguration and how it can be avoided?
3. How does edge computing expand the attack surface?
4. Why is mutual trust between IoT devices important?
5. Why trust is a dynamic entity?
6. Why devices in series tend to have weaker security?
7. How do devices in parallel improve the overall system security?
8. Do IoT devices pose a higher risk than servers in the Cloud, if so why?
9. What are the risks associated with voice recognition-based Cloud access systems?

References

1. Sehgal, N. K., Sohoni, S., Xiong, Y., Fritz, D., Mulia, W., & Acken, J. M. (2011). A cross section of the issues and research activities related to both information security and cloud computing. *IETE Technical Review, 28*(4), 279–291.
2. https://www.postscapes.com/internet-of-things-history/
3. Ashton, J. (2009). That 'Internet of Things' thing. *RFID Journal, 22*.
4. http://blog.trendmicro.com/trendlabs-security-intelligence/persirai-new-internet-things-iot-botnet-targets-ip-cameras/

5. https://www.usenix.org/legacy/event/hotbots07/tech/full_papers/provos/provos.pdf
6. https://www.owasp.org/index.php/Testing_for_Clickjacking_(OTG-CLIENT-009)
7. https://www.owasp.org/index.php/About_The_Open_Web_Application_Security_Project#The_OWASP_Foundation
8. https://www.owasp.org/index.php/Top_10_2013-Top_10
9. https://www.secureworks.com/research/wcry-ransomware-analysis
10. https://technet.microsoft.com/en-us/library/security/ms17-010.aspx
11. Beckman, P., Sankaran, R., Catlett, C., Ferrier, N., Jacob, R., & Papka, M. (2016). Waggle: An open sensor platform for edge computing. *2016 IEEE SENSORS*, Orlando, FL, pp. 1–3. https://doi.org/10.1109/ICSENS.2016.7808975.
12. https://motherboard.vice.com/en_us/article/15-million-connected-cameras-ddos-botnet-brian-krebs
13. https://www.wired.com/2016/08/jeep-hackers-return-high-speed-steering-acceleration-hacks/
14. http://abc7chicago.com/technology/home-hackers-digital-invaders-a-threat-to-your-house/515520/
15. Hamilton, W. L., et al. Loyalty in online communities. *Proceedings of the eleventh international AAAI conference on Web and Social Media (ICWSM 2017)*, pp. 540–543.
16. Deng, L. et al. (2013). Recent advances in deep learning for speech research at Microsoft. *2013 IEEE international conference on Acoustics, Speech and Signal Processing*, Vancouver, BC, pp. 8604–8608. https://doi.org/10.1109/ICASSP.2013.6639345.
17. Nelson, R. (2017, June). Smart factories leverage cloud, edge computing. *Evaluation Engineering, 56*(6).
18. https://www.kdnuggets.com/2017/01/machine-learning-cyber-security.html
19. Sommer, R., & Paxson, V. (2010, May). Outside the closed world: On using machine learning for network intrusion detection. *2010 IEEE Symposium on Security and Privacy.*
20. https://www.technologyreview.com/s/611860/ai-for-cybersecurity-is-a-hot-new-thing-and-a-dangerous-gamble/
21. https://nb.fidelity.com/public/nb/default/resourceslibrary/articles/myvoicefaq
22. Sohoni, S., Shaver, C. D., Acken, J.M., Mertz, D., Nelson, L. E., Remington, J., Sadr, B., Sundararajan, G. (2009) Evaluation criteria for biometric based identity authentication systems, *Proceedings of ISSSIS2009*, Coimbatore, 8–10 January.
23. Ortega-Garcia, J., Bigun, J., Reynolds, D., & Gonzalez-Rodriguez, J. (2004, March). Authentication gets personal with biometrics. *Signal Processing Magazine, IEEE, 21*, 50–62.
24. Acken, J. M., & Nelson, L. E. (2008). Statistical basics for testing and security of digital systems for identity authentication. *CCCT, International Institute of Informatics and systematic*, Orlando, FL, pp. 122–128, June 29th–July 2nd.
25. Bonastre, J. F., Bimbot, F., Boe, L. J., Campbell, J. P., Reynolds, D. A., & Magrin-Chagnolleau, I. (2003) Person authentication by voice: A need for caution. *Proceeding of Eurospeech, ISCA*, Geneva, pp. 33–36, 1–4 September.
26. https://en.wikipedia.org/wiki/Cyber-physical_system
27. Lee, J., Bagheri, B., & Kao, H.-A. (2015, January). A cyber-physical systems architecture for industry 4.0-based manufacturing systems. *Manufacturing Letters, 3*, 18–23. https://doi.org/10.1016/j.mfglet.2014.12.001.

Chapter 12
Analytics in the Cloud

12.1 Background and Problem Statement

The sheer volume of information available on Cloud, and the rate at which new data is being generated, is overwhelming the capacity of enterprises to manage it and people to use it in a meaningful manner. This data deluge has surpassed the capacity of existing data centers to store and process it in a timely manner. This gave rise to a new class of algorithms, such as MapReduce, which we shall study in a latter section. First, consider the following statistics from 2017 about what people do on the Internet in a typical minute [1]:

1. Google does 3.6 million searches/min.
2. Spammers send 100 million emails/min.
3. Snapchatters send 527,000 photos/min.
4. Weather channel forecasts rain or shine 18 million times/mins.
5. Netflix customers watch 69,000 hours of video/min.
6. YouTube users watch 4.1 million videos/min.

These and a few more mind-blowing statistics are shown in Fig. 12.1, representing activities of 3.7 billion people on the Internet every single minute in a day.

Americans alone use over 2.6 million gigabytes (2.6×10^{15}) of data every single minute of every single day. This is not just limited to online behavior, as Uber customers take 45,787 trips each minute and Amazon customers purchase $250 K+ of goods/min. We are still at the knee of a hockey stick curve, as only half of world's population is online and most of them on slow Internet or wireless connections. With the advent of 5G technology, the best is yet to come. Expect the data usage to grow by ten times over the next few years.

The problem this poses for Cloud Computing infrastructure is not just the storage or transportation of this data, but processing it in a meaningful way to draw inferences. An example is of a retail company that needs to order a certain color and style of sweaters in summer. It has to be done months before winter comes so merchandise

© Springer Nature Switzerland AG 2020
N. K. Sehgal et al., *Cloud Computing with Security*,
https://doi.org/10.1007/978-3-030-24612-9_12

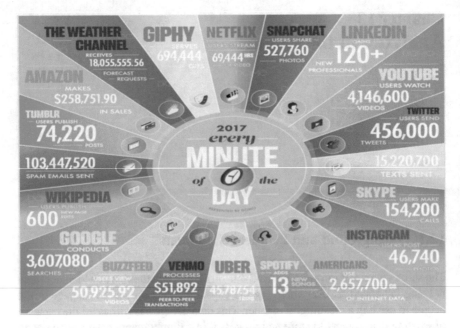

Fig. 12.1 Worldwide activity/minute of Internet users [2] in 2017

will be on the store shelves for customers to buy. In the assessment of customers' demographics, their preferences have to be guessed right, or else most of these sweaters will have to be sold in loss making clearance sales. This problem and many like it are termed as Big Data [3]. It refers to extremely large datasets that need to be analyzed computationally to reveal patterns, trends, and associations. Predictability becomes harder if it relates to human behavior and interactions vs. machines. Thus a car manufacturing company can be automated with robots, and its material ordering can be done in a predictable manner, but how those cars will sell in the marketplace is anybody's guess.

In a 2001 research report [3], Gartner defined data growth challenges and opportunities as increasing across three dimensions:

1. Volume (amount of data)
2. Velocity (speed of data in and out)
3. Variety (range of data types and sources)

Much of the industry continues to use this "3 V" model for describing Big Data. However, in this decade, 3 V model has been expanded with two other characteristics of Big Data:

4. Variability: Inconsistency of datasets can hamper processes to handle and manage it. Big Data is often a by-product of digital interactions, e.g., Facebook posts or trends of Tweeter tweets.
5. Veracity: Data quality can vary greatly affecting the accurate analysis. Machine learning can be used to detect patterns and predict future behavior.

The above two are where the analytics in the Cloud become interesting and necessary, as no single server alone is able to process Big Data sets in a reasonable manner. Businesses are starting to use Cloud analytics on Big Data to attract new customers and better serve their existing customers, in cost-effective ways. An example is online booksellers suggesting additional titles when a book is purchased, based on what other readers are reading, or sending targeted customized advertisement to Internet users based on their search patterns or social media activities.

Some prevailing uses of data analytics in the Cloud [8] are as follows:

- *Social Media*: Billions of users are active on applications such as Facebook, Instagram, Twitter, etc. sharing their stories, opinions, and preferences. This is heaven for marketers to identify potential customers as well as looking for what others are saying about their products or services online. By searching and linking activities across different online sites, it has become easier than ever before to construct a customer's profile, even without meeting that person.
- *Tracking Products*: An online business can track their inventory across warehouses and ship items to customers from the nearest location to minimize shipping time and costs. Similarly, new products can be ordered to replenish supplies, and returns can be tracked in an automated way.
- *Tracking Preferences*: Online movie and song companies log what each user watches or listens to. Then this information is used to recommend other movies or songs along similar themes, to keep the user interested. Internet has become a battleground for eyeballs and mindshare to keep user engaged, so more services and advertisements can be served in a relevant manner.
- *Keeping Records*: Cloud enables real-time recording and sharing of data regardless of location. An example is an online retailer notifying customers when goods are delivered at their home. Furthermore, a facility is offered to alert and buy the next round of supplies at home just before the previous batch finishes. This data is stored in Cloud to track patterns across regions and seasons, so business can stock their goods in an efficient manner.

An advantage of data analytics in Cloud is that entire datasets can be used instead of smaller statistical samples that may not represent the heterogeneous nature of a Big Data set. This helps to eliminate guesswork and enables identification of data patterns to minimize uncertainty. In order to process a vast amount of data for meaningful results, new techniques such as MapReduce have emerged.

12.2 Introduction to MapReduce

Every problem has a solution; we just need to find it. Sometimes the solution is not scalable, e.g., with NP (non-polynomial) class of algorithms where the run time grows exponentially with increase in the size of datasets.

   MapReduce is a possible solution for large datasets that splits the inputs into independent chunks, which are mapped to different processors and processed in a parallel manner. Then output of each mapped tasks is combined (i.e., reduced) to get the final output that is equivalent to the value as if the original data was processed sequentially. An example is the task of finding how many times the word Cloud is used in this book, which admittedly will be many. One way to find this is by writing a single-threaded program that reads each word, compares it to string "Cloud," and increases a single counter by 1 if match is true. Given the number of large number of words, although nowhere close to Big Data, our sequential algorithm will process one page at a time. However, since each page is independent of others, imagine if these pages are mapped to different processors, each of which runs our sequential algorithm in parallel and updates its own counter, and at the end all the counters are added (reduced) to produce a single integer value. That result should be identical to our first single algorithmic run through the entire book. However, the run time of our little MapReduce algorithm will be much faster because each thread has to read through only one page. The only overhead is in the split or map and final reduce tasks. Figure 12.2 illustrates both the serial and parallel nature of our search schemes.

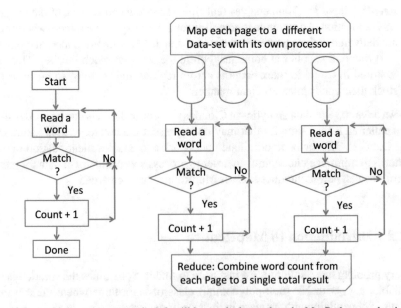

Fig. 12.2 A serial word count on the left will be much slower than the MapReduce on the right-hand side, due to the parallel nature of the latter, both giving the same final result

12.3 Introduction to Hadoop

Hadoop is an open-source software project [4] for reliable and scalable computing. The Apache Hadoop software library is a framework that allows for the distributed processing of large datasets across clusters of computers using simple programming models. It is designed to scale up from single servers to thousands of machines, each offering local computation and storage. Rather than rely on hardware to deliver high availability, the library itself is designed to detect and handle failures at the application layer, so delivering a highly available service on top of a cluster of computers, each of which may be prone to failures.

The project includes these modules, with the full software stack shown in Fig. 12.3:

1. Hadoop Common: The common utilities that support the other Hadoop modules.
2. Hadoop Distributed File System (HDFS™): A distributed file system that provides high-throughput access to application data.
3. Hadoop YARN: A framework for job scheduling and cluster resource management.

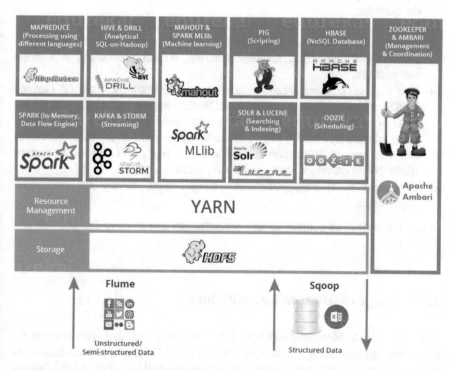

Fig. 12.3 A representation of Hadoop software stack [4]

4. Hadoop MapReduce: A YARN-based system for parallel processing of large datasets.

Other Hadoop-related projects at Apache include:

5. Ambari™: A Web-based tool for provisioning, managing, and monitoring Apache Hadoop clusters which includes support for Hadoop HDFS, Hadoop MapReduce, Hive, HCatalog, HBase, ZooKeeper, Oozie, Pig, and Sqoop. Ambari also provides a dashboard for viewing cluster health such as heat maps and ability to view MapReduce, Pig, and Hive applications visually along with features to diagnose their performance characteristics in a user-friendly manner.
6. Avro™: A data serialization system.
7. Cassandra™: A scalable multi-master database with no single points of failure.
8. Chukwa™: A data collection system for managing large distributed systems.
9. HBase™: A scalable, distributed database that supports structured data storage for large tables.
10. Hive™: A data warehouse infrastructure that provides data summarization and ad hoc querying.
11. Mahout™: A Scalable machine learning and data mining library.
12. Pig™: A high-level data-flow language and execution framework for parallel computation.
13. Spark™: A fast and general compute engine for Hadoop data. Spark provides a simple and expressive programming model that supports a wide range of applications, including ETL, machine learning, stream processing, and graph computation.
14. Tez™: A generalized data-flow programming framework, built on Hadoop YARN, which provides a powerful and flexible engine to execute an arbitrary DAG of tasks to process data for both batch and interactive use cases. Tez is being adopted by Hive™, Pig™, and other frameworks in the Hadoop ecosystem and also by other commercial software (e.g., ETL tools), to replace Hadoop™ MapReduce as the underlying execution engine.
15. ZooKeeper™: A high-performance coordination service for distributed applications.

A wide variety of companies and organizations use Hadoop for research and production.

12.4 Usage of Amazon's MapReduce

Amazon's Elastic MapReduce (EMR) is a Web service [5] that uses Hadoop for processing large amounts of data. Advantage of using EMR over generic Hadoop is that the system administration tasks, e.g., compiling, building, and maintaining the entire software framework, are handled by Amazon.

Public Cloud customers have been using EMR for log files analysis, Web indexing, data transformation, machine learning, financial analysis, scientific simulation, and bioinformatics projects. These can use data already stored in AWS databases such as Amazon Simple Storage (Amazon S3). Its security is handled by EC2 firewall and users have the following level of control:

1. Overall machines within a user's cluster
2. Root access to every instance
3. Ability to install additional applications
4. Customization of every cluster
5. Supporting multiple Hadoop open-source distributions and applications

An example of an EMR cluster with master and slave nodes is shown in Fig. 12.4. Master node handles the mapping tasks, whereas each slave node does the computation or data processing, the results of which are then presented back to the master for final reduction.

Master node manages the cluster by:

1. Running software components
2. Coordinating the distribution of data and tasks among other nodes for processing
3. Tracking status of tasks and monitors cluster health

Slave nodes do processing work as follows:

1. Core node: a slave node that has software components, which run tasks and store data in the Hadoop Distributed File System (HDFS) on the user cluster.
2. Task node: a slave node that has software components, which only run tasks. Task nodes are optional.

Life cycle of an EMR cluster, just like any other virtual server network in a Public Cloud, is ephemeral as shown in Fig. 12.5. Based on the dataset and desired parallelism, user can decide how many slave nodes are needed. Speedup will be

Fig. 12.4 An example of a master-slave cluster for MapReduce tasks

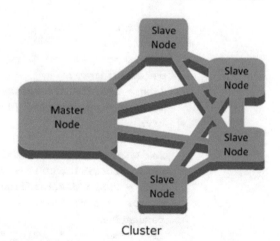

Cluster

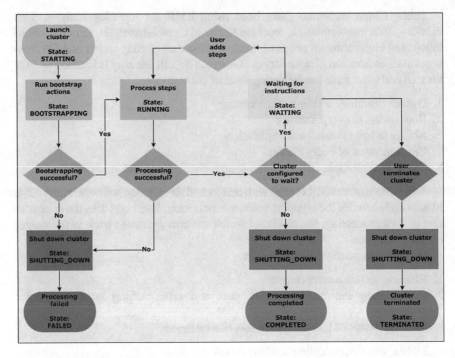

Fig. 12.5 An EMR cluster life cycle

proportional to it, barring any inherent data dependencies in the computational algorithms. The user will only pay for the uptime of cluster, so it is better to not keep any nodes idle.

A Python implementation of our previous word count using MapReduce on EMR's Hadoop is shown below. It will count the number of times a word occurs within a text collection.

```
#!/usr/bin/python
import sys
import re
def main(argv):
line = sys.stdin.readline()
pattern = re.compile("[a-zA-Z][a-zA-Z0-9]*")
try:
while line:
for word in pattern.findall(line):
print "LongValueSum:" + word.lower() + "\t" + "1"
line = sys.stdin.readline()
except "end of file":
return None
if __name__ == "__main__":
main(sys.argv)
```

This can become the basis of sentimental analysis in Cloud, e.g., reaction of Twitter users to any latest breaking news story, if their reaction can be categorized in favorable (i.e., like, happy, prefer, good, etc.) or unfavorable (i.e., do not like, angry, bad, ugly, etc.) words. As one can see, any such prediction is subjective based on the extent of sentimental analysis words used and may not capture the whole sentences or meanings. As an approximation, this has been found to be useful [6] on social media trends.

12.5   Twitter Sentimental Analysis Using Cloud

An example of studying the popularity of US presidential candidates before 2016 elections will be presented. The sole purpose of this project was to study the positive and negative sentiments of Twitter users during a specific period of time, and results were obtained a year before the actual election took place. This project was done by two graduate students (Nora Susan Kurian and Ruchika Tayal) in a Cloud Computing class that one of the authors of this book taught at Santa Clara University during the Fall 2015 quarter.

These students followed Amazon's guidelines [7] and used Amazon's EC2 and EMR services to setup their MapReduce cluster. Then on a live Twitter stream after presidential debates or any major news item, a key-value pair analysis was done to search for positive and negative sentimental words in users' tweets. The actual results, much before presidential primaries were held, are shown in Fig. 12.6. Looking at these, there should be no surprise at how the actual elections turned out then.

Further details of analyzing 10,000 users' tweets revealed that:

(a) Republicans were leading as 69% of the twitter users are talking about them.
(b) Within Republicans, Donald Trump was the leading candidate with 66% of the users mentioning him in their tweets.

Fig. 12.6 A Twitter sentimental analysis study done in Fall 2015 by students

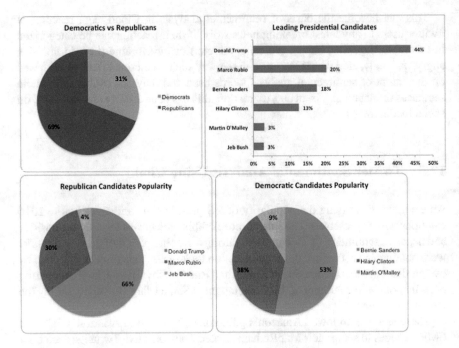

Fig. 12.7 Analysis of 10,000 users' tweets done during Fall 2015

(c) Within Democrats, Bernie Sanders was leading candidate with 53% of the users mentioning him in their tweets.
(d) Out of the six candidates, Donald Trump was leading by a large margin with 44% twitter mentions, as shown in Fig. 12.7.

While it is interesting to use Cloud-based analytics to assess users' emotions on a social network, there is more redeeming value in applying similar techniques for the benefit of the users. An example is applications of smart devices and Cloud-based applications for healthcare monitoring, as we will review in the next section.

12.6   IoT-Driven Analytics in Cloud

Internet of Things (IoT) collects data from many interconnected devices, which is then stored and analyzed in a Cloud. According to Forbes, IoT will contribute $117 billion to the healthcare and $1.9 trillion to the world economy by 2020 [9]. It is not just the monetary contribution but also potentially an estimated saving of 50,000 lives each year in the USA alone by avoiding deaths due to hospital errors [10].

Notwithstanding the legal issues of data sharing, proprietary architectures, who owns what, etc., let's consider a simple case of ECG (electrocardiogram) signals that can be captured by smart health watches or other personal wearable devices and then transported to Cloud for classification and diagnostics by health professions. It has the potential for early detection of heart-related issues, which can be treated in a timely manner instead of calling emergency services after an event occurs. In this healthcare value chain, there are following five players:

1. Patients
2. ECG signal capturing and recording service
3. Secure transmission service
4. Cloud system manager
5. Healthcare staff

Figure 12.8 depicts the value chain of information [11], starting with collection of a patient's healthcare data on the top left-hand side. ECG signals are captured via one or more devices and then flow through an Internet gateway to Cloud data centers for analysis and processing. Features such as irregular heartbeats are extracted, classified, and stored using Big Data management techniques. Then healthcare professionals can be alerted for unusual ECG patterns in case of emergencies. The stored telemetry data can also be used for delivering patient care during regular checkups, e.g., to adjusting blood pressure medications.

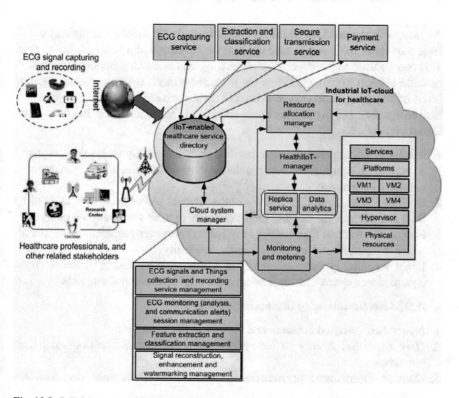

Fig. 12.8 IoT-driven medical data collection and analysis in Cloud [11]

The concerns of privacy can be addressed by data encryption with key pairs shared between patients and doctors, such as only authorized medical professionals can access an individual's data, but medical researchers can use trends across anonymous patients to detect if a particular medicine is more effective than others. The advantage of using Cloud is that many IoT devices such as health watches, other wearables, and smartphone sensors can be used to correlate heartbeats and movement activities for a given patient. However, the local data is securely transported to Cloud when connectivity is available, preventing loss of information. This also enables large dataset collections for doctors to analyze, instead of periodic ECGs performed in a doctor's office. An IoT-driven continuous monitoring system has the potential to revolutionize healthcare services in terms of delivering care, improving access to patient information, anywhere at any time. It can enable new test trails with large patient populations leading to new findings for overall health improvements. In addition to healthcare, another interesting area for IoT application is home automation, where energy consumption in a building is monitored and controlled for optimizing costs and human comfort.

12.7 Real-Time Decision-Making Support Systems

As we saw in the previous example, Big Data analytics software combined with a large number of IoT sensors have enabled the collection of a vast quantity of data. This can be used to improve decision-making and extracting important trends. A decision support system (DSS) serves the management, operations, and planning levels of an organization [12]. It is used to make decisions about the problems that may be rapidly changing and not easily specified in advance. A DSS has the following properties:

1. Relates to less structured, underspecified problem that higher-level managers of an organization care about, e.g., regulating the production of a factory in response to customer demands
2. Combines use of multiple models and analytical techniques, such as used by Wall Street traders to buy and sell in response to trade wars between nations
3. Focuses on features that make it easy for noncomputer-proficient professionals to use, e.g., medical doctors in a healthcare facility
4. Enables flexibility and adaptability to accommodate changes in environment, e.g., to steer a spacecraft for landing if the original site is not available

A DSS has the following distinct components:

1. *Inputs*: Sensor-based numbers and characteristics to analyze
2. *User Expertise*: Ability to incorporate previous manual decisions and their outcomes
3. *Outputs*: Normalized or transformed data from which new decisions are generated

4. *Decisions*: Results generated based on the user criteria

However, real-time nature of many decision-making systems imposes timing constraints that may be nontrivial to comply with the transmission and analytical latencies in a Public Cloud. There has been an implementation of a Cloud-based decision support system in a Private Cloud [13]. It uses business intelligence (BI) to evaluate effectiveness of urban development projects, for sustainable environment in local contexts. Cloud provides an advantage of elastic resources and real-time data gathering from multiple sources and access of results anywhere and at any time.

An example of using real-time data to make real-time decisions can be found in automotive industry [14]. American Axle & Manufacturing (AAM) is a multibillion-dollar supplier to automakers such as Fiat Chrysler, General Motors, Ford, and Honda. AAM designs and produces driveline and drivetrain systems for light trucks, SUVs, passenger cars, and commercial vehicles. Its management team was finding it hard to get real-time understanding of business operations, capacity planning, and financial status. They started by implementing a manufacturing Cloud by integrating a single ERP (enterprise resource planning) system across all geographies and business units. By converging on a centralized system to collect data and conduct financial reviews, AAM was able to meet its business objectives. It had to categorize data into new inputs, any data older than 3 years that was archived, and processed data used by ERP. The latter category went through a rigorous cleansing process to ensure that its data pool had more reliable information. Following benefits were observed:

1. *Inventory management*: AAM noted an improvement in inventory turns of between 5 and 10% with a positive impact on cash flow.
2. *Manufacturing output*: New ERP enabled AAM to analyze equipment efficiency and determine the root cause of any downtime. This led to continuous improvements and a 5% increase in the manufacturing output.
3. *Quality*: Using the new Cloud-based ERP, quality checks are conducted at every stage, since all of AAM's products start life as raw steel bar that requires a certain heat value to form into the final product. Now they can track these heat codes right from the mill that produced the source material to individual containers and items. This ensures detection and stoppage of any defective parts relatively early in the process.
4. *Compliance*: Using a centralized data Cloud, AAM can assign and track regulatory tariffs in compliance with ever-changing trade regulations.
5. *Workforce efficiency*: Real-time information tracking has led to better decision-making and sharing of information across the entire organization, so AAM managers can spend more time on analysis and less on data collection.
6. *Faster problem-solving*: AAM has been able to cut waste by 2% by localized decision-making, using many manufacturing devices connected to its ERP Cloud.
7. *Improved planning and decision-making*: AAM managers are able to plan, forecast, and order material in line with its 16-week window (due to its material lead

times). These decisions are refined on a real-time basis using fast, accurate, and consolidated data.

The above list can be used as a template for business improvement metrics for other Cloud implementations. AMM was pleased that it was able to launch a real-time ERP system using Cloud Computing across its entire enterprise, without suffering any downtime and while achieving continued sales growth.

12.8   Machine Learning in a Public Cloud

Machine learning (ML) refers to activities, tools, and techniques used to detect patterns and predict future behavior, based on a set of prior observations [15]. A primary goal of ML-based solutions is to perform specific tasks without using explicit instructions. The observations used as an input to ML algorithms are called training set(s). These sample datasets are used to build mathematical models and determine future decision parameters for ML algorithms.

There are broadly three types of ML algorithms:

1. *Supervised and semi-supervised learning*: In supervised learning, the training data contains the inputs and desired outputs. These are used to build a mathematical model for ML. If some training examples are missing the desired output, then semi-supervised ML algorithms are used to predict the missing output values. These have applications in ranking, recommendation systems, face and voice recognition, etc.
2. *Unsupervised learning*: In unsupervised learning, given data only specifies inputs and no desired output. ML is used to find grouping or clustering of input data points. The goal is to identify commonality, and a possible application includes density estimation in statistics.

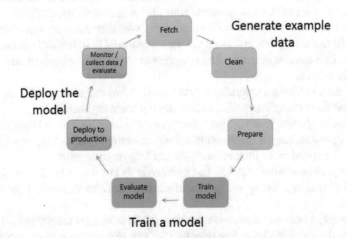

Fig. 12.9 Typical workflow for creating a ML model [16]

3. *Reinforcement learning*: if building an exact model is difficult or infeasible, then reinforcement learning is used to make decisions with a goal to maximize rewards. Applications include game theory, operations research, genetic algorithms, autonomous driving, and to interact with humans.

Without going into implementation details of specific ML techniques [15] such as neural networks, decision trees, association rules, etc., we will examine the set of services that are currently available in a Public Cloud. Sagemaker [16] is a fully managed service offered by Amazon that works in three steps, as shown in Fig. 12.9:

1. *Sample data generation*: It involves fetching the data, cleaning it for consistency, and transforming it to improve ML performance. Example of cleaning includes edits to make sure that same items are referred to by the same names, e.g., the USA vs. the United States for a country name attribute. Transformation example includes combining multiple attributes into a single one for making decisions, e.g., using temperature and humidity to determine if air-conditioning should be turned on.
2. *Building and evaluating a model*: Once the input data is ready, then one of the available algorithms can be used for model training. Then accuracy of its inferences is measured to see if the model is acceptable or not.
3. *Deploying the model*: This is the last and final step which is to use the model for actual data going beyond the initial training set. ML is a continuous improvement cycle, so inferences are often checked to minimize any drift over time. If output results are not desirable, then the model needs to be retrained with a new dataset.

Applications of ML to security are still in nascent stage and partly rely on humans to be successful [17]. The main reason for such reliance is that any ML algorithm has some false positives and false negatives. The former is acceptable, as an alarm will be raised when it should not have been, but the latter can prove costly. Imagine a face recognition system at an airport to prevent any known terrorists from boarding an airplane. Relying solely on such a system will be a mistake, since a single escape can result in fatal consequence. Hence ML plays the role of reducing the search space to focus on a few selected targets, but the broader security processes employed in an organization, such as bag and body search can't be abandoned. ML is increasingly expected to help an organization improve across all five categories of security tasks: prediction, prevention, detection, response, and monitoring. ML can continuously learn from existing data and decisions, to assist in future decisions in the domains of financial fraud detection, unusual network traffic activities, and online social media user behavior to warn the authorities. In security, there is no single silver bullet, as hackers are also using ML tools to better profile their victims and accelerate attacks [18].

12.9 Future Possibilities

As the above students' project shows, possibilities of Big Data analysis are promising, and the same can be applied to predicting financial markets. For example, before the 2008 housing crisis in the USA, financial markets were unaware of the mortgage payments being missed by many new home owners, although some data was available at the individual levels and in local markets across the country. In theory, a Big Data analysis of many small patterns could have been used to predict the larger market direction. Some individual traders saw the big crash coming and profited from their insights, as immortalized in "The Big Short" Hollywood film. A tantalizing possibility is if future terrorist attacks can also be predicted by a combination of monitoring online chatter, financial transfers, assault material purchases, etc. Training such an algorithm will be challenging. It may be acceptable to have a few false positives, as long as no large attack is missed. This is the essence of machine learning, which needs large amounts of data from multiple sources and combines it with previous occurrence to predict any future results.

12.10 Points to Ponder

1. Who owns the datasets in Big Data?
2. What is desirable for external researchers to collaborate for data analytics in Cloud?
3. Are there any vendor lock-in concerns with Big Data?
4. AWS offers several attractive services, not required by NIST, such as Elastic IP and Auto Scaling. Can you think of a few more examples?
5. What equivalent services does other Cloud service providers offer, such as Microsoft Azure and Google's GCP?
6. How do prices of different server types vary across different Cloud service providers?
7. Why Public Cloud is an attractive destination to run ML tasks?

References

1. https://www.inc.com/john-koetsier/every-minute-on-the-internet-2017-new-numbers-to-b. html
2. Domo, marketing data company's 5th annual "Data Never Sleeps" infographics, 2017.
3. https://en.wikipedia.org/wiki/Big_data
4. Hadoop: https://hadoop.apache.org
5. Amazon Elastic Map-Reduce: https://aws.amazon.com/emr/
6. Kouloumpis, E., et al. (2011). Twitter sentiment analysis: The good, the bad and the OMG! *Processings of the fifth international AAAI conference on Weblogs and Social Media*, pp. 538–541.

7. https://aws.amazon.com/blogs/big-data/building-a-near-real-time-discovery-platform-with-aws/
8. http://technologyadvice.com/wp-content/uploads/2013/05/Data-Analytics-in-Cloud-Computing_TechnologyAdvice.pdf
9. McCue, T. J. (2015, April). $117 billion market for Internet of things in healthcare by 2020, forbes, https://www.forbes.com/sites/tjmccue/2015/04/22/117-billion-market-for-internet-of-things-in-healthcare-by-2020
10. Schneider, S. (2015, January). How the industrial internet of things can save 50,000 lives a year. *Industrial Internet Consortium*, https://blog.iiconsortium.org/2015/01/how-to-industrial-internet-of-things-can-save-50000-lives-a-year.html
11. Hossain, M. S., & Muhammad, G. (2016). Cloud-assisted Industrial Internet of Things (IIoT) – Enabled framework for health monitoring. *Computer Networks, 101*, 192. http://iranarze.ir/wp-content/uploads/2016/08/4917-english.pdf
12. https://en.wikipedia.org/wiki/Decision_support_system
13. Benetia, I., et al. (2016, January). Implementing a cloud-based decision support system in a private cloud. *International Journal of Decision Support System Technology, 8*(1), 25–42. https://www.researchgate.net/publication/299499335_Implementing_a_Cloud-Based_Decision_Support_System_in_a_Private_Cloud
14. https://www.industryweek.com/Cloud-computing/real-time-data-real-time-decision-making-across-business
15. https://en.wikipedia.org/wiki/Machine_learning
16. https://docs.aws.amazon.com/sagemaker/latest/dg/how-it-works.html
17. https://www.whitehatsec.com/blog/ml-in-cybersecurity/
18. https://www.techrepublic.com/article/why-ai-and-ml-are-not-cybersecurity-solutions-yet/

Chapter 13
Future Trends in Cloud Computing

13.1 Revisiting History of Computing

The journey of computing began with a single user running a single job [1–3]. We have come a long way since. In the next phase of evolution, multiple users shared a computer system. It further evolved into a networked computer system [4], which was accessible to remote users [5]. With the emergence of PCs (personal computers) in the 1990s, we witnessed PCs being used as a gateway to networked computers, as shown in Fig. 13.1. Then Internet was developed for people to interact with far-away computers in a data center via a Web browser [6]. The twenty-first century saw the rise of mobile devices connecting to Cloud [7].

13.2 Current Limitations of Cloud Computing

Thus far we have studied the benefits of Cloud Computing including economic, elastic infrastructure, on-demand resources, pay for what you use, etc. However, there are a few unseen costs before these benefits can be realized; some are listed below:

1. *Data Movement:* Since servers are located in a remote data center, any input data needed for computation needs to be moved there, and results needs to be moved out. Such I/O (input-output) transactions cost additional money in most Public Cloud and add to the latency as compared to computing on local servers. An example is the emerging area of self-driven cars, which have a multitude of sensors including multiple cameras. There may not be sufficient time to run the image processing algorithms in a remote Cloud due to the dynamic nature of traffic for real-time decision-making while driving. Thus a self-driven car needs to have server-like computing on board. By some accounts, a

© Springer Nature Switzerland AG 2020
N. K. Sehgal et al., *Cloud Computing with Security*,
https://doi.org/10.1007/978-3-030-24612-9_13

Fig. 13.1 An example of client-server architecture, with multiple users on left side interacting with a server

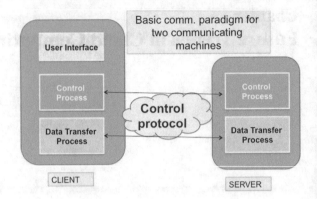

self-driven car in the future may generate up to 5 terabytes of data per day, all of which needs to be stored and processed locally, representing a mini data center on the wheels.

2. *Loss of Control:* When a user's emails are hosted in the Cloud, these are often examined by bots, which then decide on relevant advertisements to display, to generate revenue for email providers such as Google's Gmail. However, this raises a question on who owns the email content and who can access it. For example, if there is a legal case and court subpoenas the email provider to turn over the emails, it will be hard for the provider to say no. At the end, if a user wishes to own the content and keep it private, such as pictures or other business data, then it should be kept on a local computer.

3. *Perception of Cloud Security:* While multiple people can access a data center in Cloud, it may be no less safe than an enterprise data center. Due to the loss of control as mentioned previously, there is a perception of Public Cloud being less secure. This in author's opinion is a red herring, and additional steps can be performed such as to encrypt one's data in the Cloud and also any virtual machine when running on a multi-tenanted server, with keys stored separately.

4. *Uncertain Performance:* Cloud Computing operators make money by sharing same hardware infrastructure with many customers. While their virtual machines (VM) may be isolated in the memory and running on different server cores, there are other shared resources such as a memory controller and networking card that each VM's data must pass through. This creates bottlenecks similar to traffic jams in a data center at entry and exit points, as well as entry and exit to the shared servers. This causes the performance drop of a running VM without any notice. This problem has been described previously as a noisy neighbor and results in a delay in task completion.

All of the above, and a few similar issues, are causing some customers to rethink their Cloud Computing approaches, such as hybrid computing with critical tasks being performed using onsite infrastructure.

13.3 Emergence of Internet of Things (IoT)

Another emerging trend is a Cloud driven by things vs. current Cloud Computing mostly driven by people, as cameras and wireless sensors are becoming pervasive [7, 8]. Their applications include retail solutions, transportation and automotive, industrial and energy, etc. An example of retail industry is Amazon's user-facing portals where customers can visualize things and transact them. An example of transportation and automotive is a software-defined cockpit in a commercial aircraft or an autonomous vehicle. An example of manufacturing is a smart factor with robots or energy savings in a building. Lastly, additional market segment such as health, print imaging, gaming, and education are being digitized at an unprecedented rate. The phrase "Internet of Things" was first used by British technology visionary Kevin Aston in 1999. His perception was to think of "objects in physical world connected by sensors." Internet Architecture Board (IAB) RFC 7452 provides the definition of IoT, as follows:

- "Internet of Things" (IoT) denotes a trend where a large number of embedded devices employ communication services offered by Internet protocols. Many of these devices, often called "smart objects," are not directly operated by humans but exist as components in buildings or vehicles or are spread out in the environment.

Four basic communication models for IoT are:

1. Device to device
2. Device to Cloud
3. Device to gateway
4. Backend data sharing model

We are more interested in #2 and #4, as both involve Cloud services. An example is shown in Fig. 13.2, of home appliances such as a thermostat-controlled A/C connected to Cloud for better energy management.

Fig. 13.2 Cloud-based energy management, monitoring, and optimization. (Source: Tschofenig, H., et al., Architectural Considerations in Smart Object Networking. Tech. no. RFC 7452. Internet Architecutre Board, Mar. 2015. Web. https://www.rfc-editor.org/rfc/rfc7452.txt)

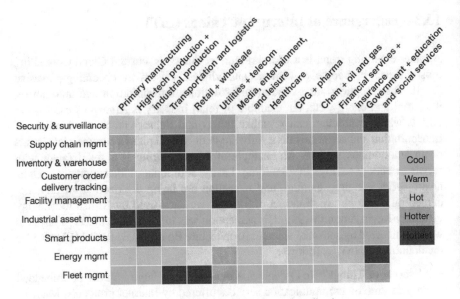

Fig. 13.3 Heat map of key IoT opportunities by industries and applications

Bains [8] predicts that by 2020 annual revenues for the IoT vendors could exceed $470B by selling hardware, software, and comprehensive solutions. Forrester [9] published a heat map in 2016, showing how opportunities vary by the industry and applications, as shown in Fig. 13.3.

As seen above, the hottest (i.e., most financially attractive) applications are in transportation, government, and retail. Further discussion of IoT business opportunities is beyond the scope of this book but can be found in [9]. These new opportunities also bring new security challenges. As an example, if these devices are connected to Internet, then a hacker can potentially gain access to read the output data or alter device configurations to yield unexpected results. We will explain the security implications and potential solutions in a later section.

13.4 Emergence of Machine Learning

Due to the preponderance of IoT data being generated, it is nearly impossible for a human to draw any meaningful comparisons. This is reviving expert systems and artificial intelligence (AI), this time aided by unprecedented compute power and self-learning systems that improve with more incoming data. Some of the use cases for IoT-based machine learning are shown in Fig. 13.4, where a smart meter and building temperature control based on when its occupants are expected to arrive or leave. Furthermore, different parts of buildings where people are present or absent can be heated or cooled at different levels, instead of a single setting for the whole floor.

At a basic level, we can add intelligence to remote devices if processing elements and storage are used to locally collect data for rule-based control decisions.

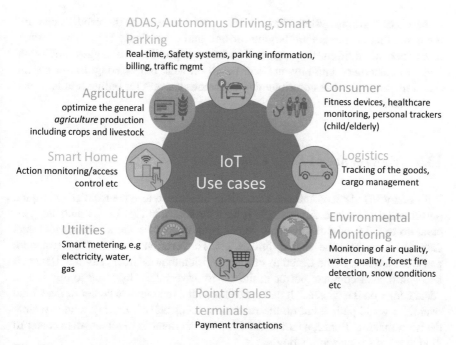

Fig. 13.4 IoT use case that needs both local and Cloud Computing

As an example, in a company cafeteria with multiple work-shifts and different number of employees served on different days, an intelligent refrigerator can check for the remaining packaged food items, including their expiration dates. If the goal is to have food ingredients for at least next 2 days in store, the cafeteria manager can be notified to replenish them as needed. Machine learning part comes from not having a predetermined supply at hand but making the solution self-learning based on consumption pattern of employees. If it is a Friday, and the company is closed for the next 2 days, then the system will look for supplies needed until the following Tuesday. Also, on different work days, menu and specific food items needed may vary, requiring a solution that can intelligently predict what needs to be ordered to minimize expenses and avoid food wastage while ensuring that essential ingredients will never run. In general, such smart devices offer desired functionality and operate in an energy-efficient manner with minimal compute power and memory while connected to a mobile app and Cloud on the back end. These connections are needed for human intervention and record keeping.

Current machine learning technology has its limitations. A case in point is an automobile accident [10] in which driver was on a self-driving system, but vehicle's camera failed to recognize a white truck reflecting a bright sky, and the car failed to brake. However, regulatory authorities absolved the automaker and blamed this accident on the dead driver. They ruled that the driver should have paid attention and not be dependent on the self-driving system. Future liability in accidents will be hotly contested.

Machine learning systems have proven useful in retail as the vendors can find the items that customers are buying, or not, and accordingly build next production order. In addition, they can build customer profiles and suggest additional items to customers who buy an item, based on what others bought after buying the same item. This has contributed to enormous success for online retailers such as Amazon.

13.5   Emergence of Edge Computing

With many IoT devices and use cases, it is imperative to have localized compute power and data storage. An example is a car, as shown in Fig. 13.5, which can generate up to 5 TB of data/day. This comes from onboard cameras, IR sensors, and data collected from the engine, brakes, etc. However, an autonomous car cannot pause for a server in the Cloud to make a decision to accelerate or brake. Hence, it needs sufficient compute power in the car to drive safely, by some to dub it as a "data center on the wheels." It can synch up with a remote data center in the Cloud overnight while parked but on the road must focus on safe driving with real-time decision-making. Hence, a part of the Cloud is migrated from remote data center to field, termed as edge computing.

Similar examples can be found in other application domains, such as smart homes with security cameras, which can decide on the spot if an intruder is a family member or a stranger and in the latter case sound an alarm.

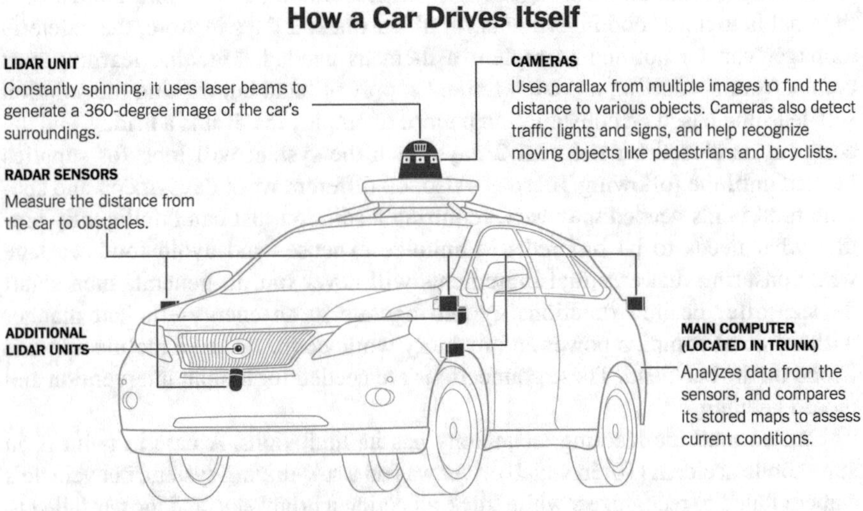

How a Car Drives Itself

LIDAR UNIT
Constantly spinning, it uses laser beams to generate a 360-degree image of the car's surroundings.

RADAR SENSORS
Measure the distance from the car to obstacles.

CAMERAS
Uses parallax from multiple images to find the distance to various objects. Cameras also detect traffic lights and signs, and help recognize moving objects like pedestrians and bicyclists.

ADDITIONAL LIDAR UNITS

MAIN COMPUTER (LOCATED IN TRUNK)
Analyzes data from the sensors, and compares its stored maps to assess current conditions.

Fig. 13.5 A car's self-driving system with multiple sensors [10]

13.6 Security Issues in Edge Computing

Security concerns abound with the emergence of edge computing. In the car example, its computers are not behind a firewall but physically accessible to many people besides the owner. When a car is taken to a mechanic for an oil change or another repair, there is a risk of someone tampering with the hardware or software components setting up a future failure of the self-driven car. It is also possible for someone to access private data stored in the car, e.g., its travel points. Vulnerabilities in other unprotected devices such as home appliances (TV, fridge) on a network can be used to launch a cyberattack. A recent DDOS (distributed denial-of-service) attack was launched using hijacked home security cameras, while in another instance private video clips were stolen and posted on the Internet.

Even for a simple home automation system, such as an intelligent door lock, needs following security features for safety:

1. A firewall to dissuade remote hackers with login authentication.
2. Authentication requires identification of phone numbers, password, or biometrics such as face recognition, thumbprint, retina scan, etc.

Note that any single biometric can be easily defeated, e.g., a pictured mask to fool a face recognition or copy of a thumbprint image, presented to the door camera. It is desirable to have a multifactored authentication system. Furthermore, a data-logging system is needed to record who opened or locked the door and when. This data is immediately backed to a remote Cloud to avoid local tampering. Machine intelligence can be used to create a regular usage pattern and flag anomaly if door is opened at unexpected hours or with unusual frequency.

We need to remember that IoT devices are constantly collecting data about an environment or individuals, which can be potentially shared with third parties compromising privacy. It can range from personal preferences of Web browsing habits, TV channel selection, or images from home security cameras. Some devices can be programmed to selectively transmit data to a Cloud service for processing, e.g., a security camera which has a buffer of 15 seconds but records and transmits a 30-second clip only if any motion is detected, for 15 seconds before and 15 seconds after the motion is detected. This reduces storage requirements but increases chances of a mistake. Such devices are designed to render service with minimal intervention, and yet they need to be directed using voice activation or image recognition. On the flip side, if there is a continuous recording dashcam, which is a forward looking recording device in a car. Purpose of this dashcam is to establish other party's guilt in case of an accident in a vehicle. It will also record voice conversations of passengers potentially violating their privacy rights. It is recommended for the vehicle driver to inform passengers and seek their consent in advance to make them aware.

For ensuring trust in edge computing, it has to start with a trusted environment, trusted protocols, and tamper-proof components. Vendors need to provide "anti-tamper" solutions to start with. Software upgrades in field are needed for any bug fixes during the lifetime of an edge computing device. A secure channel must exist to provide signed binary packets that are transmitted and installed in the field, e.g., on a car or TV at home. In our door example, vendor needs to provide an anti-tamper solution, to prevent someone locally changing the firmware or settings in an unauthorized manner. Even remote software upgrades are authenticated. Otherwise, unprotected home appliances can be used to launch cyberattacks. For example, someone can open garage doors via remote Internet attacks. Besides security, there are privacy concerns, as home sensors are collecting data about individuals that can be shared with third parties for commercial purposes.

Undesirable consequence may emerge if a third party can remotely gain control of a self-driven car causing an accident on the road or someone with malice can access the medicine drip meters in a hospital with fatal consequences for the patients. This can be avoided with a balanced approach to interoperability and access control. This needs to be addressed at different layers of architecture and within the protocol stacks between the devices. Standardization and adoption of communication protocols should specify when it is optimal to have standards. Some vendors like to create a proprietary ecosystem of compatible IoT products. This creates user lock-in to their particular ecosystem, which from a vendor's point of view is desirable because a closed ecosystem approach can offer benefits of security and reduces costs. However, from a user's point of view, such practices can create interoperability problems with solutions from other vendors, thereby limiting user's choices in case of upgrades or future system expansion.

Solution-level cost considerations involve technical factors such as limited internal processing, memory resources, or power consumption demands. Vendors try to reduce the unit cost of devices by minimizing parts and product design costs. It may be more expensive to design interoperability features into a product and test for compliance with a standards specification. A non-interoperable device may lack in standards and the documented best practices. It may limit the potential use of IoT device, and absence of these standards can result in deviant behavior by IoT devices.

13.7   Security Solutions for Edge Computing

Edge computing is the most recent inflection point in the history of computing. With the advent of edge computing, evolutionary cycles between a concentration of powerful centralized computing and an emphasis on distributed powerful computing have changed to a network made of a combination of centralized powerful computing and distribution of simple computers at the edges of the network. This network has vastly different security requirements. For example, a central system in the Cloud can send security breaches to the edge, or the edge computers can send security breaches to a server in the Cloud. A system-wide trust is difficult to achieve based upon the current start of art strategies, policies, standards, or implementations. Edge computing expands the potential threat plane and introduces the possibility of attacks from many directions.

The following classification describes the types of security issues related to the edge computing:

1. *Identity authentication:* By definition, the number of players in edge computing is large, and these may not belong to the same organization. It is infeasible to verify their identity in a foolproof manner. Trust needs to be extended, as new customers buy their devices, such as security cameras, and bring these online with a remote registration. Central authority then must depend on the ability of these remote customers to protect their own devices.

2. *Unauthorized access:* Depending on the nature of sensors at the edge, their access into data center may be bi-directional (from/to DC) in nature. If someone hacks into a remote device to impersonate a previously trusted remote device, it is nearly impossible to differentiate between genuine or fake users. Similarly, someone pretending to act as a central computer can access remote devices to get critical user data.

3. *Denial-of-service attacks:* An attack was launched by hijacking multiple remote devices and simultaneously contacting the central server. This will cause the server to be overloaded, denying access to a genuine device in a timely manner.

4. *Data theft:* Depending on where data is stored and for how long opens the possibility of it being stolen. An example is a security camera at home with local storage. In event of a theft, it may be possible for an intruder to delete the local storage, thus circumventing the purpose of a security camera. However, if camera immediately uploads an image to Cloud upon detecting a motion, then any physical tampering will still protect the image of intruders.

5. *Data integrity and falsification:* A key difference between confidentiality and integrity is that in the latter case, an attacker doesn't need to read the protected data but merely modify it. One example is with a buffer overflow, rendering it useless. This system-level attack can happen if multiple devices from different sources are writing back to a central server memory or database. This can be protected with assigning a virtual partition or container to the data coming from each distinct source and checking the address range of each access to prevent data integrity of other users on the same server.

6. *Invasion of privacy:* Since multiple players may combine their data inputs from different sources to arrive at a desired conclusion, e.g., for real-time traffic updates, their identities need to be protected. This may include an individual's location, movements, and any other aspects of personal nature.

7. *Activity monitoring:* A simple example of cell phone, which constantly pings the signal tower, is sufficient for someone to monitor the location of phone owner's movements. Furthermore, if a remote app can turn on the microphone or camera in a phone, then additional information and activities can be monitored in an illegal manner. Similar effects can be achieved with fixed cameras at commercial or public locations, e.g., in a shopping center.

It is recognized that traditional trusted compute boundary (TCB) expands with edge computing to include domains that are physically outside the control of remote device or central data center owners. The best they can do is to monitor/track a threat, identify an attacker, launch a recovery, and prevent false positives. These steps are outlined below:

1. *Monitor/track a threat:* This is possible by establishing a normal usage pattern and then looking for anomalies. Any deviation represents a potential threat.
2. *Identifying attackers:* Once a threat is detected, then attackers need to be identified. These could take the form of an IP address that is repeatedly pinging the central server to launch a denial-of-service attack.
3. *Attack recovery:* This can take the form of blocking the offending IP address. However, the situation is not always so simple as an attacker can corrupt the critical data before the attacker's presence is detected. In such a case, frequent checkpoints must be taken to do rollback to a known good state.
4. *Accidental and unintentional failures confused with security attacks:* Most detection methods suffer from the risks of false positives, e.g., mistaken flagging of a genuine access as a potential threat. An example of this is a stock market trading computer that detects unusual activity, which is genuine yet may flag a false alarm. Similar situation can happen with security alarms due to false sensor activity data, etc. This calls for a learning system that becomes smarter over time.

13.8   Example of an IoT-Based Cloud Service

A Cloud service where intelligence extends beyond a data center to the edge-based sensors is also known as fog computing [11], which is a clever name for gathering and processing data at the local computing devices. In this model, sensors and other connected devices such as cameras send data to a nearby edge computing device, which has processing power to analyze this data, make some local decisions, and then send the results to the Cloud. BI Intelligence forecasts that 5.8 billion IoT devices owned by enterprises and governments will use fog computing in 2020, up from 570 million devices in 2015 (Fig. 13.6).

An example comes from mining industry [12], where drilling equipment is working in harsh conditions, with autonomous trucks and trains and tunneling and boring machines moving at high speeds. In order to ensure worker safety and increase productivity, decisions need to be made locally. Even though mining equipment can generate terabytes of data/hour during normal operation, there may not be a reliable connection to backend Cloud given 100s of feet of underground operation, as in a coal mine. This is where fog computing can help by processing the data locally, make appropriate decisions, and, as shown in Fig. 13.7, send only small uploads to the backend Cloud every few hours or at the end of each day.

However, this also increases risk of accountability and security as different legal entities may own the local sensors, edge gateway, and backend Cloud. If something goes wrong, e.g., in the event of a mining accident, then finger pointing will begin with hard to assign liabilities. This is where companies offering end-to-end services

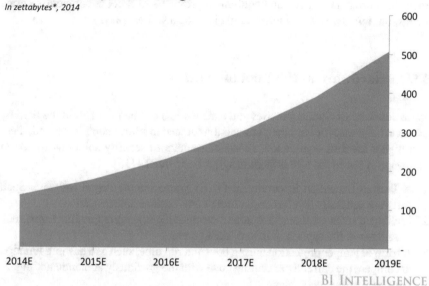

Total Amount Of Data Created Worldwide By Connected People And Things
In zettabytes*, 2014

Fig. 13.6 Growth of data being generated by IoT and Cloud together [11]. *Note: 1 Zettabyte = 1 trillion gigabytes. (Source: Cisco, Global Cloud Index 2014)

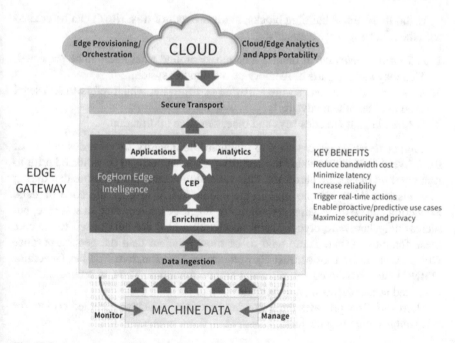

Fig. 13.7 An example of fog computing to support real-time decisions [12]

will have an advantage in edge computing deployment resulting in a business success. In the long run, as industry evolves and standards emerge, there will be a room for horizontal service and equipment providers to excel at competitive price points, but initially fog computing is likely to be a vertical play.

13.9   Hardware as the Root of Trust

Since software (including binaries and executables) can be modified in the field by hackers, but transistors and their connections etched in silicon can't be altered, there is a higher level of confidence among consumers in security solutions rooted in hardware. A hardware root of trust has four basic blocks [13]:

1. A Trusted Execution Environment (TEE) to execute privileged software. Such software generally handles critical functions, such as authentication.
2. Privileged software that at a minimal can provide cryptographic functions, such as Advanced Encryption Standard (AES) code.
3. A form of tamper protection during boot and run time, such as a hashing function to compare the currently executing code with the previously authenticated signatures of a trusted code binary.
4. An API (application programming interface) or a simple user interface for higher-level applications or users to access the underlying hardware security features.

Using these above building blocks, hardware root of trust (RoT) can be created with the following three elements:

1. *A Security Perimeter:* It defines a boundary around what needs to be protected. This may include parts of memory on a computer system.
2. *A Secure CPU:* It runs secure software or firmware, which refers to low-level code to access the hardware.
3. *Secure Data:* It includes keys and other sensitive information.

Tamper resistance is essential to maintain the integrity for a hardware root of trust. Every code from outside the security parameter needs to be validated prior to running it on the secure hardware. This validation is essential to ensure that during transmission, there was no man-in-the-middle attack to alter the code or data. Protection can be done using cryptography. It can be implemented in software, but increasingly hardware cryptography accelerators [14] are being used to execute these functions faster. These tend to be more efficient than the general-purpose CPUs. A hardware root of trust also requires a True Random Number Generator (TRNG), which is used by various security functions. Thus, a secure and un-tampered access to this module is critical.

Hardware RoT provides the following functions, which are deemed critical for many edge computing devices:

1. *Secure Monitoring:* This is to ensure that all components and interactions between the components are secure. Any attempt to insert a malicious code or data will result in a notification from the RoT to the host server.
2. *Secure Authentication:* This is responsible for cryptographically verifying the validity of code on an edge device and the data that is being sent back to host server in Cloud.
3. *Storage Protection:* Since an edge device will need to have local data, its storage needs to be protected using encryption and authentication. Such a protection can use a Device Unique Key (DUK) so that only an entity with the matching DUK can read the storage. Thus, as an example, moving a memory card from one security camera system recording to another will result in no playback.
4. *Secure Communications:* These are available between an edge device and Cloud server only after an authentication and key exchange protocol is completed.
5. *Key Management:* This refers to management of security keys, such that only an indirect access to keys is permitted. It is managed by permissions and policies on the application layer. Example of common key management systems include a Hardware Secure Module (HSM) using Public Key Cryptography Standard (PKCS)#11 interface application to manage the policies, permission, and handling of keys [15].

13.10 Security in Multi-party Cloud

Imagine a scenario where multiple entities wish to come together for a common purpose, but do not quite trust each other. An example is new drug research that needs hospitals to provide patient data, pharmaceuticals to provide their drug data, and medical researchers to explore new treatment protocols. Neither party may want to give away its crown jewels, but all are interested to know which drug protocols are effective in fighting and preventing diseases. For this type of situations, multi-party Cloud offer a viable solution. There multiple users come together on a shared hardware to accomplish a common computational goal. It allows data of each party to be kept secured from other users.

A framework for secure multi-party computation (SMPC) is proposed [16] with four entities: proxy server, Cloud server, analyzer, and parties that are taking part in the shared computations. As shown in Fig. 13.8, a Cloud server is used to perform the required computations upon the receipt of user data. A proxy server is used to hide the identity of each user while providing anonymity.

Each user can start sending private data for computations, after being authenticated into the system. Proxy server hides the traceability of messages sent from each user toward the Cloud server. This data is encrypted to provide protection and integrity against a man-in-the-middle attack. Analyzer in this model is the external party, which receives the statistical parameters of user data, decrypts it, and performs the analytics on it. There are obvious questions on the performance and efficiency of Cloud environments to deploy such a model while enforcing the security requirements of all user parties.

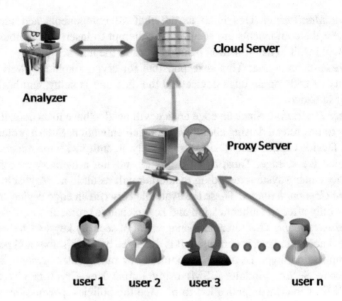

Fig. 13.8 A secure multi-party-based Cloud computing framework [16]

13.11   Privacy-Preserving Multi-party Analytics in a Cloud

Privacy-preserving algorithms for data mining tasks, which look for trends and hidden patterns, use techniques such as clustering, classification, or associate rule mining. Security assurance strategies are broadly listed as follows:

1. *Anonymization:* this approach partitions a database table in two subgroups, containing personally identifiable information (PII) and the rest. It removes the PII set, such as a person's name, date of birth, etc. These attributes are replaced by an ID for the purpose of performing analytics. However, an attacker can use some prior or external information, such age, gender, and zip codes to decipher the identity of subjects. An example is de-anonymization of Netflix dataset based on the published movie ratings and externally available datasets to identify a subscriber's records in a large dataset [17].
2. *Secure Multi-Party Computation (SMPC):* this strategy considers all attributes of a dataset as private and uses cryptographic protocols for peer-to-peer communications. These techniques tend to be secure but also slow; so they do not scale well for large datasets.
3. *Randomization:* this approach seeks the underlying data while preserving the statistical properties of the overall dataset. Examples are additive data perturbation, random subspace projections, and simple mixing of time series from different sensors within acceptable windows or slices of measurement intervals. These techniques are fast and efficient, but do not provide any security guarantees [19].

An example of a real-life project is MHMD (My Health My Data), which enables sharing of sensitive data [20] in Europe. It aims to be the first open biomedical information network, allowing hospitals to make anonymized data available for open research while prompting citizens to become the owner and controller of their health data.

As shown in Fig. 13.9, it uses a SMPC platform and provides insights into medical datasets, without compromising individual patient's privacy. This enables researchers to run medical data analytics, beneficial to patients, doctors, and researchers. These analytics include data aggregation, statistics, and classification methods. They are implemented through privacy-preserving algorithms under the secure multi-party computation scenarios.

An end-to-end query execution, as shown in Fig. 13.9, happens in the following five steps:

1. A user makes a privacy-preserving analytics request to the coordinator.
2. The coordinator server is responsible for contacting all the involved parties.
3. Data providers extract the requested data from their datasets and securely import them to the SMPC cluster applying secret sharing methods.
4. The SMPC cluster computes the privacy-preserving analytics on the requested data and returns the results to the coordinator.
5. Finally, the results are returned to the user through the coordinator.

The final results do not reveal anything except for the actual results of the secure computations performed by the computing parties. Another application for such a technique is the Department of Homeland Security (DHS) in the USA, which

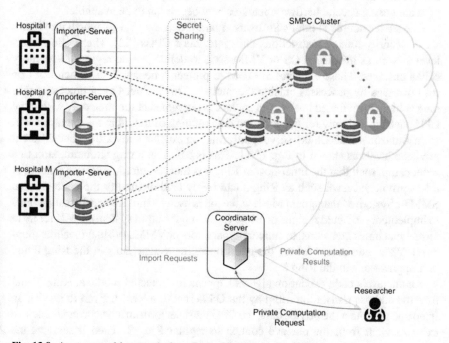

Fig. 13.9 A secure multi-party platform for health data sharing [18]

routinely wants to check the list of passengers on an incoming flight against its own list of suspected terrorists. In this case, if the flight is operated by a foreign airline, it may want to limit the details of passenger information it shares, while the DHS doesn't want to give away its list of suspected terrorists. Yet another example is in the space, where many countries operate their satellites and have risks of collision with the satellites of other countries. How can each country share its database with the other, without revealing critical details about the purpose or intent of their assets in the space? Currently, there is no such database in the world, but given the space race, one may be needed to avoid accidental collisions.

13.12   Hardware-Based Security Implementation

Trusted Computing in Fig. 13.10 refers to a setup, where users trust the manufacturer of hardware or software in a remote computer and are willing to put their sensitive data in a secure container hosted on that computer [21].

As we noted in a previous section that using hardware as a root of trust in a Cloud environment increases customers' confidence, as their VMs are running on the known and attested remote servers. Furthermore, such servers may be also running other tenants' VMs in shared pool of resources. Thus our security-conscious customers want to ensure that there is no in-memory attack or inadvertent data corruption from other tasks. Both confidentiality and integrity of sensitive data can be protected using special hardware features now beginning to be available.

One such example is Intel's Software Guard Extension (SGX), which provides a set of security-related instructions built into latest CPUs [22]. These allow user-level as well as privileged OS or VMM code to define private regions of memory called enclaves. These enclaves are used to protect code and data, which can't be read or saved by processes outside the enclave. This includes even processes running at higher privileged levels. The enclave is decrypted at run time only within the CPU package and only for the code and data running from within the enclave itself.

In Amazon's EC2 Cloud, server platforms can execute software at four different privilege levels as shown in Fig. 13.11. This is based on a ring structure, akin to a scout camp, such that the innermost or Ring 0 is most secure. A software running at a less privileged level such as Ring 3 can freely read or modify the code or data. SMM or System Management Mode at the top is used by motherboard manufactures to implement protected regions of BIOS (basic input-output system). VMX refers to virtual machine extensions, to support hypervisors or VMMs (virtual machine monitors). VMX non-root is where the guest operating system runs in the Ring 0 and user applications in the Ring 3.

An enclave's code execution always happens in protected mode, at Ring 3, and uses the address translation setup by the OS kernel or a VMM. Even to service an interrupt, to protect the private data, the CPU must perform an asynchronous enclave exit to switch from the enclave context to regular Ring 3. Then it services the

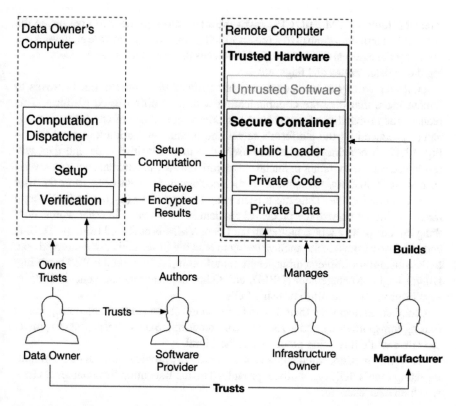

Fig. 13.10 Basis of Trusted Computing [22]

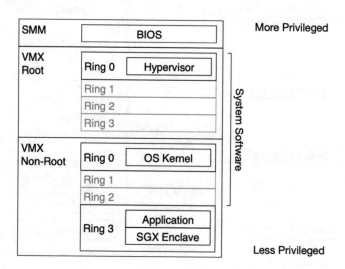

Fig. 13.11 Privilege levels in a server platform

interrupt, fault, or VM exit. CPU saves the state into a predefined area inside the enclave and transfers control to a pre-specified instruction outside the enclave. After servicing the external system call, CPU switches the state back to an enclave, restoring the register values and flags, etc.

First step in any secure computing is certification or attestation. It proves to remote users that they are communicating with a specific trusted platform. This reduces the probability of a man-in-the-middle attack. Proof of attestation is a signature produced by the platform's secret and unique attestation key, as shown in Fig. 13.12. It convinces the remote users that their sensitive code and data will reside in a secure container or enclave. Execution flow can enter an enclave only via special SGX instructions, similar to the mechanics for switching from the user mode to kernel mode. Thus the trusted compute boundary (TCB) for SGX threat model is limited only to the processes resident in an enclave. In other words, anything outside of an enclave including the OS or VMM is excluded from the TCB. A code or process outside the enclave trying to read it will see cipher or encrypted text. SGX is useful for implementing secure remote computation, secure Web browsing, Digital Rights Management (DRM), etc. Other applications can conceal security keys or proprietary algorithms using SGX.

Cloud customers want their data to be protected at rest (in storage), during transit (during transportation), and at run time (in execution). SGX enables users to protect their data while it is being processed in the Cloud.

As of this writing, Microsoft's Azure Cloud [23] provides confidential computing using Intel's SGX capability. It provides Trusted Execution Environment (TEE), which enables users to:

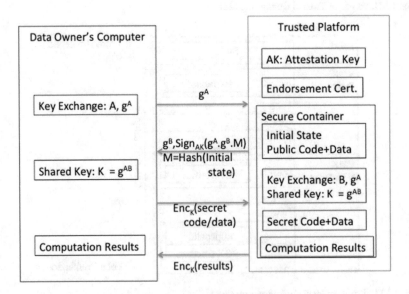

Fig. 13.12 Attesting authenticity of trusted platform to a remote user [21]

1. Safeguard information from malicious and insider threats while in use
2. Maintain control of data through its lifetime
3. Protect and validate the integrity of code in a Public Cloud
4. Ensure that data and code are opaque to the Cloud platform provider

Another US Public Cloud provider using SGX is IBM [24]. They reported that while external attacks outnumber internal incidents as causes of breaches, internal security incidents are on a rise. In 2017, 46% of attacks were malicious insider incidents. IBM Global Cloud deploys SGX-enabled non-virtualized servers, also known as bare metal servers, to provide run-time protection for users' sensitive application code and data.

13.13   Outsourced Computing Using Homomorphic Encryption

Homomorphic encryption (HE) uses a type of encryption to enable computations on ciphertext data. Its result is also encrypted, which after decryption matches the result of the same computation done on plaintext data. This is desirable if multiple parties contribute different parts of a dataset, without revealing their contents to other participants, and still be able to generate the shared results. As we observed in a previous section, this is an excellent solution for medical data privacy concerns. Another example is of a complex supply chain, with parts coming from different countries with different currency exchange rates and shipping costs. Each supplier provides its data in an encrypted format. The final cost for the final goods can be computed without exposing the confidential exchange rates or individually negotiated shipping costs.

Homomorphic encryption schemes are inherently malleable [25, 26]. Fully homomorphic encryption (FHE) supports arbitrary computation on ciphertexts. It is an excellent match for the problem of outsourcing of private computations to public or shared Cloud. An example is shown in Fig. 13.13.

Without going into the mathematical details behind FHE implementations, we will review some of the applications in a Public Cloud:

1. *Querying Encrypted Databases:* An encrypted database generally needs to be decrypted before it can be parsed. Using FHE schemes, database queries can be run on the ciphertext data directly [27]. There is of course a performance penalty as a trade-off for the added security, as the plaintext data is never exposed on the shared database.
2. *Decentralized Voting Protocols:* If a voting community wants to ensure that the vote counters are honest, then each member of the community splits its vote in pieces using homomorphic secret sharing [28]. These pieces are carved so a vote counter can't predict how altering a piece will affect the whole vote. Then each

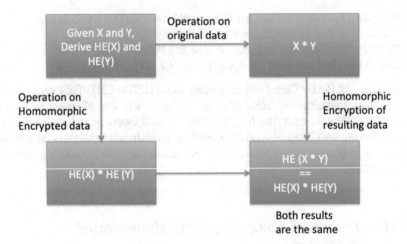

Fig. 13.13 Equivalence of homomorphic encryption for mathematical operations

piece is submitted to a different vote counter. After all the votes are cast, these pieces are gathered together to recover the aggregated election results.

3. *Private Biometrics:* Biometric security enables a better user authentication, but it also implies risks to personal privacy. While a stolen password can be replaced, compromised personally identifiable information (PII) is hard to substitute. Private biometrics addresses this concern by using a one-way FHE. It uses an IEEE open protocol 2401-2018 standard [29]. A private biometric feature vector is only 0.05% of the original biometric template but maintains the same accuracy as the original reference biometric [30].

4. *Client-Side Encryption:* Data is encrypted on the sender's side, before transmitting to a Cloud storage service. Since the encryption key is not available to the Cloud service provider, the data is secure from the man-in-the-middle or server-side admin-level attacks [31]. The data can be searched or used for computations using FHE.

Customers desire security in the Cloud, but currently it has a discernable performance penalty, as most of the FHE techniques are rooted in software. In the near future, there may well be a wave of next-generation hardware solutions to accelerate homomorphic encryption techniques.

13.14   Patching for Security

A business ought to make reasonable efforts to build a threat model for its products, study their attack surface, to validate their security before launching them in the marketplace. However, hackers are also continuously sharpening their tools and devising innovative ways to compromise the security. It is like a game between two competing forces, with no continuous winner. In order to reduce national risk of systemic cyber security and improve incident response, US Department of Homeland Security (DHS) established a national center [32]. It monitors risks and vulnerabilities, classifies all threats via a US-CERT (United States Computer Emergency Readiness Team) scoring system [33], and suggests mitigations to minimize losses. This scoring system takes into account attack complexity, privileges required, and potential impact to classify an incoming vulnerability as None, Low, Medium, High, or Critical. Such a ranking determines the urgency and priority for mitigating the threat.

One way a business deals with the ongoing threats is to issue software patches that can be applied to its products in the field. Recently, Nvidia released 13 patches for its low-end embedded computing boards [34]. These are to stem the flaws that could lead to malicious code execution, denial of service, escalation of privileges, or information disclosure. These boards are intended for AI-enabled applications such as embedded deep learning and computer vision. Products that are impacted range from drones to smart Internet facial recognition security cameras.

As an example of Common Vulnerabilities and Exposures (CVE), a vulnerability# CVE-2018-6269 was assigned Critical score in the security advisory [35]. It resulted from a flaw in the kernel driver's input-output control handling for user mode requests. This could have led to potentially malicious code execution. Other bugs in the same advisory were termed high, e.g., CVE-2017-6278 in kernel's thermal driver could allow an attacker to read or write after the end of a buffer. A CVE-2018-6267 exists in the driver for OpenMAX, which is a set of C-language programming interfaces for multimedia processing. Driver fails to validate metadata, which could allow an attacker to deny service or escalate their privileges by submitting malicious metadata. Another bug in that driver, CVE-2018-6271, improperly validates input, potentially affecting program control flow.

Above description of bugs illustrates the complexity of managing a product in the field, as no product can be secure forever and patching seems to be a way of life in the Internet and Cloud era. However, publicizing these bugs and their patches can lead to another unintended consequence as hackers also read such information. If the security community is not quick to react and apply the patches to all devices in the field promptly, hackers become aware of the new vulnerabilities. This happened in the case of Equifax breach [36], where hackers exposed confidential data for 143 million US consumers. It was done by exploiting a Web application vulnerability publicized 2 months earlier, but not all Apache Servers were patched properly. The vulnerability was Apache Struts CVE-2017-5638, and hackers used this known flaw to install rogue applications on Equifax's Web servers. The critical lesson here is that bad guys are also watching, so before any vulnerability information is made public, its mitigation needs to be in place and deployed by the relevant product users. In the case of IoT, this problem exacerbates as the attack surface is large and ownership of devices is widely spread. Sometimes it may not be possible to apply patches remotely from a central Cloud. For example in some cases, security cameras are installed at homes and may not be updated remotely due to firewalls.

13.15   Machine Learning for Security

In the previous chapters, we learned about machine learning (ML) terms and techniques. These are already being applied in many areas of life, such as social networks, retail sites, smart grids, financial sector, automotive industry, etc. Since cyber security incidents across all these domains are also growing, there is a need to apply ML techniques to curb unauthorized transactions.

In any transaction involving humans, the most common cybercrime is phishing [37]. It is an attempt to fraudulently obtain personal information such as passwords and bank or credit card details, by posing as a trustworthy entity. This also offers excellent ML-based solutions, in the following three areas:

1. *Detective Solutions:* By monitoring an account's activities and flagging any unusual transactions. Incoming phishing emails can be prevented by anti-spam detection. Web content can be filtered using anti-malware software. ML can help to strengthen these techniques by training with known vulnerabilities and then looking for some unexpected patterns.
2. *Preventive Solutions:* Incoming users and login requests can be checked using multifactor authentication. In addition to usual techniques such as sending OTP (one-time password) verification codes by email or mobile addresses, ML techniques can be used to avoid SIM hijacking attacks [38].
3. *Corrective Solutions:* After recognition of repeated attacks coming from a particular set of IP addresses, or geographies, it is possible to detect the location of phishing sites and take them down. ML techniques can be used for forensics and investigation. A comparison of the following six ML classifiers pegs their error rates between 0.075 and 0.105, the detailed discussion of which can be found here [39]:

     (i)   Logistic Regression (LR)
    (ii)   Classification and Regression Trees (CART)
   (iii)   Bayesian Additive Regression Trees (BART)
    (iv)   Support Vector Machines (SVM)
     (v)   Random Forests (RF)
    (vi)   Artificial Neural Networks (ANN)

Interestingly, hackers are also using ML techniques. An example is the usage of machine learning tools to break Human Interaction Proofs (HIP aka CAPTCHA). CAPTCHA stands for Completely Automated Public Turing test to tell computers and humans apart. It is a challenge-response test used in computing to detect if the user is a human. This involves distorted characters often hidden so it becomes a recognition challenge for a regular computer. However, ML approach trains by using the segmentation steps of hidden characters and then uses neural networks to recognize characters in new challenges [40]. Historically, the training stages were computationally expensive due to complex segmentation functions. With the advent of Cloud Computing and using elasticity of affordable resources, hackers can now overcome the difficulties in identification of valid characters. This requires the bar to be raised for security public facing Websites.

13.16 Future Work Needed

There is a growing need for interoperability between IoT devices and services in the Cloud. Internet Engineering Task Force (IETF) has identified the problem of interoperability, as many suppliers build "walled gardens" that limit users to interoperate with a curated subset of component providers, applications, and services.

Interoperability solutions between IoT devices and backend systems can exist at different layers of the architecture and at different levels within protocol stack between the devices. Key is the standardization and adoption of protocols, which should specify when and where it is optimal to use standards. More work is needed to ensure interoperability within the cost constraints for edge computing to become pervasive.

There are other regulatory and policy issues at play, such as device data being collected and stored in a Cloud may cross-jurisdictional boundaries, raising liability issues if the data leaks. This is especially important if data is of personal nature, e.g., related to shopping patterns or patient health records. ML techniques can be used to secure Cloud Computing resources, but hackers are also using ML tools to overcome barriers, so the race between good and bad actors in the cyber space is moving to the next plateau [41].

13.17 Summary

Combination of locally intelligent devices with backend Cloud-based processing is giving rise to a new class of edge or fog Cloud Computing, which offers new usage models but also raises potential of new vulnerabilities with possibility of widespread cyberattacks. There are additional concerns of user lock-ins if vendors do not follow interoperability standards in their edge-based devices in proprietary Cloud solutions. Additional issues of user data privacy and legal jurisdiction currently lag the fast evolution of edge computing domain with IoT-based solutions. This requires policy framework to be discussed by vendors and Cloud service providers with the users for avoiding any legal pitfalls.

As shown in the historic computing spiral of Fig. 13.1, industry has oscillated between large central computers and localized computing, resulting in hybrid models contributing to the spiral usage growth. This now requires large central computers to handle the distributed edge computing demand.

This trend is likely to continue as networks will become faster and machines will become more intelligent to recognize patterns of data to make decisions. In this evolution, it is important to develop standards for interoperability of computing devices on the edge and servers on the back end, to ensure a level playing field for all players. ML tools and techniques are increasingly being used by security professionals and hackers to advance their respective interests.

13.18   Points to Ponder

1. There is potential to have more devices and machines in an increasingly automated world, next wave of Cloud Computing growth is coming from IoT. Can you list additional areas to drive the growth of Cloud Computing?
2. How could one improve the Cloud's performance and support for IoT?
3. Why is edge computing needed for self-driven cars in the future?
4. Can you think of another example of edge computing devices on a road?
5. What is the trust and security model for edge devices?
6. What kinds of attacks are possible using IoT and edge devices?
7. What is the impact of vendor lock-in on edge computing devices?
8. Can hardware be the sole root of trust?
9. Who owns data in a secure multi-party Cloud?

References

1. http://www.pcmag.com/slideshow/story/348634/the-forgotten-world-of-dumb-terminals
2. https://en.wikipedia.org/wiki/Job_Control_Language
3. https://en.wikipedia.org/wiki/Personal_computer
4. https://en.wikipedia.org/wiki/Network_Computer
5. https://en.wikipedia.org/wiki/Client–server_model
6. https://en.wikipedia.org/wiki/Web_browser
7. https://en.wikipedia.org/wiki/Mobile_Cloud_computing
8. https://www.forbes.com/sites/louiscolumbus/2016/11/27/roundup-of-Internet-of-things-forecasts-and-market-estimates-2016/#634d80ab292d
9. https://www.Cloudera.com/content/dam/www/static/documents/analyst-reports/forrester-the-iot-heat-map.pdf
10. https://www.nytimes.com/2017/01/19/business/tesla-model-s-autopilot-fatal-crash.html
11. http://www.businessinsider.com/Internet-of-things-Cloud-computing-2016-10
12. http://www.nanalyze.com/2016/08/fog-computing-examples/
13. https://www.synopsys.com/designware-ip/technical-bulletin/understanding-hardware-roots-of-trust-2017q4.html
14. https://en.wikipedia.org/wiki/Cryptographic_accelerator
15. http://docs.oasis-open.org/pkcs11/pkcs11-base/v2.40/os/pkcs11-base-v2.40-os.html
16. Pussewalage, H. S. G., Ranaweera, P. S., Oleshchuk, V. A., & Balapuwaduge, I. A. M. (2013). Secure multi-party based Cloud Computing framework for statistical data analysis of encrypted data. ICIN 2016, At Paris, http://dl.ifip.org/db/conf/icin/icin2016/1570221695.pdf
17. https://www.cs.cornell.edu/~shmat/shmat_oak08netflix.pdf
18. MHMD. My Health My Data (MHMD), 2018. [Online; Accessed 2018].

19. Alkhadhr, S. B., & Alkandari, M. A. (2017). Cryptography and randomization to dispose of data and boost system security. *Cogent Engineering, 4*, 1. https://www.cogentoa.com/article/10.1080/23311916.2017.1300049. Tao Song (Reviewing Editor).
20. Giannopoulos,M.(2018,September).Privacypreservingmedicaldataanalyticsusingsecuremulti partycomputation.Anend-to-endusecase.MastersThesis,NationalandKapodistrianUniversityof Athens. https://www.researchgate.net/publication/328382220_Privacy_Preserving_Medical_Data_Analytics_using_Secure_Multi_Party_Computation_An_End-To-End_Use_Case/stats
21. Costan, V., & Devadas, S. Intel SGX explained. https://eprint.iacr.org/2016/086.pdf
22. https://en.wikipedia.org/wiki/Software_Guard_Extensions
23. https://azure.microsoft.com/en-us/solutions/confidential-compute/
24. https://www.ibm.com/blogs/bluemix/2018/05/data-use-protection-ibm-Cloud-using-intel-sgx/
25. https://en.wikipedia.org/wiki/Homomorphic_encryption
26. https://en.wikipedia.org/wiki/Malleability_(cryptography)
27. Gahi, Y., Guennoun, M., & El-Khatib, K. (2015, December). A secure database system using homomorphic encryption schemes. https://arxiv.org/abs/1512.03498
28. https://en.wikipedia.org/wiki/Homomorphic_secret_sharing
29. Biometrics Open Protocol (BOPS) III. IEEE 2401-2018, IEEE Standards Association. 2018.
30. https://en.wikipedia.org/wiki/Private_biometrics#ex1
31. http://www.infosectoday.com/Articles/Client-Side_Encryption.htm#.XJHUlxNKiuU
32. https://www.us-cert.gov/
33. https://www.first.org/cvss/user-guide
34. https://nakedsecurity.sophos.com/2019/04/05/nvidia-patches-severe-bugs-in-edge-computing-modules/
35. https://nvidia.custhelp.com/app/answers/detail/a_id/4787
36. https://arstechnica.com/information-technology/2017/09/massive-equifax-breach-caused-by-failure-to-patch-two-month-old-bug/
37. Anti-Phishing Working Group. Phishing and Fraud solutions. http://www.antiphishing.org/
38. https://www.pandasecurity.com/mediacenter/security/sim-hijacking-explained/
39. Abu-Nimeh, S., Nappa, D., Wang, X., & Nair, S. (2007, October 4–5). A comparison of machine learning techniques for phishing detection. *APWG eCrime Researchers Summit*, Pittsburg.
40. Chellapilla, K., & Simard, P. Y. (2005). Using machine learning to break visual human interaction proofs (HIPs). *Advances in Neural Information Processing Systems, 17*, 265–272.
41. Ford, V., & Siraj, A. (2014, October). Applications of machine learning in cyber security. *27th International Conference on Computer Applications in Industry and Engineering*. https://www.researchgate.net/publication/283083699_Applications_of_Machine_Learning_in_Cyber_Security

Chapter 14
A Quick Test of Your Cloud Fundamentals Grasp

14.1 Multiple Choice Questions

1. _____ describes a Cloud service that can only be accessed by a limited amount of people.

   A. Data center
   B. Private Cloud
   C. Virtualization
   D. Public Cloud

2. _____ describes a distribution model in which applications are hosted by a service provider and made available to users.

   A. Infrastructure as a Service (IaaS)
   B. Platform as a Service (PaaS)
   C. Software as a Service (SaaS)
   D. Cloud service

3. *True or False*: Access to a Cloud environment always costs more money compared to a traditional server environment.

4. _____ is the feature of Cloud Computing that allows the service to change in size or volume in order to meet a user's needs.

   A. Scalability
   B. Virtualization
   C. Security
   D. Cost savings

5. *True or False*: A Cloud environment can be accessed from anywhere in the world as long as the user has access to the Internet.

© Springer Nature Switzerland AG 2020
N. K. Sehgal et al., *Cloud Computing with Security*,
https://doi.org/10.1007/978-3-030-24612-9_14

6. Which Cloud characteristic refers to the ability of a subscriber to increase or decrease its computing requirements as needed without having to contact a human representative of the Cloud provider?

   A. Rapid elasticity
   B. On-demand self service
   C. Broad network access
   D. Resource pooling

7. Which customer scenario is best suited to maximize the benefits gained from using a virtual Private Cloud?

   A. A small start-up business focused primarily on short-term projects and has minimal security policies
   B. An enterprise whose IT infrastructure is underutilized on average and the system load is fairly consistent
   C. An enterprise that requires minimal security over their data and has a large existing infrastructure that is capable of handling future needs
   D. An enterprise that does not want to sacrifice security or make changes to their management practices but needs additional resources for test and development of new solutions

8. Which of the following statements are true about the Public Cloud model?

   A. It meets security and auditing requirements for highly regulated industries.
   B. Resources and infrastructure are managed and maintained by the enterprise IT operations staff.
   C. It shifts the bulk of the costs from capital expenditures and IT infrastructure investment to a utility operating expense model.
   D. It shifts the bulk of the costs from capital expenditures to creating a virtualized and elastic infrastructure within the enterprise data center.
   E. Resources are dynamically provisioned on a self-service basis from an off-site third-party provider who shares resources in a multi-tenanted infrastructure

9. A company would like to leverage Cloud Computing to provide advanced collaboration services (i.e., video, chat, and Web conferences) for its employees but does not have the IT resources to deploy such an infrastructure. Which Cloud Computing model would best fit the company needs? For interactive and self-paced preparation of exam 000-032, try our practice exams. Practice exams also include self-assessment and reporting features!

   A. Hybrid Cloud
   B. Public Cloud
   C. Private Cloud
   D. Virtual Private Cloud

10. A company has decided to leverage the Web conferencing services provided by a Cloud provider and to pay for those services as they are used. The Cloud provider manages the infrastructure and any application upgrades. This is an example of what type of Cloud delivery model?

   A. Platform as a Service
   B. Software as a Service
   C. Application as a Service
   D. Infrastructure as a Service

11. Which statement best describes the Software as a Service Cloud delivery model?

   A. A virtual machine provisioned and provided from the Cloud which allows the customer to deploy custom applications
   B. A multi-tenant storage service provisioned from the Cloud which allows the customer to leverage the Cloud for storing software data
   C. A solution stack or set of middleware delivered to the client from the Cloud which provides services for the design, development, and testing of industry aligned applications
   D. An application delivered to the client from the Cloud which eliminates the need to install and run the application on the customer's own computers and simplifying maintenance and support.

12. Which statement is true about the Platform as a Service Cloud Computing delivery model?

   A. It provides a virtual machine and storage so that computing platforms can be created.
   B. It provides a run-time environment for applications and includes a set of basic services such as storage and databases.
   C. It provides the entire infrastructure along with a completed application that is accessible using a Web-based front end.
   D. It is required by the Infrastructure as a Service delivery model so that end user applications can be delivered on the Cloud.

13. What is one benefit of Cloud Computing?

   A. Computer resources can be quickly provisioned.
   B. A workload can quickly move to a Cloud Computing environment.
   C. There is no operational cost for a Cloud Computing environment.
   D. The resources can quickly move from one Cloud environment to another.

14. What is one benefit of a Cloud Computing environment?

   A. It improves server performance.
   B. It minimizes network traffic to the virtual machines.
   C. It automatically transforms physical servers into virtual machines.
   D. It maximizes server utilization by implementing automated provisioning

15. Which statement is true about the maintenance of a Cloud Computing environment?

    A. In a Software as a Service (SaaS) environment, patches are automatically installed on the clients.
    B. In a SaaS environment, customers do not need to worry about installing patches in the virtual instances.
    C. In an Infrastructure as a Service (IaaS) environment, patches are automatically installed on the clients.
    D. In an IaaS environment, customers do not need to worry about installing patches in the virtual instances.

16. What are two common areas of concern often expressed by customers when proposing a multi-tenancy software solution compared to a single tenancy solution? (Choose two.)

    A. Data privacy
    B. Higher costs
    C. Greater network latency
    D. Complexity to customize solution
    E. Less efficient use of computing resources

17. What is the role of virtualization in Cloud Computing?

    A. It removes operating system inefficiencies.
    B. It improves the performance of Web applications.
    C. It optimizes the utilization of computing resources.
    D. It adds extra load to the underlying physical infrastructure and has no role in Cloud Computing.

18. A company currently experiences 7–10% utilization of its development and test computing resources. The company would like to consolidate to reduce the number of total resources in their data center and decrease energy costs. Which feature of Cloud Computing allows resource consolidation?

    A. Automation
    B. Elasticity
    C. Provisioning
    D. Virtualization

19. A company operates data centers in two different regions. Energy costs for one of the data centers increases during the warmer, summer months. The company already uses server virtualization techniques in order to consolidate the total number of required resources. How might the company further reduce operating costs at this data center?

    A. The company can shut down the data center in the summer season.
    B. The company does not need to do anything because they are already using server virtualization techniques.

C. The company can leverage provisioning to optimize the availability of their environments in the summer season.

D. The company can further leverage virtualization to easily and quickly move as many assets from the data center in the warmer region to the data center in the cooler region during the summer season.

20. A company has peak customer demand for its IT services in the month of April. It has enough IT resources to handle off peak demand but not peak load. What is the best approach to handle this situation?

A. Outsource all of IT services to a Cloud-based provider.

B. Virtualize its servers and implement a Cloud-based system.

C. Utilize an external Cloud service provider to handle the peak load.

D. Acquire additional IT resources and design the system to handle the peak load.

14.2 Detailed Questions

1. An ELB is deployed in AWS, which has a public IP of 98.24.20.10. Two EC2 instances are behind this ELB, each with public IP of 98.20.10.10 and 98.10.10.12. You have traffic from West Coast and East Coast coming to your ELB. Add comments for each of the following options if they will work and if not, why not?

A. You decide to move one EC2 in one of the East Coast regions and another one in a West Coast region and put both the regions behind the ELB. (Circle one). Works Doesn't Work.

B. You decide to create two availability zones and migrate an EC2 to each availability zone with the ELB front ending it. (Circle one). Works Doesn't Work.

C. You decide to deploy an EC2 instance and install your own load balancer (like an open source load balancer) and deploy EC2 to each availability zones, thus getting rid of the ELB from AWS. (Circle one 2 pts). Works Doesn't Work.

D. What service model does ELB fall under and why?

2. What is the availability of a Cloud system (Web+DB+LB+Firewall+DataCenter +ISP), where the availability of each component is 99.9%/year?

• How much downtime can you expect in a year with such a system?
• Describe the steps and a diagram of your new setup that gets you 99.95% availability a year.

3. Draw the block diagram of ELB, 4 EC2 instances, and S3, where the application running on EC2 requires 8 GB setup minimum.

The S3 instance needs a setup of 20 GB to handle all the external sources. Assume the cost of EC2 instance to be $0.10 an hour for 5 GB and $0.15 for 10 GB, S3 to be $0.10 an hour for 10 GB, and ELB to be $0.10 an hour.

Cost the monthly price for ELB + 4 EC2 (8 GB) instances + S3 (4 points)

4. How will you make the above setup more reliable and redundant in the most economical way, and how much extra will it cost? AS group is not an option.
5. What's a noisy neighbor problem in Cloud Computing?

14.3   Answer Key for Multiple Choice Questions

1. *B. Private Cloud*
2. *C. Software as a Service (SaaS)*
3. *False*
4. *A. Scalability*
5. *True*
6. *C. On-demand self service*
7. *D. An enterprise that does not want to sacrifice security or make changes to their management practices but needs additional resources for test and development of new solutions.*
8. *C. It shifts the bulk of the costs from capital expenditures and IT infrastructure investment to a utility operating expense model.*
   E. Resources are dynamically provisioned on a self-service basis from an off-site third-party provider who shares resources in a multi-tenanted infrastructure.
9. *B. Public Cloud*
10. *B. Software as a Service*
11. *D. An application delivered to the client from the Cloud which eliminates the need to install and run the application on the customer's own computers and simplifying maintenance and support.*
12. *B. It provides a run-time environment for applications and includes a set of basic services such as storage and databases.*
13. *A. Computer resources can be quickly provisioned.*
14. *D. It maximizes server utilization by implementing automated provisioning.*
15. *B. In a SaaS environment, customers do not need to worry about installing patches in the virtual instances.*
16. *A. Data privacy*
    D. Complexity to customize solution
17. *C. It optimizes the utilization of computing resources.*
18. *D. Virtualization*
19. *D. The company can further leverage virtualization to easily and quickly move as many assets from the data center in the warmer region to the data center in the cooler region during the summer season*
20. *C. Utilize an external Cloud service provider to handle the peak load.*

14.4   Answer Key for Detailed Questions

1. *A. Doesn't Work*

   Key here is to know if an ELB (Elastic Load Balancer) can cover across regions, and answer is NO. It can cover across AZs (availability zones) but not regions as they may be thousands of miles away.

   B. Works

   Yes, for the above reason, an ELB can manage instances within an AZ and across AZs that are located within a Region.

   C. Works

   Yes, but you will have a lot of work to do, for doing health checks and making sure that your own load balancer knows how to distribute traffic across other instances.

   D. IaaS: because it is still giving access to one instance or another but though a layer of indirection.

2.

$$\text{Total Availability} = 0.999*0.999*0.999*0.999*0.999*0.999 = 0.994 = 99.4\%$$

$$(100-99.4)*365.25*24 = 52.46469 \, \text{hours} \, (2 \, \text{Days and} \, 4.56 \, \text{hours})$$

Double up on everything except the data center, so the availability of each step becomes six 9s, and then their multiplication will give you 99.95% total availability.

Web	0.999		0.999999
DB	0.999		0.999999
LB	0.999		0.999999
Firewall	0.999		0.999999
DC	0.999		0.999
ISP	0.999		0.999999
Total	99.401%	52.46469	99.900%

3.

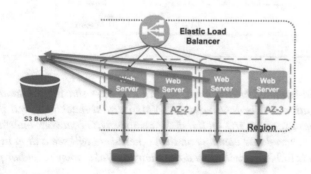

Elastic Load Balancer

S3 Bucket

AZ-2   AZ-3

Region

EBS 8 GB each

4.

	5 GB	10 GB	1 hour	Month
EC2	$0.10	$0.15	$0.60	
S3		$0.10	$0.20	
ELB		$0.10	$0.10	
Total			$0.90	$648.00

If no auto scaling, then extra instances need to be up and running all the time. S3 already has a lot of redundancy, so no need, but you do need four extra instances, preferably in a different AZ, doubling the cost of EC2 line item. Rest remains the same. Extra cost = $1080 − $684 = $396/month

	5 GB	10 GB	1 hour	Month
EC2	$0.10	$0.15	$1.20	
S3		$0.10	$0.20	
ELB		$0.10	$0.10	
Total			$1.50	$1080.00

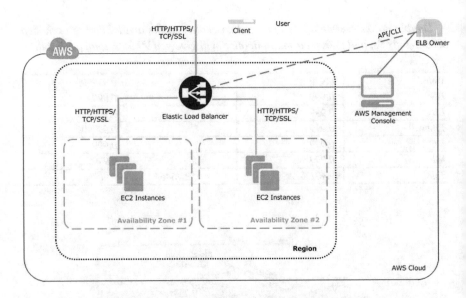

5. *In a multi-tenanted environment, several users are sharing the same hardware. Virtualization can isolate individual VMs (virtual machines), but if one of the users does excessive I/O on the disk, memory, NIC (network interface card) or any other shared HW component, then other users will see a drop in their available bandwidth. This will slow down their jobs, aka noisy neighbor problem.*

14.5   Additional Challenging Questions

6. What are the PaaS or SaaS components that Cloud vendors provide, and how do they compare with similar offerings from AWS? (For instance, VMware's Cloud is big on DR (Disaster Recovery), but what does AWS provide?). Pick one of the PaaS or SaaS offering from an alternative vendor, and compare it to AWS, in terms of features and price.
7. What would be the cost of the following configuration for the Cloud Alternative vendor when run 24/7 for a whole year? Compare the cost to what AWS provides using Reserved Instances or spot prices, with upfront payments. For the following configuration, what would be the lowest cost possible on AWS and on the Cloud Alternative vendor and how?
   1 Load balancer
   4 EC2 instances
   2 DB instances
8. Compare the security offering of the Cloud Alternative vendor to AWS. Compare VPC and other ways that AWS provides security protection to what Azure, Google, or VMware provide. Which one is more secure and why? Give references, screenshots, etc. to support your answer.
9. How is Docker supported in the Alternative Cloud vendor? If so compare to AWS support for Docker. Share some pros and cons between what AWS has and what the vendor provides?
10. Research and compare the uptime (in 4 or 5 nines) and SLA offered by AWS with your Cloud Alternative vendor. How does the Cloud Alternative vendor report downtimes? If the Alternative Cloud Vendor has data centers around the globe, compare the global data center presence and the uptime for the various data centers around the globe.
11. If you had to describe one really cool feature of the Cloud Alternative vendor, what would that be, and does AWS have anything like that or not?
12. How does an on-premise computing stack help customers with multiple and distinct business divisions?

Chapter 15
Hands-On Project to Use Cloud Service Provider

Open an account on Amazon EC2 or Microsoft Azure or Google's GCP Cloud. For the first time account holder, these Cloud service providers (CSP) will give you free access to use their facilities for a limited time.

(a) *Project 1:* Install LAMP stack (consisting of Linux, Apache server, MySQL, and PHP) on your CSP account.
(b) *Project 2:* Set up WordPress and PHP services for other users to access your blog.
(c) *Project 3:* Enhance security of your Cloud server.
(d) *Project 4:* Set up load balancer to manage multiple users.
(e) *Project 5:* Use Elastic IP for your Cloud instances, so if multiple servers are spawned, users can access any of them with the same IP address.
(f) *Project 6:* Evaluation of encryption strength of key size for end-to-end security for user data between multiple clients and a Cloud server by measuring factoring effort vs key size.
(g) *Project 7:* Evaluation of encryption performance vs strength for end-to-end security for user data between multiple clients and a Cloud server as a function of key size.
(h) *Bonus:* Set up auto scaling to increase or decrease the number of servers as a number of users vary.

Steps below are shown for illustration purposes only:

15.1 Project 1: Install Lamp Stack on Amazon EC2

15.1.1 *Installing Lamp Web Server on AWS via EC2*

(Instance used for this, Instance 1 – i-9adc516c)

© Springer Nature Switzerland AG 2020
N. K. Sehgal et al., *Cloud Computing with Security*,
https://doi.org/10.1007/978-3-030-24612-9_15

1. Launch the instance.

```
                                                        ec2-user@ip-172-31-38-180:~ [
 File  Edit  Settings  Plugins  Tunnels  Help
[ec2-user@ip-172-31-38-180 ~]$ █
```

2. Perform a software update.

```
                                                        root@ip-172-31-38-180:
 File  Edit  Settings  Plugins  Tunnels  Help
[ec2-user@ip-172-31-38-180 ~]$ sudo su
[root@ip-172-31-38-180 ec2-user]# yum update -y█
```

3. Group install the Web server, PHP support, and MySQL database using the yum
   groupinstall command. Check to see if they have been properly installed.

```
 File  Edit  Settings  Plugins  Tunnels  Help
[root@ip-172-31-38-180 ec2-user]# yum grouplist
Loaded plugins: priorities, update-motd, upgrade-helper
Installed groups:
    MySQL Database
    PHP Support
    Web Server
Available Groups:
    Console internet tools
    DNS Name Server
    Development Libraries
    Development tools
    Editors
    FTP Server
    Java Development
    Legacy UNIX compatibility
    Mail Server
```

4. Start the Apache Web server.

```
[root@ip-172-31-38-180 ec2-user]# yum grouplist
Loaded plugins: priorities, update-motd, upgrade-helper
Installed groups:
   MySQL Database
   PHP Support
   Web Server
Available Groups:
   Console internet tools
   DNS Name Server
   Development Libraries
   Development tools
   Editors
   FTP Server
   Java Development
   Legacy UNIX compatibility
   Mail Server
   MySQL Database client
   NFS file server
   Network Servers
   Networking Tools
   Performance Tools
   Perl Support
   PostgreSQL Database client (version 8)
   PostgreSQL Database server (version 8)
   Scientific support
   System Tools
   TeX support
   Technical Writing
   Web Servlet Engine
Done
[root@ip-172-31-38-180 ec2-user]# service httpd start
Starting httpd:                                             [  OK  ]
[root@ip-172-31-38-180 ec2-user]# service httpd status
httpd (pid  2470) is running...
[root@ip-172-31-38-180 ec2-user]#
```

5. Ensure that the Apache server remains switched on with every reboot.

```
[root@ip-172-31-38-180 ec2-user]# chkconfig --list httpd
httpd           0:off   1:off   2:on    3:on    4:on    5:on    6:off
[root@ip-172-31-38-180 ec2-user]#
```

6. Copy the public DNS link for this instance from the AWS console onto a browser tab, and click enter.

   We should be able to view the Amazon Linux AMI test page.

7. Add the www group to the current instance.

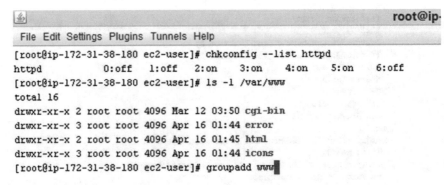

8. Add the ec2-user to this group.

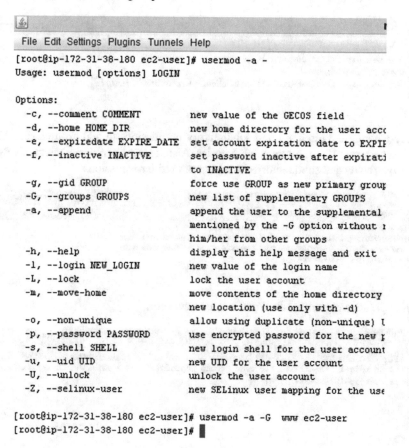

```
[root@ip-172-31-38-180 ec2-user]# usermod -a -
Usage: usermod [options] LOGIN

Options:
  -c, --comment COMMENT        new value of the GECOS field
  -d, --home HOME_DIR          new home directory for the user acc
  -e, --expiredate EXPIRE_DATE set account expiration date to EXPIF
  -f, --inactive INACTIVE      set password inactive after expirati
                               to INACTIVE
  -g, --gid GROUP              force use GROUP as new primary group
  -G, --groups GROUPS          new list of supplementary GROUPS
  -a, --append                 append the user to the supplemental
                               mentioned by the -G option without r
                               him/her from other groups
  -h, --help                   display this help message and exit
  -l, --login NEW_LOGIN        new value of the login name
  -L, --lock                   lock the user account
  -m, --move-home              move contents of the home directory
                               new location (use only with -d)
  -o, --non-unique             allow using duplicate (non-unique) l
  -p, --password PASSWORD      use encrypted password for the new p
  -s, --shell SHELL            new login shell for the user account
  -u, --uid UID                new UID for the user account
  -U, --unlock                 unlock the user account
  -Z, --selinux-user           new SELinux user mapping for the use

[root@ip-172-31-38-180 ec2-user]# usermod -a -G www ec2-user
[root@ip-172-31-38-180 ec2-user]#
```

9. Log out and log in again to view membership.

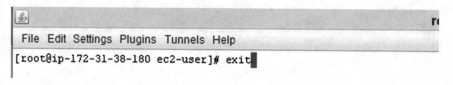

```
[root@ip-172-31-38-180 ec2-user]# exit
```

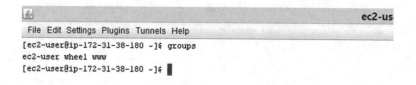

```
[ec2-user@ip-172-31-38-180 ~]$ groups
ec2-user wheel www
[ec2-user@ip-172-31-38-180 ~]$
```

10. Changing group ownership of the /var/www folder to www group.

```
                                                                          ec2-user@ip
File  Edit  Settings  Plugins  Tunnels  Help
[ec2-user@ip-172-31-38-180 ~]$ groups
ec2-user wheel www
[ec2-user@ip-172-31-38-180 ~]$ sudo chown -R root:www /var/www
```

11. Change the directory and subdirectory permissions to add group write. Also,
    set a group ID on future subdirectories. Also, change permissions recursively of
    /var/www and subdirectories to add group write permissions.

```
                                                          ec2-user@ip-172-31-38-180:
File  Edit  Settings  Plugins  Tunnels  Help
[ec2-user@ip-172-31-38-180 ~]$ sudo chmod 2775 /var/www
[ec2-user@ip-172-31-38-180 ~]$ find /var/www -type d -exec sudo chmod 2775 {} +
[ec2-user@ip-172-31-38-180 ~]$ find /var/www -type f -exec sudo chmod 0664 {} +
[ec2-user@ip-172-31-38-180 ~]$ 
```

12. Create a php file in the Apache document root.

```
                                                                       ec2-user@ip-
File  Edit  Settings  Plugins  Tunnels  Help
[ec2-user@ip-172-31-38-180 ~]$ echo "<?php phpinfo(); ?>" > /var/www/html/phpinfo.php
[ec2-user@ip-172-31-38-180 ~]$ 
```

13. Type the public DNS name followed by /phpinfo.php to view the following through a browser.

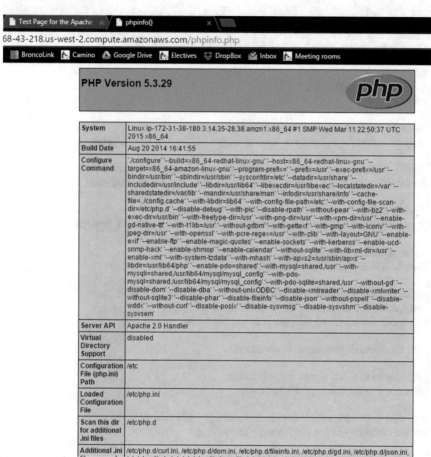

14. Delete the phpinfp.php file for security reasons.

```
[root@ip-172-31-38-180 ec2-user]# rm /var/www/html/phpinfo.php
rm: remove regular file â ~/var/www/html/phpinfo.phpâ ™? y
[root@ip-172-31-38-180 ec2-user]# ls /var/www/html
health.html
[root@ip-172-31-38-180 ec2-user]#
```

15. Start the MySQL server to run mysql_secure_installation.

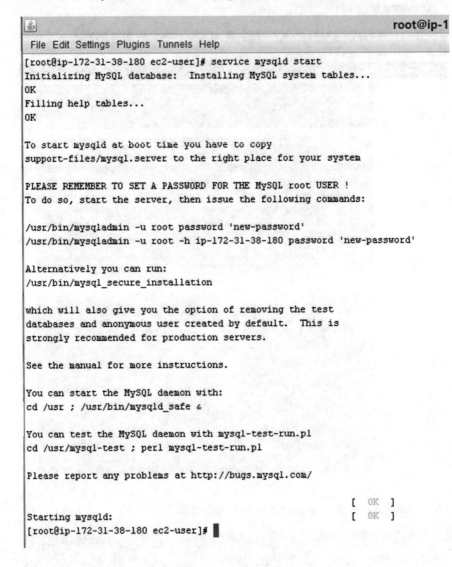

16. Set a strong password.

```
[root@ip-172-31-38-180 ec2-user]# mysql_secure_installation



NOTE: RUNNING ALL PARTS OF THIS SCRIPT IS RECOMMENDED FOR ALL MySQL
      SERVERS IN PRODUCTION USE!  PLEASE READ EACH STEP CAREFULLY!


In order to log into MySQL to secure it, we'll need the current
password for the root user.  If you've just installed MySQL, and
you haven't set the root password yet, the password will be blank,
so you should just press enter here.

Enter current password for root (enter for none):
OK, successfully used password, moving on...

Setting the root password ensures that nobody can log into the MySQL
root user without the proper authorisation.

Set root password? [Y/n] y
New password:
Re-enter new password:
Password updated successfully!
Reloading privilege tables..
 ... Success!


By default, a MySQL installation has an anonymous user, allowing anyone
to log into MySQL without having to have a user account created for
them.  This is intended only for testing, and to make the installation
go a bit smoother.  You should remove them before moving into a
production environment.

Remove anonymous users? [Y/n]
```

```
                                                                    root@i

 File  Edit  Settings  Plugins  Tunnels  Help

By default, a MySQL installation has an anonymous user, allowing anyone
to log into MySQL without having to have a user account created for
them.  This is intended only for testing, and to make the installation
go a bit smoother.  You should remove them before moving into a
production environment.

Remove anonymous users? [Y/n] y
 ... Success!

Normally, root should only be allowed to connect from 'localhost'.  This
ensures that someone cannot guess at the root password from the network.

Disallow root login remotely? [Y/n] y
 ... Success!

By default, MySQL comes with a database named 'test' that anyone can
access.  This is also intended only for testing, and should be removed
before moving into a production environment.

Remove test database and access to it? [Y/n] y
 - Dropping test database...
 ... Success!
 - Removing privileges on test database...
 ... Success!

Reloading the privilege tables will ensure that all changes made so far
will take effect immediately.

Reload privilege tables now? [Y/n] y
 ... Success!

Cleaning up...


All done!  If you've completed all of the above steps, your MySQL
installation should now be secure.

Thanks for using MySQL!


[root@ip-172-31-38-180 ec2-user]# █
```

17. Stop the MySQL server as it will not be used for now.

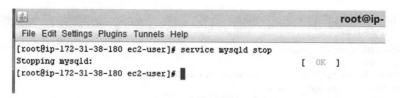

15.1.2 Installing WordPress

1. Download the WordPress.tar file.

```
[ec2-user@ip-172-31-38-180 ~]$ wget https://wordpress.org/latest.tar.gz
--2015-04-18 02:52:59--  https://wordpress.org/latest.tar.gz
Resolving wordpress.org (wordpress.org)... 66.155.40.250, 66.155.40.249
Connecting to wordpress.org (wordpress.org)|66.155.40.250|:443... connected.
HTTP request sent, awaiting response... 200 OK
Length: 6186275 (5.9M) [application/octet-stream]
Saving to: â "latest.tar.gzâ "

latest.tar.gz                    100%[=====================================================

2015-04-18 02:53:00 (17.1 MB/s) - â "latest.tar.gzâ " saved [6186275/6186275]

[ec2-user@ip-172-31-38-180 ~]$
```

2. Unzip this tar file.

```
[ec2-user@ip-172-31-38-180 ~]$ wget https://wordpress.org/latest.tar.gz
--2015-04-18 02:52:59--  https://wordpress.org/latest.tar.gz
Resolving wordpress.org (wordpress.org)... 66.155.40.250, 66.155.40.249
Connecting to wordpress.org (wordpress.org)|66.155.40.250|:443... connected.
HTTP request sent, awaiting response... 200 OK
Length: 6186275 (5.9M) [application/octet-stream]
Saving to: â "latest.tar.gzâ "

latest.tar.gz                              100%[============================:

2015-04-18 02:53:00 (17.1 MB/s) - â "latest.tar.gzâ " saved [6186275/6186275]

[ec2-user@ip-172-31-38-180 ~]$ tar -xzf latest.tar.gz
[ec2-user@ip-172-31-38-180 ~]$ ls
latest.tar.gz  wordpress
[ec2-user@ip-172-31-38-180 ~]$
```

3. Create a MySQL database and user.

```
ec2-user@ip-17
 File Edit Settings Plugins Tunnels Help
[ec2-user@ip-172-31-38-180 ~]$ mysql -u root -p
Enter password:
Welcome to the MySQL monitor.  Commands end with ; or \g.
Your MySQL connection id is 2
Server version: 5.5.42 MySQL Community Server (GPL)

Copyright (c) 2000, 2015, Oracle and/or its affiliates. All rights reserved.

Oracle is a registered trademark of Oracle Corporation and/or its
affiliates. Other names may be trademarks of their respective
owners.

Type 'help;' or '\h' for help. Type '\c' to clear the current input statement.

mysql>
```

```
ec2-user@ip
 File Edit Settings Plugins Tunnels Help
[ec2-user@ip-172-31-38-180 ~]$ mysql -u root -p
Enter password:
Welcome to the MySQL monitor.  Commands end with ; or \g.
Your MySQL connection id is 10
Server version: 5.5.42 MySQL Community Server (GPL)

Copyright (c) 2000, 2015, Oracle and/or its affiliates. All rights reserved.

Oracle is a registered trademark of Oracle Corporation and/or its
affiliates. Other names may be trademarks of their respective
owners.

Type 'help;' or '\h' for help. Type '\c' to clear the current input statement.

mysql> CREATE DATABASE `wordpress-db`;
Query OK, 1 row affected (0.00 sec)

mysql>
```

```
[ec2-user@ip-172-31-38-180 ~]$ mysql -u root -p
Enter password:
Welcome to the MySQL monitor.  Commands end with ; or \g.
Your MySQL connection id is 8
Server version: 5.5.42 MySQL Community Server (GPL)

Copyright (c) 2000, 2015, Oracle and/or its affiliates. All rights reserved.

Oracle is a registered trademark of Oracle Corporation and/or its
affiliates. Other names may be trademarks of their respective
owners.

Type 'help;' or '\h' for help. Type '\c' to clear the current input statement.

mysql> CREATE USER ' user1        '@'localhost' IDENTIFIED BY 'mysql_password@123';
Query OK, 0 rows affected (0.00 sec)

mysql> █
```

4. Grant privileges to the database for the WordPress user.

	ec2-user@ip-172
File Edit Settings Plugins Tunnels Help	

```
[ec2-user@ip-172-31-38-180 ~]$ mysql -u root -p
Enter password:
Welcome to the MySQL monitor.  Commands end with ; or \g.
Your MySQL connection id is 10
Server version: 5.5.42 MySQL Community Server (GPL)

Copyright (c) 2000, 2015, Oracle and/or its affiliates. All rights reserved.

Oracle is a registered trademark of Oracle Corporation and/or its
affiliates. Other names may be trademarks of their respective
owners.

Type 'help;' or '\h' for help. Type '\c' to clear the current input statement.

mysql> CREATE DATABASE `wordpress-db`;
Query OK, 1 row affected (0.00 sec)

mysql> GRANT ALL PRIVILEGES ON `wordpress-db`.* TO " User1        "."@"localhost";
Query OK, 0 rows affected (0.00 sec)

mysql> █
```

5. Screenshots of list of databases, user, and privileges on the database that would be used for WordPress.

• View all users.

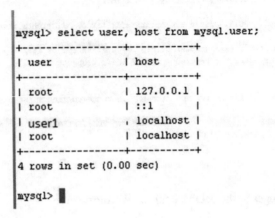

• View contents of the mysql.user table.

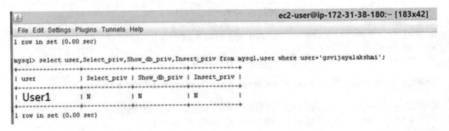

• View all the databases created.

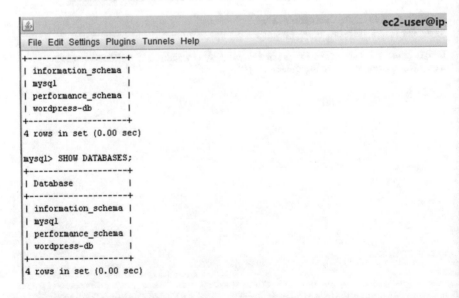

- View the privileges on the newly created database.

```
mysql> SHOW GRANTS FOR "User1" "@"localhost";
+--------------------------------------------------------------------------------------------------------------------+
| Grants for gsvijayalakshmi@localhost                                                                               |
+--------------------------------------------------------------------------------------------------------------------+
| GRANT USAGE ON *.* TO 'User1' @'localhost' IDENTIFIED BY PASSWORD '*1DEE070DFADEB327EA19138FD6DCA6EA9104A04A' |
| GRANT ALL PRIVILEGES ON 'wordpress-db'.* TO 'gsvijayalakshmi'@'localhost'                                          |
+--------------------------------------------------------------------------------------------------------------------+
2 rows in set (0.00 sec)

mysql>
```

6. Exit mysql.

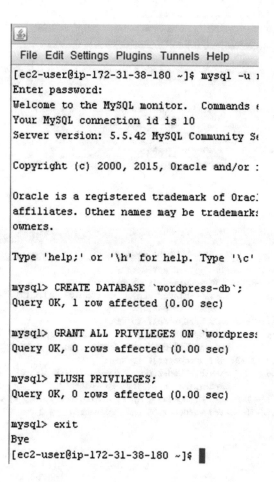

7. Create and edit the wp-config.php file.

```
ec2-user@ip-172-31-38-180:~/wordpress [183x42]
File Edit Settings Plugins Tunnels Help
GNU nano 2.3.1                               File: wp-config.php


/** MySQL database username */
define('DB_USER', 'User1          ');

/** MySQL database password */
define('DB_PASSWORD', 'mysql_password@123');

/** MySQL hostname */
define('DB_HOST', 'localhost');

/** Database Charset to use in creating database tables. */
define('DB_CHARSET', 'utf8');

/** The Database Collate type. Don't change this if in doubt. */
define('DB_COLLATE', '');

/**#@+
 * Authentication Unique Keys and Salts.
 *
 * Change these to different unique phrases!
 * You can generate these using the (@link https://api.wordpress.org/secret-key/1.1/salt/ WordPress.org secret-key service)
 * You can change these at any point in time to invalidate all existing cookies. This will force all users to have to log in again.
 *
 * @since 2.6.0
 */
define('AUTH_KEY',         'FK1YdFFB{!ee$P_tc}8jJ3m!*(h4D]k/T<9dwURttd)(z'#>'tspT}Gaab'URca3');
define('SECURE_AUTH_KEY',  'Kt+&K}-Bdo*'+^M-OI6l(4t&N}/z/wY}2RS ]xW;G1j6G@_Oxx0.J:v)]Xah+&<1');
define('LOGGED_IN_KEY',    'Fe@BOcJ}Ru+'&]anY_Wo#x0!(^igOz7}\Fsi]z:iwoAZ P{CI3Y(^JP!C\Wq+&MzH');
define('NONCE_KEY',        '!/c4&+GNYzz4fAM:%+QRHJz:Tw% L_i2TCz44B(`d(`}x/.mXU}^.C]+YVE(vIdt');
define('AUTH_SALT',        ')80!-*4Rs`sY[z4]}KRPv+*MZP:;}hiP?4Zt2AD)%6+4>IaV-z C:NCkL/7d:2Pn');
define('SECURE_AUTH_SALT', '*< I6{Z6)}ap-_zt@?f3zUYJxG;*Nsz}WK{1-;}QL3Cxg;FAP5qY /74q9FZ%Z,(');
define('LOGGED_IN_SALT',   '/[4_wOy#&8;MMFh =]g>*l8Lq.2S[-+E-mC1*n+786zxtz90*.8HGzZzqOp0]OTx');
define('NONCE_SALT',       'u_gJ+ 1P3B+me*]Iq3CLBOK-M<dzGq=M-Xxp04@xL}qhyC:l[9!!Gp$_-2/dD2/3');

/**#@-*/
```

```
ec2-user@ip-17
File Edit Settings Plugins Tunnels Help
[ec2-user@ip-172-31-38-180 ~]$ cd wordpress/
[ec2-user@ip-172-31-38-180 wordpress]$ cp wp-
wp-activate.php       wp-blog-header.php      wp-config-sample.php   wp-cron.php
wp-admin/             wp-comments-post.php    wp-content/            wp-includes/
[ec2-user@ip-172-31-38-180 wordpress]$ cp wp-
wp-activate.php       wp-blog-header.php      wp-config-sample.php   wp-cron.php
wp-admin/             wp-comments-post.php    wp-content/            wp-includes/
[ec2-user@ip-172-31-38-180 wordpress]$ cp wp-
wp-activate.php       wp-blog-header.php      wp-config-sample.php   wp-cron.php
wp-admin/             wp-comments-post.php    wp-content/            wp-includes/
[ec2-user@ip-172-31-38-180 wordpress]$ cp wp-config-sample.php wp-config.php
[ec2-user@ip-172-31-38-180 wordpress]$ nano wp-config.php █
```

8. Creating a directory called blog and moving all WordPress content into that directory.

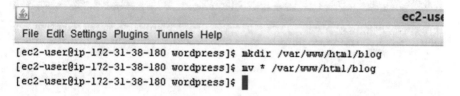

```
[ec2-user@ip-172-31-38-180 wordpress]$ mkdir /var/www/html/blog
[ec2-user@ip-172-31-38-180 wordpress]$ mv * /var/www/html/blog
[ec2-user@ip-172-31-38-180 wordpress]$ ▮
```

9. Fix the file permissions for the Apache Web server.

```
ec2-user@ip-172-31-38-
File  Edit  Settings  Plugins  Tunnels  Help
[ec2-user@ip-172-31-38-180 wordpress]$ mkdir /var/www/html/blog
[ec2-user@ip-172-31-38-180 wordpress]$ mv * /var/www/html/blog
[ec2-user@ip-172-31-38-180 wordpress]$ sudo usermod -a -G www apache
[ec2-user@ip-172-31-38-180 wordpress]$ sudo chown -R apache /var/ww
chown: cannot access â ˜/var/wwâ ™: No such file or directory
[ec2-user@ip-172-31-38-180 wordpress]$ sudo chown -R apache /var/www
[ec2-user@ip-172-31-38-180 wordpress]$ sudo chgrp -R www /var/www
[ec2-user@ip-172-31-38-180 wordpress]$ sudo chmod 2775 /var/www
[ec2-user@ip-172-31-38-180 wordpress]$ find /var/www -type d -exec sudo chmod 2775 {} +
[ec2-user@ip-172-31-38-180 wordpress]$ find /var/www -type f -exec sudo chmod 0664 {} +
[ec2-user@ip-172-31-38-180 wordpress]$ sudo service httpd restart
Stopping httpd:                                        [  OK  ]
Starting httpd:                                        [  OK  ]
[ec2-user@ip-172-31-38-180 wordpress]$ ▮
```

10. Check if all the required services are running.

```
                                                          e
File  Edit  Settings  Plugins  Tunnels  Help
[ec2-user@ip-172-31-38-180 wordpress]$ sudo chkconfig httpd on
[ec2-user@ip-172-31-38-180 wordpress]$ sudo chkconfig mysqld on
[ec2-user@ip-172-31-38-180 wordpress]$ sudo service mysqld status
mysqld (pid 3308) is running...
[ec2-user@ip-172-31-38-180 wordpress]$ sudo service httpd status
httpd (pid 24596) is running...
[ec2-user@ip-172-31-38-180 wordpress]$ ▮
```

11. Copy paste the public DNS link followed by/blog to access the WP config page.

12. Install WordPress.

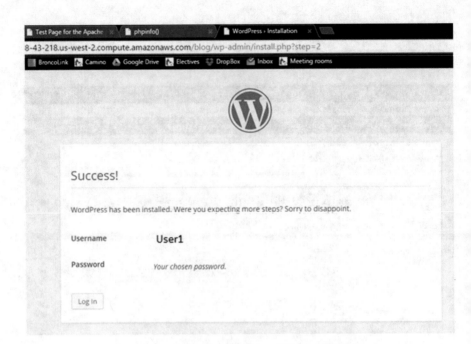

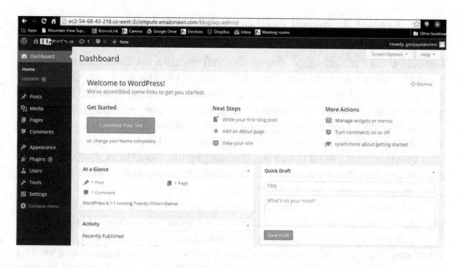

15.1.3   WordPress URL

WordPress Link
http://ec2-54-186-40-103.us-west-2.compute.amazonaws.com/blog/

Uname
xyz

Password
abc

15.2   Project 2: Install PHP on Your AWS Instance

Generally, the "phpinfo.php" file is disabled because it consists of information about the server where our site is currently running on. This file is useful for debugging purposes, but its removal makes the server more secure. It would tell hackers what versions of software are running in the server, thereby enabling hack attacks.

The phpinfo() function is majorly used to check for configuration settings and available system variables and to test if the PHP installation was successful or not. This function is contained within the phpinfo.php file. Sometimes, owners forget to remove this file, thereby exposing information about physical paths in the system, environment variables, and the PHP settings.

The best work-around for ensuring our servers do not get hacked via the phpinfo. php file is to ensure that this file is not used on a production server. It is okay to use it on a development server that has a restricted member access. Also, as an additional step, we can disallow other IPs from accessing it. It is important that we give an obscure name to this file or maybe just delete it after use (e.g., delete this file after Website development and testing are done).

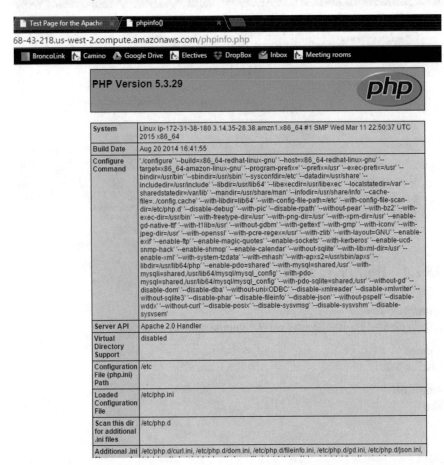

```
                                                                        ec2-user@ip-
File  Edit  Settings  Plugins  Tunnels  Help
[ec2-user@ip-172-31-38-180 ~]$ echo "<?php phpinfo(); ?>" > /var/www/html/phpinfo.php
[ec2-user@ip-172-31-38-180 ~]$ █
```

```
                                                                        root@ip-172-3
File  Edit  Settings  Plugins  Tunnels  Help
[root@ip-172-31-38-180 ec2-user]# rm /var/www/html/phpinfo.php
rm: remove regular file â ~/var/www/html/phpinfo.phpâ ™? y
[root@ip-172-31-38-180 ec2-user]# ls /var/www/html
health.html
[root@ip-172-31-38-180 ec2-user]# █
```

15.3   Project 3: Enhance Security of Your Aws Instance

Amazon security groups are used to provide functionality similar to a "firewall" to safeguard the EC2 servers. These security groups can be used to filter the incoming traffic, also called ingress, on the basis of the chosen protocol (TCP, ICMP, UDP), an IP address or a range of IP addresses, and even ports. It is important that every instance or server configured through AWS is linked to at least one security group.

By default, all the incoming traffic to a server is rejected till the time that server is attached to a security group and that security group should allow traffic requests of that kind. EC2 security group only handles incoming traffic and does not have control over a server's outgoing data (requests/responses).

Security groups are not specific to availability zones but sync with AWS regions. Any user can have more than one security group in their account to a maximum to 500. In the event that a user fails/forgets to attach a security group to their instance, AWS attaches a default security group to that instance.

The default security group opens ports and protocols only to servers that are in the default group.

In the case of an EC2 instance acting as a Web server, the security group must allow the HTTP and the HTTPS protocols.

Every server must be mandatorily attached to a security group.

Once the server is launched, security groups cannot be added or removed from it.

Changes made to the ingress rules are applied immediately on the running servers.

It is a best practice to select different security groups for different levels or sections of a project, unless they have similar functionalities.

It is also a good practice to remove unused security groups or to keep them clean to avoid any untoward incidents.

To allow incoming traffic from the same region (EC2 instance), we need to create a rule with source port = 11.0.0.0/8.

Servers that are connected to different security groups can communicate with one other either via private or public IPs, as long as they are in the same AWS region. Servers belonging to other AWS regions communicate through public IPs.

A security group can be added to itself. In this feature, servers use their private IPs for connecting with other servers through any port and protocol. However, in this case, the rules of the security group will apply only to the servers that have a private IP.

When we add one security group (A) to another (B), servers of A can connect with servers of B through their private IPs, through a specific set of ports and protocols.

When a user makes use of EIP (Elastic IP), security groups can be used to allow requests only from this IP to be sent to another server.

VPC: Virtual Private Cloud are dedicated to the user's AWS account. They cater to a single region but span over multiple availability zones. However, the scope of the subnet in use is just one availability zone.

The user has control over the methods used by instances to access information that are outside of the VPC.

The VPC security group can be used for controlling both incoming (ingress) and outgoing (egress) network traffic.

We can have separate rules set for the ingress and egress traffic.

It is always a best practice to reserve IP addresses for both instances and subnets for further expansion across availability regions.

It is always better to place subnets alongside the tiers (e.g., ELB, Web/app).

To have all the data in private subnets by default is another good practice, keeping only the ELB or other filters in public subnets.

IAM should be used for assigning users and also for setting their access privileges.

It is also important to define subnet routes and tables and VPC security groups.

To sum up, VPC provides extra security and isolates your network.

Differences between security groups and firewall:

- Firewalls control network traffic between subnets of networks or between different networks.
- They are provided by either vendors or open sourced.
- AWS security groups belong to Amazon. They are easier to manage than firewalls.
- Firewalls require manual updates whenever the IP address of a server changes or when a new set of servers are added as a cluster so that the users will be able to receive traffic from all these servers.
- AWS security groups make use of policies wherein more than one server can reference the same set of policies. This makes managing updates easier as it lowers the configuration error rates.
- For most part, firewalls and AWS security groups provide the same functionalities.

15.4   Project 4: Set Up a Load Balancer for Your AWS Instance

15.4.1   Elastic Load Balancer Setup

1. Enable two instances for the load balancer.
2. Ensure that the security groups of the instances must be covered by the load balancer.

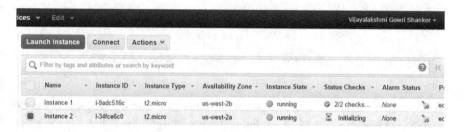

3. Click on load balancer on the left panel.

4. Define a name for the load balancer. Also define the http port.

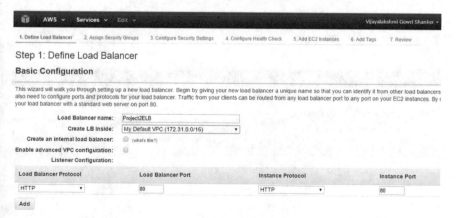

5. Choose all those security groups that are associated with the instances that would be serviced by this load balancer.

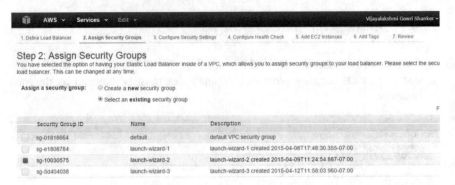

6. Configure the health check parameters. Make sure that the ping path (health. html) is actually present inside the /var/www/html folder. If not, the instances would be deemed out of service.

7.

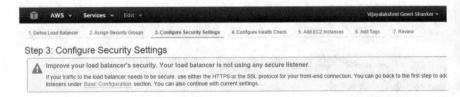

8. Add the required instances to this LB.

9. Add tags.

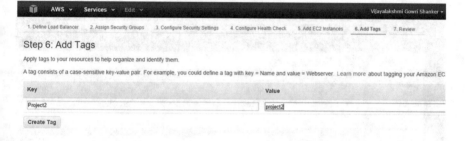

10. Review the specifications.

11. View LB creation message.

12. The two instances are initially listed as Out of Service.

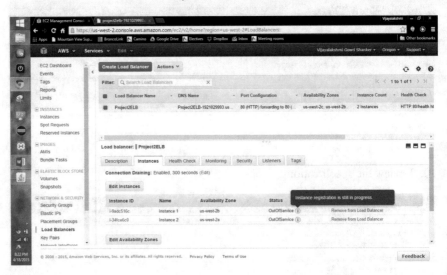

13. Wait for some time. Once the health.html file is accessed across both the instances, we get the two instances to be in service.

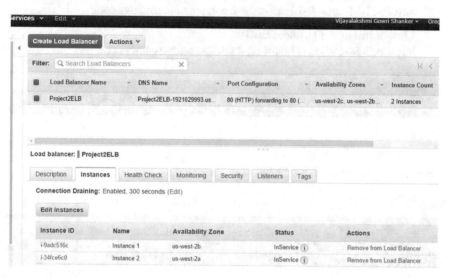

14. URL of the project with the load balancer:
    http://project2elb-1921029993.us-west-2.elb.amazonaws.com/

15. URL of the project without ELB:
Link of Instance 1 (without ELB):
http://ec2-54-186-40-103.us-west-2.compute.amazonaws.com/

Link of Instance 2 (without ELB):
http://ec2-54-186-51-77.us-west-2.compute.amazonaws.com/

Hi

Hey. I am from Instance 1. My health is fine. Welcome!

Hi

Hey. I am from Instance 2. My health is fine. Welcome!

15.4.2   Unique Features of AWS Load Balancer

- ELBs distribute traffic from the Web across all the tagged EC2 instances that reside on one or more than one availability region.
- ELB makes use of the DNS link to distribute traffic.
- *Cross-zone load balancing*: Prevents some instances from receiving a higher than average amount of inbound request traffic. This feature ensures that the ELB distributes requests to all instances equally irrespective of the zones they are located in.
- *Connection draining*: This feature serves existing connections on a deregistered instance for the configured timeout period. If an instance is removed from the ELB network due to faulty behavior or maintenance, the end user will not experience any glitch on the services rendered by the ELB.
- *ELB access logs*: This feature was made available recently and can go a long way in debugging. Users can therefore identify and rectify any issues they may face at all levels: instance, ELB, zones, etc.
- *Security:* Employs SSL termination in addition to the security groups provided by Amazon.
- *Sticky session or session affinity*: Helps connect the application to a user's session. As a result, all of the user's requests are sent to the same application.

15.5 Project 5: Use Elastic IP for Your AWS Instance

15.5.1 How to Make an Instance Elastic

Auto scaling is generally used to make our instances in the Cloud structure elastic. Typically, elasticity of an instance or a structure means its ability to expand or contract (in terms of resources) depending on the current network demand. And, this process should be automatic without the user having to intervene to either scale up or scale down.

Auto scaling has two major parts. One is launch configuration: this tells us how to create new instances. Here, we get to choose the name and type of AMI, storage, etc. Second is the scaling groups: this tells us what the rules and regulations for the scaling procedure would be. AWS provides many functions which can be used to scale our infrastructure on the basis of network traffic or other parameters.

Auto scaling is also used to determine the health of the instances created and define checkpoints based on which we can have more/less number of instances (notifications or alarms). Once an instance is deemed out of service (not working), auto-scaling function will take that instance out of its group and will create a new instance in its place.

If we use the Elastic Load Balancer to distribute traffic among the various instances tied to it, ELB also functions as an elastic device. But this does not work in conditions that have huge traffic spikes.

Auto scaling works in that under this, no instance can be classified as indispensable. This will help us to avoid storing information on any of the instances. Also, the launch or the termination of any new or existing instance is automated, thereby saving a lot of time for owners of the infrastructure.

15.5.2 Extra: Elastic IP

An EC2 instance currently has a public DNS that is used to access or view the Website content. However, the DNS setting changes with every reboot. These public IPs are that not available in abundance, thereby making it difficult to allocate a dedicated IP address for every instance. This is where Elastic IPs figure in. The running instance can be easily accessed by these elastic IPs.

1. Ensure that the instance that is to be made elastic is running.
2. Go to the Amazon EC2 console. Click on "Elastic IPs" under Network & Security.
3. Click on "Allocate new address" that can be used in a VPC.
4. Select EC2-VPC under the network platform list. Click on "Yes, Allocate."
5. Then click on "Associate address" to link your running instance with this IP address.

6. To change the association of an Elastic IP to another instance, we have to first "Disassociate Address" from the current instance and then tag this IP to the new instance.
7. When switching between AMIs, we have to stop the current instance, which would automatically disassociate the Elastic IP from this instance. We then start a new instance and then associate the Elastic IP to the latest instance.
8. The new Elastic IP address becomes the public DNS for that instance. To access the Web page, we need to copy this link onto URL tab.

The command/syntax to associate an EC2 instance to an elastic IP is:
ec2-associate-address [-i *instance_id*| -n *interface_id*] [*ip_address*| -a *allocation_id*] [--private-ip-address *private_ip_address*] [--allow-reassociation]

This command outputs a table that has:

(a) An address identifier
(b) Elastic IP (EIP)
(c) Instance to which the EIP has been associated
(d) An allocation ID
(e) A private IP linked to the EIP (if specified)

15.5.3   *Bonus*

Auto scaling is generally done when the current load on the existing instances/servers goes above their thresholds. This may lead to delays in processing. To avoid this, we can deploy an auto-scaling mechanism wherein we can scale out (add instances) or scale in (remove instances) depending on our current needs. This process of scaling is automated.

We can set the conditions as to when to scale in or scale out, the configurations of the new instances that will be added, get notifications to your email id whenever a new instance is created, or if an existing instance gets terminated.

We can also specify the minimum, maximum, and the desired number of instances that are required in a group. Also, we have the option of tagging a load balancer to this auto-scaling group.

Some screenshots are included to depict the steps followed to set up a functioning auto-scaling group.

We can either have a load balancer initially set up, or we can create a load balancer once auto-scaling instances have been created and then add these instances to the LB (but this requires manual addition).

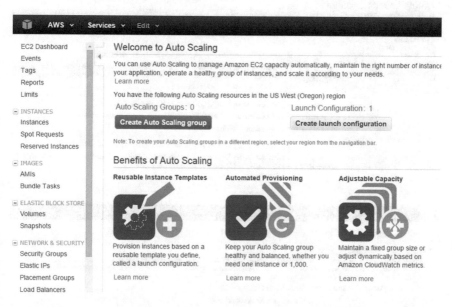

Create Auto Scaling Group

To create an Auto Scaling group, you will first need to choose a template that your Auto Scaling group will use when it launches instances for you, called a launch configuration. Choose a launch configuration or create a new one, and then apply it to your group.

Later, if you want to use a different template, you can create another launch configuration and apply it to this group, even if you already have instances running in it. Using this method, you can update the software that your group uses when it launches new instances.

- ◉ **Create a new launch configuration**

- ○ **Create an Auto Scaling group from an existing launch configuration**

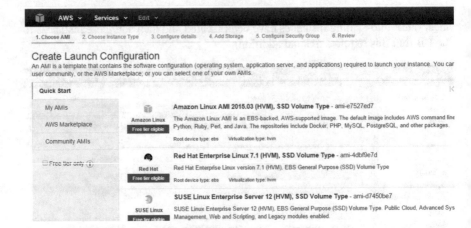

Create Launch Configuration

An AMI is a template that contains the software configuration (operating system, application server, and applications) required to launch your instance. You can select an AMI provided by AWS, our user community, or the AWS Marketplace; or you can select one of your own AMIs.

Quick Start		
My AMIs	Amazon Linux Free tier eligible	**Amazon Linux AMI 2015.03 (HVM), SSD Volume Type** - ami-e7527ed7 The Amazon Linux AMI is an EBS-backed, AWS-supported image. The default image includes AWS command line Python, Ruby, Perl, and Java. The repositories include Docker, PHP, MySQL, PostgreSQL, and other packages. Root device type: ebs Virtualization type: hvm
AWS Marketplace		
Community AMIs	Red Hat Free tier eligible	**Red Hat Enterprise Linux 7.1 (HVM), SSD Volume Type** - ami-4dbf9e7d Red Hat Enterprise Linux version 7.1 (HVM), EBS General Purpose (SSD) Volume Type Root device type: ebs Virtualization type: hvm
Free tier only		
	SUSE Linux Free tier eligible	**SUSE Linux Enterprise Server 12 (HVM), SSD Volume Type** - ami-d7450be7 SUSE Linux Enterprise Server 12 (HVM), EBS General Purpose (SSD) Volume Type. Public Cloud, Advanced Sys Management, Web and Scripting, and Legacy modules enabled.

Create Launch Configuration

A security group is a set of firewall rules that control the traffic for your instance. On this page, you can add rules to allow specific traffic to reach your instance. For example, if you want to set up a web server and allow Internet traffic to reach your instance, add rules that allow unrestricted access to the HTTP and HTTPS ports. You can create a new security group below. Learn more about Amazon EC2 security groups.

Assign a security group: ○ Create a **new** security group

● Select an **existing** security group

	Security Group ID	Name	VPC ID	Description
	sg-dd5f53b8	default	vpc-2451eb41	default VPC security group
■	sg-fa2d219f	launch-wizard-1	vpc-2451eb41	launch-wizard-1 created 2015-04-19T12:35:59.549-07:00

Inbound rules for sg-fa2d219f Selected security groups: sg-fa2d219f.

Type ⓘ	Protocol ⓘ	Port Range ⓘ	Source
SSH	TCP	22	0.0.0.0

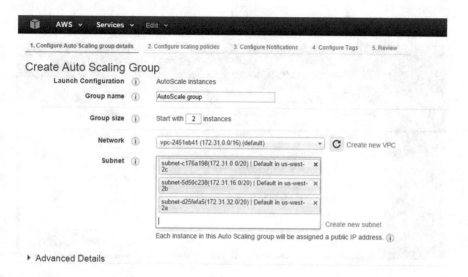

Creating alarm for increasing group size:

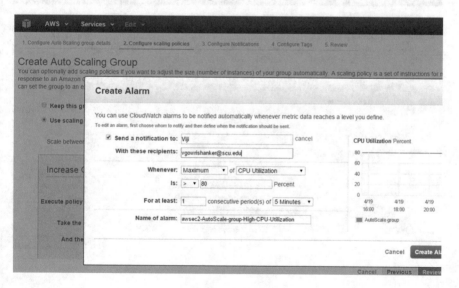

Creating alarm for decreasing group size

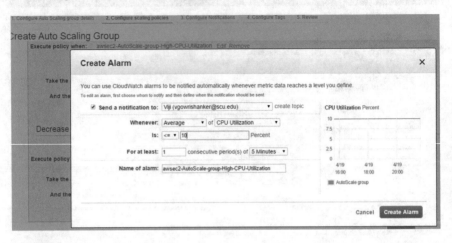

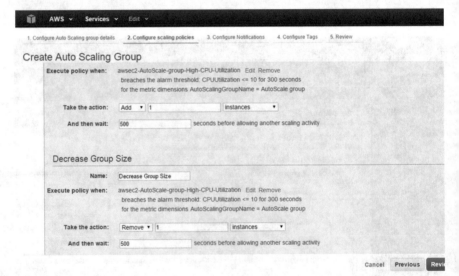

AWS ∨   Services ∨   Edit ∨

1. Configure Auto Scaling group details   2. Configure scaling policies   **3. Configure Notifications**   4. Configure Tags   5. Review

Create Auto Scaling Group

Configure your Auto Scaling group to send notifications to a specified endpoint, such as an email address, whenever a specified event takes place, including: su
instance launch, instance termination, and failed instance termination.

If you created a new topic, check your email for a confirmation message and click the included link to confirm your subscription. Notifications can only be sent to

Send a notification to: Viji (vgowrishanker@scu.edu)   ▼   create topic

Whenever instances: ☑ launch
                       ☑ terminate
                       ☑ fail to launch
                       ☑ fail to terminate

Add notification

New instances getting automatically created due to auto scaling

We can also connect these newly created instances.

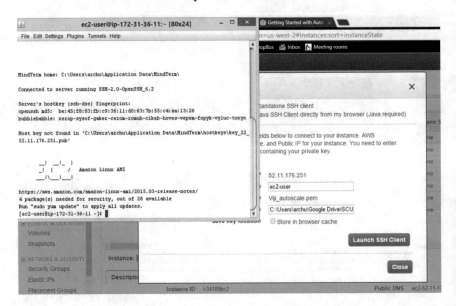

We can also add a load balancer to this group.

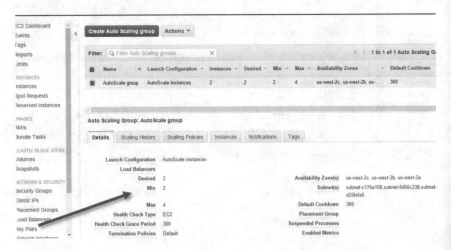

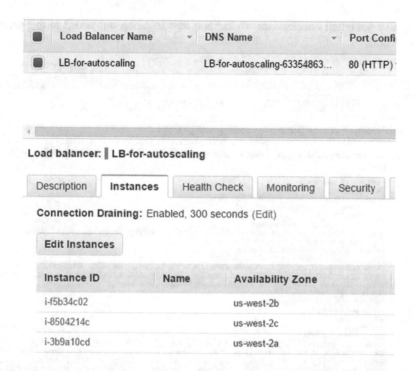

We can also see that when the load on the CPU increases, a new instance gets created automatically. The maximum limit of this group has been set to four instances.

I also got email notifications whenever a new instance was added to this group (scale out):

Auto Scaling: launch for group "AutoScale group"   Inbox   x

AWS Notifications no-reply@sns.amazonaws.com via amazonses.com   2:56 PM (6 minutes ago)
to me

Service: AWS Auto Scaling
Time: 2015-04-19T21:56:39.951Z
RequestId: dc2c1c53-d995-4d00-9337-7feb07dcbc05
Event: autoscaling:EC2_INSTANCE_LAUNCH
AccountId: 927490047934
AutoScalingGroupName: AutoScale group
AutoScalingGroupARN: arn:aws:autoscaling:us-west-2:927490047934:autoScalingGroup:f84d1b33-7605-439b-b6c0-
26a89008c10b:autoScalingGroupName/AutoScale group
ActivityId: dc2c1c53-d995-4d00-9337-7feb07dcbc05
Description: Launching a new EC2 instance: i-9a735653
Cause: At 2015-04-19T21:56:04Z a monitor alarm awsec2-AutoScale-group-High-CPU-Utilization in state ALARM triggered
Increase Group Size changing the desired capacity from 3 to 4. At 2015-04-19T21:56:06Z an instance was started in respon
a difference between desired and actual capacity, increasing the capacity from 3 to 4.
StartTime: 2015-04-19T21:56:08.038Z
EndTime: 2015-04-19T21:56:39.951Z
StatusCode: InProgress
StatusMessage:
Progress: 50
EC2InstanceId: i-9a735653
Details: {"InvokingAlarms":[{"Region":"US-West-2","AWSAccountId":"927490047934","OldStateValue":
"ALARM","AlarmName":"awsec2-AutoScale-group-High-CPU-Utilization","AlarmDescription":null,"NewStateReason":"Thresh
Crossed: 1 datapoint (0.4713333333333333) was less than or equal to the threshold (10.0).","StateChangeTime":
1429479844437,"NewStateValue":"ALARM","Trigger":{"Period":300,"Statistic":"AVERAGE","MetricName":"CPUUtilization",
"Threshold":10,"EvaluationPeriods":1,"Dimensions":[{"name":"AutoScalingGroupName","value":"AutoScale
group"}],"Namespace":"AWS/EC2","ComparisonOperator":"LessThanOrEqualToThreshold","Unit":null}}],"Availability Zone":"
west-2c","Subnet ID":"subnet-c176a198"}

Email notification when an instance is terminated from this group (scale in):

Service: AWS Auto Scaling
Time: 2015-04-19T22:05:50.650Z
RequestId: 191473fd-917b-4591-9914-a85f9989da69
Event: autoscaling:EC2_INSTANCE_TERMINATE
AccountId: 927490047934
AutoScalingGroupName: AutoScale group
AutoScalingGroupARN: arn:aws:autoscaling:us-west-2:927490047934:autoScalingGroup:f84d1b33-7605-439b-b6c0-
26a89008c10b:autoScalingGroupName/AutoScale group
ActivityId: 191473fd-917b-4591-9914-a85f9989da69
Description: Terminating EC2 instance: i-4676538f
Cause: At 2015-04-19T22:05:04Z a monitor alarm awsec2-AutoScale-group-High-CPU-Utilization in state ALARM triggered policy
Decrease Group Size changing the desired capacity from 4 to 3. At 2015-04-19T22:05:07Z an instance was taken out of service
in response to a difference between desired and actual capacity, shrinking the capacity from 4 to 3. At 2015-04-19T22:05:07Z
instance i-4676538f was selected for termination.
StartTime: 2015-04-19T22:05:07.698Z
EndTime: 2015-04-19T22:05:50.650Z
StatusCode: InProgress
StatusMessage:
Progress: 50
EC2InstanceId: i-4676538f
Details: {"InvokingAlarms":[{"Region":"US-West-2","AWSAccountId":"927490047934","OldStateValue":
"ALARM","AlarmName":"awsec2-AutoScale-group-High-CPU-Utilization","AlarmDescription":null,"NewStateReason":"Threshold
Crossed: 1 datapoint (0.866111111111111) was less than or equal to the threshold (10.0).","StateChangeTime":
1429479844437,"NewStateValue":"ALARM","Trigger":{"Period":300,"Statistic":"AVERAGE","MetricName":"CPUUtilization","
Threshold":10,"EvaluationPeriods":1,"Dimensions":[{"name":"AutoScalingGroupName","value":"AutoScale
group"}],"Namespace":"AWS/EC2","ComparisonOperator":"LessThanOrEqualToThreshold","Unit":null}}],"Availability Zone":"us-
west-2c","Subnet ID":"subnet-c176a198"}

If you wish to stop receiving notifications from this topic, please click or visit the link below to unsubscribe:

If we delete the auto-scale group, all the instances that were created as a result of
this group also get terminated (not just stopped).

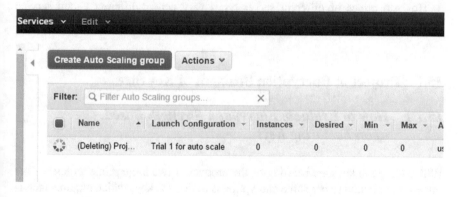

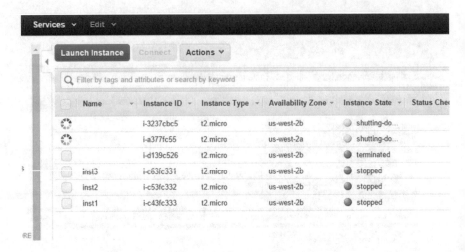

We can also view the scaling history to see the list of instances that were either launched or terminated:

15.6   Points to Ponder

1. AWS offers several attractive services, not required by NIST, such as Elastic IP and auto scaling. Can you think of a few more examples?
2. What equivalent services are offered by other Cloud service providers, such as Microsoft Azure and Google's GCP?
3. How do prices of different server types vary across different Cloud service providers?

15.7   Project 6: Encryption Strength of Key Size

15.7.1   How to Evaluate Encryption Strength Based upon Factoring Keys

RSA encryption keys are based upon the product of two large prime numbers. One brute force method to break the encryption is to find the keys which requires factoring the large number that is part of the key pair. Evaluation of encryption strength of key size for end-to-end security for user data between multiple clients and a Cloud server can be done by measuring factoring effort vs. key size.

Using your favorite programming language environment, implement any of the factoring algorithms found on the Web. Also, find a table of large prime numbers. Multiply two of the prime numbers, and then apply your chosen factoring algorithm.

Of course, you already have the answer because you created the number in the first place. Plot the run time vs. the size of the keys to evaluate how strong a given key is against the brute force factoring attack.

Example Websites for factoring:

http://mathforum.org/library/drmath/view/65801.html
https://projecteuclid.org/download/pdf_1/euclid.bams/1183495051
https://www.alpertron.com.ar/ECM.HTM

Example Websites for large prime numbers:

https://www.bigprimes.net/archive/prime/
https://primes.utm.edu/largest.html

Appendix A

Appendix A: Points to Ponder

Each of the previous chapters has a list of points to ponder, as an additional food for thought. Students are expected to come up with their own answers first, after reading each chapter. Then these can be compared with authors' proposed responses.

Chapter #1: Points to Ponder

1. *Cloud Computing has enjoyed double-digit yearly growth over the last decade, driven mainly by economics of Public Cloud vs. enterprise IT operations. Can you think of three economic reasons why Public Cloud Computing is cheaper than an enterprise DC?*

    (a) Three reasons are capital expenditure reduction, operational expenditure reduction, and compute elasticity.

    (b) Cloud Computing growth has been driven mainly due to reduced CapEx (capital expenditure, e.g., cost of servers and other data center equipment). Servers are shared among different customers, and hardware costs are amortized over multiple tenants.

    (c) Cloud also reduces OpEx (operational expenditure, e.g., power to run the servers and cool them, salaries for IT resources), due to multi-tenancy.

    (d) Lastly, compute elasticity refers to the flexibility and fast reaction time to provision additional capacity in the Cloud (of the order of seconds or minutes). This is due to higher available capacity than placing orders to buy new servers (hours), waiting for their arrival (days), and then deploying them in a private IT data center.

© Springer Nature Switzerland AG 2020
N. K. Sehgal et al., *Cloud Computing with Security*,
https://doi.org/10.1007/978-3-030-24612-9

2. *Concentration of data from various sources in a few data centers increases the risk of data corruption or security breaches. How can this possibility be minimized?*

   (a) Having lots of data at one place makes it an attractive target of malicious hackers. So additional care should be taken to ensure the confidentiality of data with encryption.

   (b) Integrity of data needs bit-level error detection with frequent checkpoints. This will help to identify data corruption and restore it to a previously known good state.

   (c) The above steps are needed in addition to restricting physical and electronics access to the location of this shared data. It should be accessible to authorized personnel only.

3. *Under what circumstances an enterprise should opt to use a Public Cloud vs. its own Private Cloud facilities?*

   (a) Cost is clearly a major consideration to use any Public Cloud. It imposes little to no capital expenditure burden on the users and only charges them what they use.

   (b) Second consideration is if the data being placed in a Public Cloud is not mission critical, e.g., its access time and performance have some latitude, then it is okay to rely on shared facilities.

   (c) Lastly, resource availability within an enterprise, e.g., if the data and compute requirements vary greatly over time then elastic nature of Public Cloud is attractive.

4. *Why does a Public Cloud provider want to support a Hybrid Cloud, instead of asking enterprises to use its public facilities?*

   (a) Any Public Cloud provider, who wants to attract new business from an established enterprise, recognizes the hesitation of an IT manager due to loss of control.

   (b) Thus, offering an on-premise compute service owned and remotely managed by the Cloud provider forms an attractive way to try a new usage model.

   (c) An option to extend into a Public Cloud for backups is attractive due to its low cost and overheads, making hybrid model a success.

5. *Is there a scenario when Private Cloud may be cheaper than the Public Cloud?*

   (a) Yes, in a rare circumstance if a private data center can be kept fully loaded 24×7 (i.e., 24 hours a day for all 7 days a week), then its total cost of ownership may be less than using a Public Cloud.

6. *Can you think of scenarios where a Cloud user favors (a) PaaS over IaaS PaaS, and (b) SaaS over PaaS?*

   (a) Depending on the nature of workloads and applications, one or more of Public Cloud usage models are useful.

(b) Customers who want to work with a bare bone server or closer to hardware prefer IaaS. They can specify number of CPU cores, size of memory, etc.

(c) Users that want to use operating system facilities prefer PaaS, such as different storage systems and databases.

(d) Companies that want to focus on their end-user applications often go with SaaS, e.g., Netflix and Salesforce.

7. *Think of ways in which you can detect if a DOS attack is underway?*

(a) A denial-of-service attack creates an overcrowding effect on a victim's network or server, resulting in the victim not being able to serve its legitimate users.

(b) User activities are regularly monitored and a normal behavior pattern is documented. Any uncharacteristic usage or traffic is flagged.

(c) Such algorithms are often termed as anomaly detection, as an attack represents a deviation from the normal usage patterns. An example is too many Website requests coming in from the same IP address.

Chapter #2: Points to Ponder

1. *What led to the desire of people with PCs to connect with one another?*

(a) Professionals wanted to share data and programs with other users.

(b) People wanted to share music with their friends.

(c) Socially motivated sharing of content, e.g., pictures, stories, and news between family and friends, gave rise to social networks. This in turn gave rise to sites such as Facebook. This phenomenon brought many new consumers and devices to the Cloud connected networks and databases.

2. *What led to the growth of thin clients? How thick should be a thick client, and how thin should be thin clients?*

(a) Previously mentioned trend of people wanting to stay connected at all times through various means, such as while travelling on vacation with a smartphone but not carrying a PC, meant that phones evolved to have a larger screen and mobile applications. However, phones have limited compute capability due to smaller form factors and users' desire to make the phone charge last a full day. This is the essence of a thin client. Another example of thin clients is an airport display terminal with little or no local processing or storage capability.

(b) A laptop PC is a thick client due to its larger form factor, as compared to a smartphone. However, a laptop should be comfortable to carry around, have a reasonable battery life, and not get overly hot during its operation.

(c) Thickness of a thin client depends on the usage model for each category (e.g., an information panel at the airport needs a larger display vs. an enterprise handheld computer in the hands of an Amazon parcel delivery driver needs more storage and larger battery vs. a smartphone to fit in the vacation going tourist's pocket). Besides usage, software update frequency and security requirements are also a consideration. Enterprises typically need encryption on their mobile handheld devices to protect confidential data, which needs more compute and battery power.

3. *How SOA has helped with the evolution of Cloud Computing?*

   (a) If a service is well defined, then a series of services can be combined. Service-oriented architecture defines inputs, outputs, and transformations carried for each service, a series of which can be chained to offer a compounded service. An example is requesting a virtual machine to run on a certain type of server, then provisioning it with certain OS and apps, and lastly deploying it to accomplish desired results. This is the essence of IaaS in a Public Cloud, upon which then PaaS and SaaS service models are constructed using SoA as a guiding principle.

4. *Cloud Computing is a natural evolutionary step as communication links grew in capacity and reliability*

5. *Even though cookies pose security risks, why do browsers allow them, and what's the downside for a user to not accept cookies?*

   (a) Cookies are like a dual-edged sword. They store valuable information to give the appearance of continuity for repeat visitors of a public Website, by storing passwords, shopping carts, and other user preference data on a user's local machine. However, in case someone else can access the cookies, he/she will be able to see this data. Not accepting any cookies will mean that browser will not have any memory, causing some inconvenience to the users. So they are a necessary evil, but cookies need to be cleaned up via a browser to save storage space.

6. *What's the minimal precaution a public Website's user should take?*

   (a) A user needs to ensure that any secure Website starts with https://, wherein the last "s" means secure. Also, if user arrives at a Website by clicking on a link through an email or text message, then the spellings need to be checked to ensure that it is the correct service provider's site instead of a phishing attack to steal financial information.

7. *What are the trade-offs of securing information during transmission?*

   (a) During transmission, any sensitive information such as user's passwords, credit card information, or any sensitive data needs to be protected. This information may be travelling over public nodes, which are susceptible to a man-in-the-middle attack. It can be protected with encryption. However, the process of encryption by the sender, and decryption by the receiver, requires time and compute resources. Hence, an appropriate trade-off must be done on the need and level of encryption requirements, e.g., 128 bits vs. 256 bits.

Chapter #3: Points to Ponder

1. *Small and Medium Business (SMB) users lack financial muscle to negotiate individual SLAs with Cloud service providers. So what are their options?*

   (a) SMB users should analyze their needs and look for a better fit before moving their entire datasets to a Public Cloud. Reason is that vendor lock-ins are expensive and users may lack in-house capability to juggle data between multiple Cloud service providers. SMB customers want to maintain a backup copy of their data in-house, to deal with immediate business needs, so hybrid computing is a good option to consider.

2. *Some potential Cloud users are on the fence due to vendor lock-in concerns. How can these be addressed?*

   (a) Some potential users may be on the fence, as they are apprehensive about making a choice. Also, they are concerned about not being able to migrate to another Cloud service provider in future. A Cloud service provider needs to gain the trust of new users to win their business. Perhaps this can be done by offering to co-locate customers' data in multiple data centers and giving them an ability to convert from proprietary Cloud data formats to other industry-acceptable standards, etc. This will assure customers that their data is not locked in any specific Cloud service provider location or format.

3. *What are the new business opportunities that enable movement of users and data between different Cloud providers?*

   (a) OpenStack was started with such an aim to use only open-source software components to build a complete Cloud solution, so users can move between different providers with similar offerings. However, the difficulty of implementing and maintaining such a stack, with no single software vendor to support, has made its industry acceptance harder. This presents opportunities for new solution providers to help customers migrate their data from one Public Cloud vendor to another.

4. *What are some key differences between the concerns of a Private vs. Public Cloud's IT managers?*

   (a) An enterprise's IT manager is generally aware of the users' workloads and thus can place them optimally. In contrast, a Public Cloud's IT manager may not be able to look into users' workload content due to privacy concerns.

   (b) Secondly, the amount of trust is higher within an enterprise, since all employees work for the same company, whereas in a Public Cloud, an incoming user may be a hacker, so extra caution is warranted in terms of resource usage, workloads' behavior, etc.

5. *Why is virtualization important for Public Cloud? Would it also help in Private Cloud?*

(a) By definition, Public Cloud's servers are shared between different users, so virtualization provides a convenient solution to isolate these users.

(b) In a Private Cloud, all users may belong to the same enterprise, so it is acceptable to mix their workloads on the same machine without paying any performance overhead due to virtualization. However, this requires the users to use the same operating system.

(c) An alternative to virtualization is containers, which allow for process isolation between different programs and their datasets on the same operating system.

6. *Is workload migration a viable solution for a large Public Cloud service provider?*

(a) Workload migration is an excellent way to do load balancing between servers.

(b) However, in a Public Cloud, the number of servers is very large, and their workloads have an uncertain lifetime; migration is of questionable value unless done at a local level, e.g., in a rack of servers.

7. *Why is NIST cyber security framework applicable to Public Cloud?*

(a) Even though users of a public Cloud may be more worried about the security of their data, the Cloud managers care about the security of their infrastructure as well as the security of their customers' data. This situation is akin to the operators of a public airline, who can't assume that all passengers have good intentions. Thus security checking is done for anyone wishing to fly on a commercial plan. Similarly, the Public Cloud managers are on a constant looking to avert any threats in their data center, potentially coming from insiders, hackers, or their own customers.

Chapter #4: Points to Ponder

1. *Multi-tenancy drives load variations in a Public Cloud. Different users running different types of workloads on the same server can cause performance variations. How can you predict and minimize the undesirable effects of performance variability?*

(a) Analytics can be used to predict periodic patterns of workloads, e.g., for enterprise users, weekdays may be busy, while for recreational users weekdays may use more compute on the weekends. For example, Netflix customers watch more movies on the weekend using Amazon's data centers. Their jobs can be scheduled to run at complementary times, to balance the overall server utilization in a Cloud.

2. *How can you counter the undesirable effects of performance variability?*

(a) A user can submit jobs when the systems are less loaded and more available. Also, user needs to monitor long-running jobs to ensure that they are getting the compute resources necessary or else may need to check point (i.e., save the state of a running program) and migrate the job to another server or Cloud.

3. *How is performance variability addressed in enterprise data centers?*

(a) Enterprises can earmark different server types for running different types of jobs, e.g., a pool of graphics engines for video rendering applications.
(b) Also, an enterprise can restrict when certain type of jobs can be run, e.g., Software QA only during the nights or weekends with fewer interactive users. This may not be always possible in a Public Cloud.

4. *What is a good way to assess if a noisy neighbor problem exists?*

(a) It is nontrivial for a Public Cloud user to know if her job is slowing down due to a noisy neighbor problem, because IT manager isolates each customer's tasks. However, when these tasks use any shared resource, such as a memory controller, then tasks may slow down due to resource contention. This can be detected by careful monitoring and logging the time it takes for individual tasks at different times. A comparison will reveal if the same tasks, such as a memory access time, is increasing or decreasing in a substantial manner. This change can be attributed to noisy neighbors.

5. *How can a noisy neighbor problem be prevented?*

(a) In a Public Cloud, once a persistent noisy neighbor presence is detected, it is beyond the scope of an individual user to avoid it. Reason is that one user has no control over another user's tasks. However, it can be avoided by stopping the task and requesting a different machine. This may interrupt the service, so a better way is to start another server in parallel and migrate the task in a seamless manner, if possible.

6. *How can understanding a Public Cloud's usage patterns help?*

(a) Observing the performance of a server over time, via a monitoring agent, can often reveal its loading patterns that may be cyclic. An example is to find out when other users on the same server in a Public Cloud are running their jobs and if possible to avoid delays by scheduling one's own tasks at a different time. Time is money, and in a pay per use model, one can save both time and money by scheduling tasks at a time when Cloud servers are lightly loaded.

7. *What operating systems-related strategies can a user consider to optimize Cloud usage?*

(a) A Cloud customer needs to know the isolation technique used by the service provider. If Hyper-V is used as a VMM, then it is better suited for running Windows-based virtual machines. If Xen or KVM is used, then Linux-based VMs will run faster. Else, putting a Linux VM on top of Windows-based

VMM will result in a multilayered call stack that may cause unexpected latencies. In any case, user should do their own benchmarking in a Cloud environment before committing production workloads and making any long-term migration plans.

Chapter #5: Points to Ponder

1. *Many applications combine different workload types, can you give an example?*

    (a) Yes, sometimes an application exhibits different behaviors during its different stages. For example, an online geographically mapping service used by mobile GPS users needs to read one or more large files with map data. This involves accessing storage servers. Then an in-memory graph is built on which different users can search for their traffic routes. This involves compute servers. Lastly, users can query GPS, which needs I/O and networking capabilities to support thousands of users in the Cloud.

    (b) Similar scenario can play for other social networking sites too, as users upload or watch pictures or video, post comments, or hold a chat etc.

2. *Different applications in a Public Cloud can offer opportunities to load balance and optimize resource usage, can you give some examples?*

    (a) Compare workload posed by a social networking application vs. a video playing site. Each requires different kinds of computational resources.

    (b) A video movie playback involves transfer of a large file over a few hours, possibly doing real-time decompression, needing some compute but a large networking bandwidth.

    (c) A social networking site has a large number of customers, engaging in small data interactions via text messages or photos being shared.

3. *What is the best way to improve fault tolerance with multitier architectures?*

    (a) By having multiple sources of data and compute at each tier in a solution stack, the probability of a system-level failure due to any component-level failure is reduced. An example is of database servers, where the same information can be duplicated making any single server redundant.

4. *What is the advantage of characterizing your workload in a Cloud?*

    (a) From a Public Cloud user's point of view, if you understand the nature of your workload, it will help you to pick the right configuration of server to rent. For example, if it is CPU-intensive, then pick a stronger compute machine vs. a server with larger memory for a memory-intensive workload such as for Big Data.

5. *From a Cloud service provider's perspective, how the knowledge of a workload's characterization may help?*

(a) As an IT manager for a Cloud data center, one can use the knowledge of customers' workloads to properly mix them on shared hardware, such that a balanced resource loading can be achieved. An example is to put CPU-intensive job from a customer with memory-intensive task from another. This will ensure that both the CPU and memory controller can be kept busy, instead of having multiple tasks wait for CPU to free up.

6. *Can machine learning play a role to improve the efficiency of a Cloud data center?*

(a) A Public Cloud service provider may agree to not look into customers' workloads for preserving confidentiality. However, one may look at the resource usage of individual customers' tasks to profile them and predict in advance the future needs. This will help to optimally place them on the right type of servers and other hardware accelerators, such as graphic engines or faster numerical machines etc.

Chapter #6: Points to Ponder

1. *Distance causes network latency, so AWS has set up various regional data centers to be close to its customers and reduce delays. This is one way to mitigate network latency. Are there other methods you can think of?*

(a) Last mile bandwidth is often the cause of latency, where traffic within a neighborhood causes the congestion over short distances.
(b) Depending on the nature of data, a service provider can use mirror sites or edge-based servers at locations closer to the consumers.
(c) Netflix uses such a technology, known as CDN (content delivery network), to serve its movies. This works well if the data being delivered is read-only in nature; otherwise synchronization problems will arise.

2. *In an AWS region, there are different availability zones (AZs) to isolate any electrical fault or unintentional downtime. These could be different buildings or different floors within a data center building. What are other ways to build a failure-safe strategy?*

(a) Another method is to replicate the content across different servers located in different geographies and then use network switches to route the incoming traffic for both load balancing and fault tolerance to a lightly loaded site.

3. *What kind of metrics need be monitored to ensure the health of a data center?*

(a) Metrics are of two types, a DOA (dead or alive) type of test, which involves regular pings to the compute and storage servers, and secondly performance tests which are needed to ensure that all critical components are optimally loaded.

4. *What are the trade-offs of frequent vs. infrequent monitoring of a server's performance?*

   (a) There is no free lunch in life, so any monitoring agent will need some CPU time and memory to run. This causes an additional loading of a server, which, if done too often, may result in a perceptible slowdown.

5. *How long the monitoring data should be kept and for what purposes?*

   (a) Assuming that Cloud server monitoring has a value beyond instantaneous decision, such as to immediate load balancing, then the result of monitoring should be saved in a log file for later offline analysis. That will help to reveal patterns of usage and enable cost optimizations with future scheduling decisions.

6. *What's the role of monitoring if a SLA is in place?*

   (a) Even if a performance-based SLA (service-level agreement) is in place, its compliance needs to be ensured by the CSP's (Cloud server provider) IT managers and Public Cloud users. This can only happen with regular monitoring and comparing actuals with the promised performance, such as the CPU cycles, memory bandwidth, etc. These may translate to an application's observed performance, such as database read-write transactions per second, etc. Any gaps will result in renegotiation of a SLA.

7. *What are the security concerns of Follow-Me Cloud?*

   (a) Connectivity for a Follow-Me Cloud (FMC) user migrates from one data center to another, in response to physical movements of that user's equipment. Backend implementation involves moving user's information between different data centers, so this information needs to be protected. If user has any encrypted data in the Cloud, then appropriate care should be taken to migrate the keys separately, so that the confidentiality and integrity of data are maintained. Secondly, the infrastructure of Cloud data center is accessible by the middle-layer protocols. IT managers need to ensure that the middle-tier gateways are not used to launch denial-of-service (DOS) attacks on the Cloud.

Chapter #7: Points to Ponder

1. *Networking gear, such as Ethernet cards, represents the first line of defense in case of a cyberattack. How can it be strengthened?*

   (a) When a denial-of-service (DOS) type of attack happens, a lot of packets of the same type, often from a same or smaller group of IP addresses, arrive at a data center. Such a pattern can be detected via packet sniffing algorithm, for which a watchdog agent can be running as an observer at the network entry point. If a pattern is detected, then the offending IP addresses can be blocked, and system administrator is notified.

2. *Recent Storage crash in an AWS Public Cloud was caused by the scripting error of internal admins, how could it have been prevented?*

   (a) Human errors are generally the cause of most crashes, including those in a data center. These can be minimized with an input validator. A recent AWS crash was caused by a scripting error. Such an issue can be detected in advance, if the script is run through a simulator or filter first, and if a wild card is found, then ask the IT person if they intend to launch a broader operation. If the answer is affirmative, then they are allowed to proceed. This will not totally eliminate the errors but will reduce them.

3. *Does having multiple customers in a Public Cloud increase its risk profile? What operating system strategies you would recommend to prevent business impact?*

   (a) Yes, with multiple customers in a data center, impact of any downtime increases, as any failure will hit multiple businesses simultaneously.
   (b) By having fault tolerance in operating systems, or by using isolation technologies such as virtual machines, failure from any customer can be contained and not be allowed to spread to other users' virtual machines.

4. *Explain the role of attack surface management to minimize security risks? How would you reduce an attack surface?*

   (a) An attack surface represents domains in the Cloud that hackers can gain access to, for making unauthorized changes. This may lead to a DOS (denial-of-service) attack for other users. An IT manager must minimize the attack surface available to hackers. An example is the recent Equifax breach, where hackers gained access to valuable personal information of 143 million customers using vulnerability in the Apache server software. Interesting fact is that the vulnerability was publicly known and Equifax was informed months before but failed to apply the patch.
   (b) Reducing the number of access ports, adding authentication and encryption on the remaining few access points can shrink an attack surface. An example is servers at a bank with no USB ports and verifying any new software upgrades coming from legitimate sources before uploading and installing new drivers, firmware, etc. Authentication keys should be stored in a secure place.

5. *How the regular monitoring of a server's usage activity may help to detect a security attack?*

   (a) An IT manager should know the regular usage patterns of servers in a data center and ought to monitor for any irregular activities, e.g., unusual amount of traffic coming on certain network ports or a high amount of read-write activities on certain storage servers. Then an alarm can be raised for manual checks to verify if such an activity is legitimate and coming from authorized sources. In a Public Cloud, this means checking the IP address of incoming traffic and if necessary blocking it to prevent an attack in progress.

6. *What are the security concerns due to a residual footprint?*

(a) Whenever a virtual machine, a container, or a program ceases to run, it may leave behind some memory and disk footprint. This is in the form of data or code being used by that task. The operating system doesn't zero out the address space, but puts it on the available list for the next task to use. That next task may start reading the binary content of its memory or disk allocation, thereby getting information about the previous task, which is equivalent to unauthorized access. The previous program can prevent it by zeroing out its memory before exiting.

7. *What type of applications will not benefit from block chain?*

(a) Block chain is best for applications with many distributed participants. They may benefit from transparency and a distributed ledger. It is used to avoid attackers changing a single master copy of the records. However, applications that represent a single or few users, such as the three co-authors of this book collaborating to write the book, would not have benefitted from block chain. Other applications involving people taking online exams on a one-on-one basis will not benefit.

Chapter #8: Points to Ponder

1. *Migrating to a Public Cloud may result in IT savings, but are there any potential pitfalls?*

(a) Yes, as an old saying goes, there is no free lunch in life. Public Cloud savings are as a result of many customers' tasks consolidation on a fewer physical servers. One impact is that loading by other customers will slow down the machine such that a customer may experience higher or lower performance independent of its own loading of that machine. This in turn will determine how well that one customer's jobs can run, and the response time for its customers, impacting their business.

2. *Business continuity is critical for many mission-critical operations in an enterprise. Can multiple Cloud providers be used to improve BCP (Business Continuity Practices)?*

(a) If business continuity is important, then that business can't afford any single point of failure, including a Public Cloud. Such businesses can mitigate their risks by spreading their data and jobs, sometimes with redundant servers in different Cloud. Alternatively, they can keep a minimal set of critical data in-house with a local server, which will keep the business running even if the external Cloud goes down.

3. *If local computing is needed, e.g., when Internet connections are not reliable, then what are the other options?*

   (a) A Hybrid Cloud usage model is desirable for on premise applications and data.
   (b) Other options are check pointing to save the state of databases and virtual machines. Take their backups often, and rehearse disaster recovery before it is needed.

4. *Under the following scenario, would an IT provider prefer a Private or Public Cloud?*

   (a) Should a B2C IT service provider prefer a Private or Public Cloud?

   A B2C (business to customers) generally maintains an internal data center in a Private Cloud, e.g., eBay, but may benefit from a Public Cloud. A part of the data is already open to many customers, such as items being auctioned in the case of eBay, but other data such as customers' credit card information is strictly private. Later may be the motivation for eBay to maintain a private data center. However, a fluctuation in the compute load during weekends and holidays may call for a Public Cloud with a hybrid usage model.

   (b) Should a B2B IT service provider prefer a Private or Public Cloud?

   A B2B (business to business) supports interactions between partners, which are not open to public. Due to its private nature of these transactions, generally a Private Cloud would be preferred. However, there may be a special case to use the Public Cloud hosting based on economic factors, but security considerations should be enumerated and agreed upon in advance via a SLA (service-level agreement).

   (c) Should a C2C IT service provider prefer a Private or Public Cloud?

   C2C (consumer to consumer) is the best case for using a Public Cloud, but the most prevalent site such as Facebook maintains its own data center to keep a control on the data and performance aspects. However, other C2C users may prefer a Public Cloud to maintain a focus on their application and let someone else take care of their IT needs.

5. *Can you think of a case where a business tried a Cloud but decided to not adopt it?*

   (a) Yes, several solution providers in EDA (electronic design automation) industry considered using a Cloud but, as we will explain in the next chapter, decided to not adopt it.

6. *What factors may lead a business to consider a hybrid model?*

   (a) If a business is located in a location where Internet connections are not reliable, it may lead to a loss of access to the data and programs in a Public Cloud. In such a situation, the business owners will prefer on-premise computing but may still wish to use the Cloud for backing up their data or using on-demand compute. Besides reliability, other factors such as latency of long-distance networks and desire to maintain control with some local compute may lead a business to consider Hybrid Cloud models.

7. *What kinds of security attacks are possible in an IoT-based Cloud?*

   (a) Unlike a Private or Public Cloud infrastructure located behind a firewall, an IoT-based Cloud has many devices exposed in potentially hostile environment.
   (b) Many IoT networks have wireless or over-the-air (OTA) access points.
   (c) IoT devices can be often easily accessed and modified to launch a Distributed Denial of Service (DDOS) attack on any unsuspecting server. There have been recent examples of attacks with Mirai bots.
   (d) Mirai is derived from a Japanese word, meaning "future." It also refers to malware that repurposes networked devices running Linux into remotely controlled "bots", which can be used as a part of a botnet to launch a large-scale network attacks.
   (e) Recent targets of such attacks were consumer devices, e.g., IP cameras and home routers.
   (f) Mirai botnet infection code uses common factory usernames and passwords for the victim devices, such as admin and admin, which many users may carelessly forget to change.
   (g) Other attacks can be on the server side if the backend site or its master database is not sufficiently protected.

Chapter #9: Points to Ponder

1. *For mission-critical application, such as EDA or health data, what precautions need to be taken? EDA companies want to recoup R&D investments faster by selling private licenses, while their customers are hesitant to put confidential design data in a Public Cloud. Given this situation, would it ever make sense to migrate EDA workloads to Public Cloud?*

   (a) Mission-critical applications need timely access to the data and must protect its confidentiality and integrity.
   (b) A similar business situation existed when most semiconductor companies owned their own fabrication plants (aka fabs) and manufactured their own silicon chips. As the cost of fabs kept rising, only a few design companies could afford to have their own fabs. This drove most semiconductor companies to go fabless and send their designs to third-party fabrication plants. Their concerns of confidentiality were superseded by economic needs. Similarly, the cost of EDA tools and need to maintain one's own data centers may cause design companies to adopt third-party data centers. Of course, their business confidentiality concerns will need to be addressed first.

2. *How do IP concerns influence the decision to move to Cloud? What precaution-ary measures need to be considered?*

(a) Intellectual property protection is often cited as the top reason for EDA com-panies to avoid Public Cloud. This can be addressed with strict regulations providing confidentiality and privacy safeguards, similar to what healthcare industry is adopting before using Public Cloud.

3. *Are there any other enterprises or Private Cloud customers who faced similar dilemmas but economics enabled their migration to Public Cloud?*

(a) Consider Salesforce customers, or medical records management, or health-care providers.

(b) Most companies treat their customer's business relationships as a top secret, but high cost of in-house customer relationship management (CRM) tools drove many companies to adopt an online, Public Cloud-based solution by Salesforce.com. Currently, this Cloud company's business is growing and sustainable, causing in-house commercial tool providers to adopt Cloud too.

4. *What can Public Cloud service providers do to accelerate migration of EDA workloads?*

(a) Public Cloud vendors can offer economic incentives, well-defined SLAs, and above all some sort of insurance against data leakage accidents.

5. *What benefits can an EDA solution provider get by migrating to a Public Cloud?*

(a) fdAn EDA solution provider can focus more on their domain-specific appli-cation and leave the IT management to the CSP (Cloud service provider). Furthermore, they can pass on or share any cost savings with their custom-ers, lowering the overall cost of doing business similar to many other Cloud-based application providers.

6. *What benefits can EDA customers get by migrating to a Public Cloud?*

(a) EDA customers are looking for potential cost savings and faster software updates, both of which are possible in the Cloud. However, EDA customers and software will need to migrate their design data and code to a Cloud. Considerations for which were discussed in the Chap. 9.

7. *What are the challenges in block chain adoption for EDA industry?*

(a) EDA industry is very sensitive about IP (intellectual property) rights, whether it is for the tools or design domain. Transparency of BC solutions is #1 challenge for EDA adoption. While the content of blocks is encrypted and protected, all users on a network can see the occurrence of transactions. Many participants in EDA may not want others to see which design library cells they are using.

(b) One way the EDA industry may find BC useful is in a Private Cloud setting. A large silicon design company has multiple sites located in different geog-

raphies, with many different design teams. BC record keeping will be helpful to keep a record of changes and track evolution of design database.

Chapter #10: Points to Ponder

1. *NIST has well-defined Cloud Standards; compare how Amazon, Google, and
   Microsoft Public Cloud offerings comply with NIST standards?*

   (a) NIST refers to Cloud Computing elasticity, cost, and automation provided
       by the service providers. Since Public Cloud solutions are evolving rapidly,
       readers can compare various vendors along these lines.

2. *Amazon is innovating many additional services to attract and keep its EC2 customers. Give some examples.*

   (a) Two recent examples are AMR (Amazon's MapReduce) solution and analyt-
       ics services. More details can be found on AWS Website.

3. *OpenStack has been using open-source code to build interoperable Cloud solutions, but its adoption has been slow, why?*

   (a) OpenStack is a combination of several mix-n-match open-source-based
       tools, with no single supplier or entity to support the whole stack. Thus
       installation, deployment, and maintenance tasks are left to the users, who
       may instead want to focus more on running their own business. They may
       find OpenStack hard to use and use turnkey Public Cloud solutions.

4. *A startup wants to save money, and their tasks require 4 GB memory. Should it
   rent a small-size VM (virtual machine) that costs less, but supports only 2 GB, or
   a medium size that costs double?*

   (a) Even though a small-size VM is cheaper, using it to run tasks that need extra
       memory will slow it down. This is due to page faults in an operating system,
       which causes existing memory to be swapped out, written to a disk, and new
       data to be brought in. This is the only way a VM with 2 GB size can run tasks
       that require 4 GB. Constant swapping in and out will take extra time, and if
       the rental cost of a VM is time-based, then over time it will be more expen-
       sive to rent a smaller VM than the one that can run the tasks without page
       faults. Note that swapping will also use system resources, which will impact
       other VMs too causing a noisy neighbor behavior. An IT manager can detect
       this and evict the culprit VM.

5. *What's the value of a constant IP address for a Website hosted in the Cloud?*

   (a) A Cloud user's VM may be moved to different machines, each having a dif-
       ferent IP address. This may be confusing for the customers of that VM as
       DNS (Domain Naming Server) mapping for the public Website will be to the

old IP address, Amazon Web service offers Elastic IP facility to maintain a constant IP address, which can be used to reach a VM independent of the physical server it is hosted on.

6. *Under what circumstance will you prefer to use a load balancer vs. an auto scaler in a Cloud?*

   (a) A load balancer acts as a reverse proxy and distributes network or application traffic across a number of servers, all of which must be up and always running, while an auto scaler is a service to start a new server when the existing ones are getting overloaded. It typically works by monitoring the compute usage, and then it takes a few minutes for new server to get ready. While this saves money, because one pays for a new server only when needed, it can't handle rapid load fluctuations. That will require additional servers to be up and ready, to pick the incoming load almost instantaneously. Load balancer is generally more expensive than using an auto scaler.

7. *What criterion you would use to compare and contrast different Public Cloud service providers?*

   (a) A comparison between different Public CSPs (Cloud service providers) may start with a cost-benefit analysis, e.g., what kind of service they offer and how much they charge for it. However, making a decision solely on the basis of ROI (Return on Investment) will lead the users to later surprises, so be sure to look at a CSP's latency, reliability, customer support, and above all reputation before making a decision. It is easy to get into a Public Cloud but hard to migrate between CSPs due to their different tool sets and data formats. The migration should also be considered for the sake of BCP (Business Continuity Practices).

Chapter #11: Points to Ponder

1. *How can Web-browser cookies be both helpful and harmful at the same time?*

   (a) Cookies are used to store state on a user's local machine. These help the browser to maintain persistent context across Web sessions. Cookies can contain information such as user authentication, personalization, usage tracking, etc. However, cookies do not provide integrity protection, e.g., a network attacker can rewrite secure cookies and alter their content.

2. *What is security misconfiguration and how can it be avoided?*

   (a) Security of an organization is as good as the weakest link in the organization. For example, an employee losing an unencrypted work laptop, or someone logging in from outside hacking a weak password, can get access to the critical business data. Thus, it is important for an organization to have a good security policy, which needs to be adhered to in its implementation. It needs to be defined and deployed for the applications, frameworks, application server, Web server, database server, and platforms. Secure settings should be defined, implemented, and maintained, as defaults are often insecure. Additionally, software versions should be kept up to date with latest patches.

3. *How does edge computing expand the attack surface?*

   (a) A typical data center has a firewall that prevents outside attackers from getting in. However, IoT devices in the field may not have the same level of protection. Since these devices often have a direct connection back to the data center, someone can compromise a device to launch a denial-of-service attack on the server. Hackers using home security surveillance cameras have overwhelmed data center servers with Web-based requests. This represents an expansion of attack surface, which needs to be mitigated.

4. *Why is mutual trust between IoT devices important?*

   (a) One way to mitigate the expanded attack surface is by devices in the field checking up on themselves and each other. It can be as simple as a built-in self-test (BIST) to authenticate the versions of various software drivers loaded on the machine during each boot cycle. A device may exchange predetermined hash keys with its neighbors, which can be compared with the stored values to determine if any device has been compromised. Then such a device can be isolated or reported back to the central server for further action.

5. *Why trust is a dynamic entity?*

   (a) As in real life, trust must be earned and kept. It may take a long time to earn and still be lost easily with a single action. This is so because a hacker can compromise a device anytime, and even though it was okay in the past, it should be isolated and reported as soon as the attack is detected.

6. *Why devices in series tend to have weaker security?*

   (a) Security is all about the weakest link in a chain; thus any single compromised element in a series may cause the whole chain to fail. An example is of a communication passing through public nodes. The network packets can be sniffed and used to decipher the messages being passed. Thus, not only the end points but also the entire communication chain must be protected. Messages can be secured by protecting each node, or by encrypting the message part of the packets.

7. *How do devices in parallel improve the overall system security?*

   (a) Parallel devices represent redundancy, e.g., in a storage system, two copies of the same file may be stored. If one is corrupted or accidentally deleted, then other copies may be used. This gives integrity protection, but not the confidentiality.

8. *Do IoT devices pose a higher risk than servers in the Cloud, if so why?*

   (a) IoT devices are often located in the field with easier physical access, as compared to the servers in Cloud located behind a firewall with controlled access. Furthermore, a weak compute power on an IoT device limits the security checks that can be performed. Many times, consumers tend to leave the easy to guess default passwords unchanged on IoT devices, such as home-based surveillance cameras, which makes it easier for an attacker to hijack a weakly protected device.

9. *What are the risks associated with voice authentication-based Cloud access systems?*

   (a) Voice can be easily recorded and synthesized to match the intended speaker. Hence using voice recognition as a sole measure to grant access is risk prone. Multifactor authentication using voice recognition, a password, and the incoming device's authentication (such as the phone number used to call in) reduces the possibility of a successful security attack.

Chapter #12: Points to Ponder

1. *Who owns the datasets in Big Data?*

   (a) This is a tricky question, with no clear answer. In case of retail, a consumer owns the choice and financial transactions, but a retailer can combine data from different consumers and remove any personalization information to use it as an aggregated trend. Thus, Amazon is able to suggest what other books customers view or buy after purchasing a certain book. Similarly, a patient owns the medical records, but a hospital or doctor can use them for deriving generalized trends. Then, in turn, these can be sold to a pharmaceutical company to decide which drugs to develop. In all these cases, if data is collected and computed in a Cloud, then clearly the Cloud service provider (CSP) doesn't own this data.

2. *What is desirable for external researchers to collaborate for data analytics in Cloud?*

   (a) Different parties can get together and collaborate in a Cloud, each bringing a unique value to the partnership. As an extension of medical example in the previous question, a hospital can provide medical records devoid of patient's personal identification information, a pharmaceutical company can bring the results of latest drug trials, and multiple medical researchers can bring their ideas and algorithms for future drug or medical device inventions. Any such collaboration needs to assure that no party can copy or take out the data belonging to another party, only use and learn from it. This is called privacy-preserving analytics and is still an evolving field in the Cloud Computing.

3. *Are there any vendor lock-in concerns with Big Data?*

   (a) Yes, since each Public Cloud service provider may use a proprietary format and tools to manage the Big Data, it will be hard, if not impossible, for data owners to quickly migrate their compute to another Cloud. One way to manage this is to start with a multi-Cloud strategy or maintain copies locally.

4. *AWS offer several attractive services, not required by NIST, such as Elastic IP and auto scaling. Can you think of a few more examples?*

   (a) Two recent examples are AMR (Amazon's MapReduce) solution and analytics services. Details can be found on AWS Website.

5. *What equivalent services do other Cloud service providers, such as Microsoft Azure and Google's GCP, offer?*

   (a) Left for reader to discover as the Cloud industry is rapidly evolving.

6. *How do prices of different server types vary across different Cloud service providers?*

   (a) Left for reader to discover as the Cloud industry is rapidly evolving.

7. *Why Public Cloud is an attractive destination to run ML tasks?*

   (a) Machine learning task of training is a CPU-intensive activity, while prediction usually is smaller workload. Since Public Cloud represents zero initial capital expenditure, and customers pay as they use, it forms an attractive location to run the ML training workloads. Detection generally needs to be done locally, so a Public Cloud is not suitable, but prediction models can be run in an offline mode on a Public Cloud.

Chapter #13: Points to Ponder

1. *There is potential to have more devices and machines in an increasingly automated world, and next wave of Cloud Computing growth is coming from IoT. Can you list additional areas to drive the growth of Cloud Computing?*

   (a) Yes, self-driving cars represent a large growth area for Cloud Computing to share data and analytics and update software or instructions in real time.
   (b) Home and industrial surveillance using security cameras offer growth potential for Cloud, with any intrusion data stored in a remote Cloud.
   (c) Pollution monitoring and environmental sensors located in the field with central data collection will help to fuel the growth of Cloud.

2. *How could one improve the Cloud's performance and support for IoT?*

   (a) By having distributed and redundant systems for a failsafe solution.
   (b) Avoid having a single point of failure.

(c) Backend Cloud Services are needed to log data and results for audit and machine learning inferences.

(d) Sensors can generate enormous data requiring Cloud storage and compute power. However, moving data in and out of Cloud is slow and expensive. So input-output considerations will require local compute and storage power.

3. *Why is edge computing needed for self-driven cars in the future?*

(a) Sensors in a moving car can generate enormous data requiring Cloud storage and compute power. Examples of this are forward-looking and side-view cameras. However, moving data in and out of Cloud is slow and expensive. A car may need to react quickly due to changing road conditions. So input-output considerations for the sensor data will require local compute and storage power. However, any learning and performance data can be reconciled with backend servers during night or when the car is safely parked.

4. *Can you think of another example of edge computing devices on a road?*

(a) A network of traffic lights can communicate and coordinate between them, for adjusting to an accident that may be sending scores of cars to other roads. Normally, such a scenario can cause bottlenecks, while other roads may be empty. Using a combination of artificial intelligence and machine learning methods, in the future a network of traffic lights may be able to adapt their sequence of red-yellow-green sequence to ease the wait times on critical junctions.

5. *What is the trust and security model for edge devices?*

(a) Edge devices in a Cloud need backend Cloud services that are needed to log data and results for audit and machine learning inferences. However, devices need to trust the Cloud servers, and Cloud needs to trust the incoming device data.

6. *What kinds of attacks are possible using IoT and edge devices?*

(a) It has been shown that an army of botnets (a term used for devices on the Internet) can be hijacked by hackers and used for launching DDOS (Distributed Denial of Service) attacks on unsuspecting Cloud servers. An example is of home surveillance cameras that had unsecured IP addresses used for bringing down a security journalist's blog site.

7. *What is the impact of vendor lock-in on edge computing devices?*

(a) Vendor lock-in represents difficulty of migrating from one device maker or service provider to another. This can lead to higher costs or inefficiencies. An example is security cameras. If a business has initially installed surveillance cameras from a vendor, then as the business needs expand, it often must buy additional equipment from the same vendor. However, later on there may be another provider with better or cheaper cameras, but these may not work with the existing system. So the business must spend more money or settle for older models just to ensure that the system works in a homogenous manner. This can be avoided if industry standards exist, and only equipment complying with these standards is deployed to ensure a plug-n-play approach. An example of this is in home entertainment systems, where TVs, DVDs, and set-top boxes can come from different manufacturers yet can work together well.

8. *Can hardware be the sole root of trust?*

   (a) As we previously noted, multifactor authentication offers a better defense strategy. Having any single piece of hardware or software as the sole root of trust is risky. One possible solution is mutual attestation by various devices that are not located at the same place or are not under the same control. Thus an attacker will need to simultaneously compromise multiple devices, which is harder to accomplish than altering any single root of trust.

9. *Who owns data in a Secure Multi Party Cloud (SMPC)?*

   (a) No single party owns the entire dataset in a SMPC environment, as each contributes a subset for the common good. All participants have right to use others' data for their computations and can only extract results in an agreed-upon output format. Any personally identifiable information (PII) is removed from the output.

Appendix B: Additional Considerations for Cloud Computing

Since Cloud Computing field is wide and rapidly evolving, not every topic of relevance could be covered in the previous chapters. We picked seven important topics and describe them here with references for further reading: backups, Cloud Accelerators, checkpointing, disaster recovery, fault tolerance, Linux Containers, watchdog, Encryption, keys-based security, and Network Function Virtualization.

1. *Backups:* Backup refers to the copying of physical or virtual files or databases to a secondary site for preservation in case of equipment failure or other catastrophe [1]. Process of backing up data is pivotal to a successful disaster recovery (DR).

2. *Checkpointing:* Checkpointing is a technique [2] to add fault tolerance into computing systems. It basically consists of saving a snapshot of the application's state, so that it can restart from that point in case of failure. This is particularly important for long-running applications that are executed in failure-prone computing systems. Checkpointing enables the system to be restored to a known good configuration.

3. *Cloud Accelerators:* As Cloud applications become prevalent and specialized such as face recognition, or real-time traffic management, run-time efficiency becomes even more important. When using a VM or container, a stack of drivers and operating system services may add to the workload latency. Some Public Cloud service providers have started to use FPGAs (field-programmable gate arrays) to accelerate specialized applications. Both of these technologies allow mapping an algorithm to a hardware implementation, which enables a higher execution speed as compared to running the same application in software. Amazon's AWS is offering F1 as compute instances with FPGAs to create custom hardware acceleration [3]. Similarly, Microsoft has been deploying FPGAs in Azure servers [4], to create a Cloud that can be reconfigured to optimize a set of applications and functions. Unlike a CPU that can be rapidly programmed and run-time binaries that can be switched instantly, an FPGA takes time to be programmed for a specific application. Thus it needs to be running for a longer time to amortize the time and cost of setting up FPGAs, but in

© Springer Nature Switzerland AG 2020
N. K. Sehgal et al., *Cloud Computing with Security*,
https://doi.org/10.1007/978-3-030-24612-9

return the FPGA will run faster than the software-only implementation of an application. In contrast, Google has been using TPUs (tensor processing units) with ASICs (application-specific integrated circuits) for machine learning (ML) workloads [5]. This field is still evolving.

4. *Development and Operations (DevOps):* Deploying applications in an enterprise or Cloud is nontrivial. It may involve configuration changes in server and network, setting up environment variables, monitoring and collecting performance logs, and finally a method to roll out new releases. When development engineers (in short Dev) work closely with IT operation engineers (in short Ops), then the automation and repeatability of the following steps are called DevOps:

   1. Planning
   2. Coding
   3. Building
   4. Testing
   5. Releasing
   6. Deployment
   7. Operating
   8. Monitoring
   9. A loop back to step back for planning next release

   A successful DevOps cycle requires the following steps, for which several tools, such as Puppet [6] and Chef [7], are available:

   1. Infrastructure management
   2. Configuration management
   3. Deployment automation
   4. Log management
   5. Performance management
   6. Monitoring and corrective actions

   Finally, DevOps is a set of principles and practices, both technical and cultural, which can help deploy better software faster.

5. *Disaster Recovery:* Disaster recovery involves a set of policies, procedures, and tools to enable the recovery or continuation of vital technology infrastructure and systems following a natural or human-induced disaster. Disaster recovery focuses on the IT or technology systems supporting critical business functions and is a subset of business continuity [8].

6. *Function as a Service (FaaS):* A new category of Cloud Computing service is growing fast that allows customers to develop and run applications as functions [9], without worrying about the underlying server architecture. Examples of this are AWS Lambda, Google Cloud Functions, and Microsoft Azure Functions. In a FaaS model, functions are event-driven and expected to start in milliseconds, while customer only pays for the execution time of a function. An example of a function is image loading in response to a Website click, or an event triggered by a sensor reading, such as storing a video footage in Cloud upon motion detection by a security camera.

7. *Fault Tolerance:* Fault tolerance is the property that enables a system to continue operating properly in the event of the failure of (or one or more faults within) some of its components. If its operating quality decreases at all, the decrease depends on the severity of the failure, as compared to a naively designed system in which even a small failure can cause total breakdown. Fault tolerance is particularly sought after in high-availability or life-critical systems. The ability of maintaining functionality when portions of a system break down is referred to as graceful degradation [10].

8. *Linux Containers:* Linux Containers enable packaging of applications, including their entire run-time files necessary, in an isolated environment. This enables it easy to move the contained application from development to test and finally to production while retaining full functionality. Containers [11] are especially helpful with distribution of self-contained application packages, as described above in DevOps. Each container has code, a run-time environment, system tools, and libraries, all of which are packaged for a complete distribution and installation on a server. A container has short lifetime, if used as elastic compute resources, as a disposable unit to be spun up or shut down as the demand grows or wanes. Cloud providers are increasingly exploring containers, instead of or in addition to VMs, for deploying individual customers' applications on a shared server. One clear advantage is that containers are considered lightweight in terms of memory or run-time requirements as compared to a VM. This allows more containers, as compared to fewer VMs, to run on a given server platform. Putting an application in a container enables it to be secure from other malwares, as it has its own address space that an outside application cannot access. A possible downside is that the same operating system is shared between multiple containers, increasing the possibility of noisy neighbors. In contrast, as shown in Fig. B.1, virtualization allows a mix of different operating systems on the same server and better control of resource allocation between the VMs.

Linux Containers are based on open-source technology [12], such as CRI-O, Kubernetes, and Docker, making it easier for teams to develop and deploy applications. These technologies are also referred to as an orchestrator, which manages multiple application containers across a cluster of servers and keeps

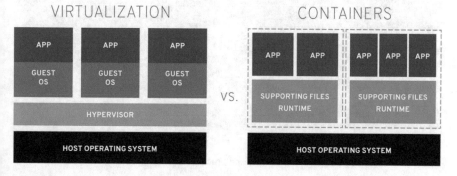

Fig. B.1 A comparison of VMs vs. containers [11]

them running. Docker container is a widely used orchestrator to ensure that the code written by programmers will be portable and have the same functionality independent of the computing environment. A Docker enables multiple containers to run on the same machine and share the operating system kernel, each one of which could run as isolated processes in user space, thereby providing additional security and isolation, as shown in Fig. B.2.

9. *Watchdog:* A watchdog [13] is a device used to protect a system from specific software or hardware failures that may cause the system to stop responding. The application is first registered with the watchdog device. Once the watchdog is installed and running on your system, the application must periodically send information to the watchdog device. If the device doesn't receive this signal within a set period of time, then it would reboot the machine or restart the application. Watchdog services are also available for Internet Websites. In these instances the watchdog may be set to monitor your own Website by attempting to access it from several different cities or country locations at regular intervals. You could then view an online report of the connectivity or have the watchdog system email you should it become inaccessible. Some Websites also offer readers a watchdog service where they receive instant email notification when the Website offering the service has been updated.

10. *Encryption:* In a Cloud there are many levels of security for a variety of data elements. Most data requires a minimal, if any, security. Data in transmission, short-term storage, long-term storage, and during execution all require security. Encryption is an underlying technology needed for systems providing protection of data. However, these four situations require different applications of encryption to support the security of the system. The issues of generating and

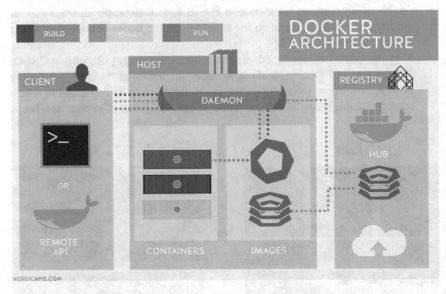

Fig. B.2 Docker Orchestration architecture and process [12]

distributing keys are discussed following this definition. First, consider the situation of long-term storage, which we define as longer than the current run time of an application or greater than 1 day. Applying encryption to disk storage in the Cloud is exacerbated by the fact that data will be stored in a distributed locations (whether different disk drives at one or at different sites). Now, consistency of the encryption is an added issue for distributed databases. Consistency must ensure that the data from a variety of locations and a variety of times can be decrypted. Also, a Public Cloud's attack surface is much greater than other types of Cloud, which leads to a higher required level of strength for the encryption used. For example, data stored on a desktop disk for a year might require a 1024-bit level because there is a small attack port, and re-encrypting the data each month or week can further reduce attack time window. Whereas in the Cloud one would require a 4096-bit level of security because there are many undefined ports of attack, and time attack window cannot as easily be reduced by frequent re-encryption of the distributed data. Cloud Computing adds a new dimension to the data transmission. Specifically, in addition to the straightforward transmission of getting data to and from an application as initiated by a user, there is the transmission of data between sites and devices initiated by the administrator for load balancing, or as part of a distributed application. Encryption of any secure messages to or from the user is the default and straightforward. However, when to apply encryption between sub-processes of distributed applications is a nontrivial decision. For short-term or temporary storage, adding encryption also adds significant performance penalty. Finally, and the toughest case of all, is protection of data while processing it. For example, as has been described before in this book, there are examples of attacks on caches to access secure data while a process is doing its computation. Also, other shared resources are susceptible to attacks while computation is being performed, and the encryption activity itself is susceptible to side channel attacks. Also, the performance effort of encryption can actually be used to the advantage of a service attack by repeatedly overloading the data for encryption or checking beyond the system capability. In summary, Cloud Computing needs to use encryption for data security, but its system implementation poses added difficulties, not the least of which is related to the keys management as described in the next section.

11. *Keys-Based Security:* Any information security system requires keys to lock and unlock the data. This is true whether it includes encryption, secure hashing, authentication, access control, or repudiation. The issue of key management includes both key generation and key distribution. Key generation requires a method that creates keys in an irreversible manner. It should also not be possible to counterfeit the keys. In some weak system implementations, the successful attacks are based upon accessing a weak key generation scheme. For example, some keys are based on prime numbers, and a weak key generation scheme would use a small subset of possible prime numbers, resulting in a small search space for the attacker. Also, a weak method might produce keys in a repeatable and identifiable order allowing an attacker to reverse the order to

identify previous keys. The key generation attacks are not significantly different in a Cloud environment than in other computing environments. A Public Cloud has the multiplier effect, which needed a high number of new keys, due to multi-tenancy among a large customer population, with almost no restriction on each customer having multiple accounts. Also, the key distribution problem is immensely exacerbated in a Cloud environment. One way this is seen is the variety of encryption needs described in the previous section. Another is the fact that keys must be distributed to many different sites. The standard approach in a communication system is to use an Encrypted Key Exchange (EKE) to generate a session-specific key. This is very cumbersome in a Cloud environment with many participating processes that must secure the data. As with security, the weakest link limits the strength of the overall system, so key management and its storage vaults need to be protected well.

12. *Network Function Virtualization (NFV):* It refers to running network services, such as routing, load balancing, and firewalls, in virtual machines (VMs) on regular servers instead of on dedicated proprietary hardware [14]. An example of NFV architecture stack is shown in Fig. B.3.

NFV is driven by economic considerations and flexibility, as it is easier to provision new servers and start VMs than investing in dedicated hardware. NFV is different than the software-defined networking (SDN). The latter refers to network management and changing network configuration through software to improve performance and monitoring. NFV moves networking services to a virtual environment but precludes policies to automate the environment.

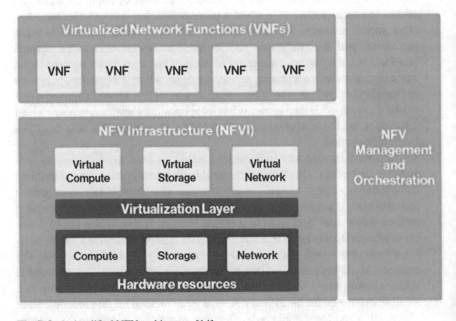

Fig. B.3 A simplified NFV architecture [14]

References

1. http://searchdatabackup.techtarget.com/definition/backup
2. https://en.wikipedia.org/wiki/Application_checkpointing
3. https://aws.amazon.com/ec2/instance-types/f1/
4. https://azure.microsoft.com/en-us/resources/videos/build-2017-inside-the-microsoft-fpga-based-configurable-Cloud/
5. https://Cloud.google.com/tpu/
6. https://puppet.com/solutions/devops
7. https://www.chef.io/devops-tools/
8. https://en.wikipedia.org/wiki/Disaster_recovery
9. https://en.wikipedia.org/wiki/Function_as_a_service
10. http://scholarworks.umass.edu/cgi/viewcontent.cgi?article=1186&context=cs_faculty_pubs
11. https://www.redhat.com/en/topics/containers/whats-a-linux-container
12. http://apachebooster.com/kb/what-is-a-docker-container-for-beginners/
13. http://www.Webopedia.com/TERM/W/watchdog.html
14. https://searchnetworking.techtarget.com/definition/network-functions-virtualization-NFV

Appendix C: Suggested List of Additional Cloud Projects

This section extends beyond the projects presented in Chap. 14, for the advanced readers, and challenges them to attempt a few real-life Cloud usage scenarios which can be deployed in a production setting.

1. AWS has made Genomes Project data publicly available to the community free of charge:

    (a) https://aws.amazon.com/public-datasets/
    (b) AWS provide a centralized repository of public data hosted on Amazon Simple Storage Service (Amazon S3).
    (c) This data can be seamlessly accessed from AWS services such Amazon Elastic Compute Cloud (Amazon EC2) and Amazon Elastic MapReduce (Amazon EMR).
    (d) AWS is storing the public datasets at no charge to the community.
        Project: Researchers can use the Amazon EC2 utility computing service to dive into this data without the usual capital investment required to work with data at this scale. Show how you can search and assemble a particular gene sequence, such as for e-coli? More details are given here: http://ged.msu.edu/angus/tutorials-2013/files/lecture3-assembly.pptx.pdf

2. *Analyze a live Twitter stream* to indicate emotions of crowd for a particular event, using MapReduce to search for keywords to determine the sentiments of each tweet.
3. *Use a popular real-estate service*, such as Zillow's APIs, to search for households that are more likely to donate money for a symphony, and heuristics to look for owners who have a certain property value, such as greater than $1 million USD, or have already paid off their houses with no mortgage payments due.
4. *Compare AWS to Google Cloud.* Technical and business case study of how the two compare. Sign up for the Google Cloud free trial. Cost-benefit analysis of the two. Identify IaaS, PaaS offerings, their cost and benefits, and which one you

© Springer Nature Switzerland AG 2020
N. K. Sehgal et al., *Cloud Computing with Security*,
https://doi.org/10.1007/978-3-030-24612-9

would use. Are there any unique offerings on the Google Cloud, if so go into the details, and how are they different and better than AWS?

5. *Compare AWS to Azure*: Technical and business case study of how the two compare. Sign up for the Azure free trial. Cost-benefit analysis of the two. Identify IaaS, PaaS offerings, their cost and benefits and which one you would use. Are there any unique offerings on the Microsoft Cloud, if so go into the details, and how are they different and better than AWS?

6. *Projects related to Online Libraries*: Imagine a community of users, such as in an apartment complex, which wishes to share books within the community or neighboring communities with like-minded readers. Create an online database of such users, with a list of books that each user is interested in, and another list of books that the user has read, to share with others. Using an online blog, such as WordPress, each user can comment on a specific book.

7. *Projects related to Online Communities:* Extend the previous project to include movies, music, or any other items that a community of users would like to discuss using a Cloud-based database and a virtual network of users.

Appendix D: Trust Models for IoT Devices

We look at the hardware and software stack of a simple home surveillance camera system, to analyze its attack surface and threat model. Then a novel method is presented to apply series-parallel reliability calculations to propose a system security scoring computation. We close this section and the book with an analysis of methods to improve the system-level reliability.

Attacks have often exploited a component-level vulnerability. Most security systems are designed using capability models. A capability model usually takes into account how various services are utilized. An example of such a model starts with a multidimensional representation composed of:

1. Hardware: an ASIC or programmable microcontroller
2. Operating system: Windows, Linux, Android, etc.
3. Applications: nature of application and its privilege level.
4. Manner in which various components, services, and utilities are deployed:

    (a) Kernel, library services, files accesses, etc.
    (b) Manner in which objects such as username, application, and function get authenticated
    (c) The kind of cryptography used: MD5 vs. SHA256

We propose to evaluate components of a given HW and SW solution of one or more Cloud-connected IoT devices based on the robustness and trustworthiness of their entire solution stack, with a multiplicative serialized model in the following order:

1. Native compiled code is trusted more than interpreted code.
2. Code that uses external libraries.
3. Third-party SW attempting to integrate with the platform.

Using the above method, it is possible for us to evaluate trust of different operating systems with applications from diverse fields. Our goal is to create a framework for evaluating and assigning a security score to each layer, which is used to compute

© Springer Nature Switzerland AG 2020
N. K. Sehgal et al., *Cloud Computing with Security*,
https://doi.org/10.1007/978-3-030-24612-9

a composite score. An application can be disassembled to see whether it uses a kernel service, a utility in the user space, or a built-in library, etc.

For each component in the stack, a list of orthogonal properties is established followed by an objective scoring system for each property. The numerical score for a utility function depends on the manner in which it is accessed, e.g., read (as a call by value) or a write (call by reference). A security score can be computed by answering a set of questions by a user or automatically computed by a testing tool. Examples of questions include:

- Whether a salt is used hash passwords?
- Which algorithm is used for hashing: MD5 or SHA256?
- Does the communication channel use SSL, and which version of TLS is being used?
- What is the version of MYSQL in operation?

Another security score determination method is whether port 3306 used by MySQL is open to the world or just to the application servers that use the MySQL database. This score can be continuously updated during the operations. More importantly, it needs to be updated after a maintenance or upgrade action is completed.

Security score questionnaires may also focus on the best practices during development. Automated score calculation focuses on the system operations hygiene: an OS without the latest patch can be at a security risk.

Security score computations have two outputs:

1. *Probability of a successful attack:* What is the probability that an attack on this device will succeed?
2. *Probable impact of a successful attack:* What is the probable impact if the attack succeeds?

The security score (S) can be computed as follows:

$$S = 1 - Pa * Pi,$$

where:

Pa: probability of the attack from 0 to 1
Pi: probable impact of a successful attack from 0 to 1
Pa * Pi: the expected loss

This score is for a single component. By describing the security-wise relationship among the different components and their individual security scores, the whole system security score can be computed.

The factors that affect probability of an attack include:

- Presence of an existing vulnerability that is known to attackers
- Hackers' focus on products of this type
- History of exploitation of this type of product

Probable impact of a security attack is defined as the sum of any regulatory fines, reputational damage, and operational losses, representing loss of trust in the product and services. This needs constant monitoring for security breaches and policy updating [1, 2].

The first step in the modeling of probable impact is to describe the whole system in terms of its components, hierarchically organization, and security-wise connections between the components, which could be in series or parallel.

The directional lines in the figure below represent the security-wise relationship between different blocks in a system. In Fig. D.1, all the blocks should be secure for system to be secure and provide the required functionality. In Fig. D.2, any one block should be available for the system to provide the required functionality. Composite Security Score can be computed by applying series-parallel reliability rules [1], as shown in Figs. D.1 and D.2:

An Example of Trust-Based Scoring

Raspberry PI is the de facto choice and starting point for many IoT devices. This choice is driven by its ubiquity and low price, making it a popular controller for many home and entry-level appliances. A higher installed base also makes it an attractive target for hackers, making it a good evaluation choice for our IoT trust model. For our sample system, we restricted probability values to high (0.9), medium (0.6), and low (0.3). Similarly, the impact values were also high (0.9), medium (0.6), and low (0.3).

We took an implementation of a Raspberry Pi Model 3B with Raspbian OS Ver 4.14, released on 2018-4-18 as a reference system for trust scoring [3]. The base Raspberry Pi system comes with a microSD card, which holds the OS and can be used to install additional software. The factory settings and factory-shipped software packages for the OS were used for trust scoring. No packages were updated. Once the basic model trust scoring was complete, we proceeded to complete the Raspberry Pi-based security camera setup [4]. We added the software components listed below, and the data flow for the system is depicted in Fig. D.3.

Fig. D.1 Risk levels of series systems

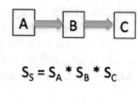

$$S_S = S_A * S_B * S_C$$

Fig. D.2 Risk levels of parallel systems

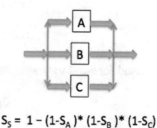

$$S_S = 1 - (1-S_A) * (1-S_B) * (1-S_C)$$

Fig. D.3 Series-parallel
implementation of our
prototype

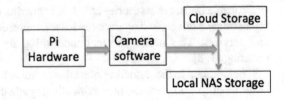

1. MongoDB
2. Rabbit MQ
3. AWS IoT client
4. MotionPie software

We use MongoDB to store images on a NAS (network-attached storage) drive. Cloud storage used Amazon's backend services and is accessed by an AWS IoT client. MotionPie image processing software is used to detect motion and decide which video clips are saved or discarded.

A problem with this security camera prototype, as shown in Fig. D.4, is that someone with physical access to the local system can easily switch off the system or alter the system software. There is no authentication of system software at boot time, so the base hardware setup has a high probability (0.9) of an attack. The impact probability of such attack is also high (0.9) as the base system can be fully compromised.

In the default setup, the username is "admin," and the password is blank. It is easy for someone to remotely hijack the system and use its camera in a Mirai botnet attack [6]. Once the password has been changed, and the camera is moved behind a secure firewall, the probability of such an attack is lowered to medium (0.6). However, the impact probability remains high (0.9). Our proposed system uses the AWS IoT security model [2], compliant with X.509 certification with asymmetric keys [4]. The backend environment used to store images is protected by high security, so the probability of an attack is low (0.3). The impact probability is also low (0.3) because the images are stored in two physically separate places: local storage and Cloud-based storage. As a consequence, the higher attack (0.9) and impact (0.6) probabilities of the local system do not sway the overall assessment.

Overall, we have the Raspberry hardware and software components in series security-wise, which itself is in series with two parallel storage systems security-wise. At component level, here is what we have so far:

$$Sc = 1 - (0.9 * 0.9) = 1 - 0.81 = 0.19$$

$$Ss = 1 - (0.6 * 0.9) = 1 - 0.54 = 0.46$$

And for the storage systems,

$$Sgc = 1 - (0.3 * 0.3) = 0.91$$

Fig. D.4 A simple
Raspberry Pi-based camera
system [5]

$$Sgl = 1 - (0.9 * 0.6) = 0.46$$

where Sc is the security of camera, Ss is the security of software, Sgc is the security of Cloud storage, and Sgl is the security of local storage. As Cloud storage and local storage are in parallel and provide redundant functionality, the security score can be computed using reliability parallel chaining rule:

$$
\begin{aligned}
Sg &= 1 - (1 - Sgc) * (1 - Sgl) \\
&= 1 - (1 - 0.91) * (1 - 0.46) \\
&= 1 - (0.09 * 0.54) = 0.9514
\end{aligned}
$$

Finally, the end-to-end system-level security protection score for an attack is a composite of three scores = 0.19 * 0.46 * 0.9514 = 0.07 which is only 7% or very low. This means that the entire camera system is prone to attacks. However, we can still use it due to our added security measures of a strengthened password, dual storage on local HW and the Cloud, and moving the camera behind a secure firewall. Thus, one of the two paths needs to be secured to continue the required functionality:

$$Path\ A : Camera \rightarrow Software \rightarrow Local\ NAS$$

$$Path\ B : Camera \rightarrow Software \rightarrow Cloud\ Storage$$

Note that all past images will still be preserved even if the system is compromised up to the point of intrusion, say if someone physically removes the microSD card on a security camera. If a home or business uses such a system, it may need multiple cameras, so if one of them is compromised, others will continue the surveillance. An example setup would consist of system with 5 Pi cameras, shared local NAS, and common Cloud storage. The security score for the camera and software part is computed as $1 - (1 - 0.19*0.46)$ ^5 = 0.36. The entire system security will be 0.36*0.9514 = 0.34 or 34%, improving the total system security by almost 5×.

Another way to achieve better security is by making it harder to compromise a single camera system, say by putting it in a cage with a backup battery, so its microSD card can't be easily replaced and the system remains powered on. This action drives the probability of a physical attack lower, from high to low, such that $Sc = 1 - 0.3 * 0.9 = 1 - 0.27 = 0.73$. The overall score for such a single camera system would be $0.73*0.46*0.9514 = 0.32$, or 32%, which is almost same as our five parallel camera systems, and at a much lower cost. A system with one camera also presents a single point of failure, making a combination of multiple cameras with physical security a better approach. Both redundancy and cost control can be achieved by using just two physically secure camera systems in parallel instead of five cameras.

References

1. Sandhu, R., Sohal, A. S., & Sood, S. K. (2017). Identification of malicious edge devices in fog computing environments. *Information Security Journal: A Global Perspective, 26*(5), 213–228.
2. https://aws.amazon.com/blogs/iot/understanding-the-aws-iot-security-model/
3. https://www.raspberrypi.org/downloads/raspbian/
4. https://pimylifeup.com/raspberry-pi-security-camera/
5. http://Web4.uwindsor.ca/users/f/fbaki/85-222.nsf/0/b7d85091e772a10185256f8 4007be5c1/$FILE/Lecture_07_ch6_222_w05_s5.pdf
6. https://www.sentinelone.com/blog/iot-botnet-attacks-mirai-source-code/

Index

© Springer Nature Switzerland AG 2020
N. K. Sehgal et al., *Cloud Computing with Security*,
https://doi.org/10.1007/978-3-030-24612-9